Current Topics in Microbiology and Immunology

213/I

Editors

Springer
Berlin
Heidelberg
New York
Barcelona
Budapest
Hong Kong
London
Milan
Paris
Santa Clara
Singapore
Tokyo

Attempts to Understand Metastasis Formation I

Metastasis-Related Molecules

Edited by U. Günthert and W. Birchmeier

With 35 Figures

Springer

URSULA GÜNTHERT, Ph.D., P.D.
Basel Institute for Immunology
Grenzacherstraße 487
CH-4005 Basel
Switzerland

Professor WALTER BIRCHMEIER, Ph.D.
Max-Delbrück-Centrum für Molekulare Medizin
Robert-Rössle-Straße 10
D-13122 Berlin
Germany

Cover illustration: Tumor cells, like lymphocytes, interact with their surface receptors with corresponding receptors on endothelial cells during intra- and extravasation. Metastasizing tumor cells extravasate from the blood or lymphatic vessels through the endothelial lining into the extracellular matrix of the underlying tissue. The requirement of specific cell surface structures and of secreted proteolytic enzymes for this process, as depicted on the cover, is described in the chapters of this volume.

Cover design: Künkel+Lopka, Ilvesheim
ISSN 0070-217X
ISBN-13:978-3-642-64697-3   e-ISBN-13:978-3-642-61107-0
DOI: 10.1007/978-3-642-61107-0

Softcover reprint of the hardcover 1st edition 1996

Library of Congress Catalog Card Number 15-12910

Typesetting: Thomson Press (India) Ltd., Madras

SPIN: 10495469   27/3020/SPS – 5 4 3 2 1 0 – Printed on acid-free paper

Preface

In metastasis, tumor cells disseminate from the primary lesion and home to secondary organs where they may remain dormant for a long time. Metastasis formation is still the most feared manifestation for tumor patients and clinicians. Although improvements have been made concerning earlier detection and specific therapy, most of the cancer patients still die of distant metastases. The purpose of these three volumes is to review the recent progress in molecular metastasis research and to attempt to further understand the biology of this multifocal process.

With respect to present day molecular biology, the pioneers of metastasis research established the basic concepts of metastasis formation in the 1970s and 1980s, namely, clonal selection of metastatic cells, heterogeneity of metastatic subpopulations, organ specificity of metastasis and the importance of angiogenesis (Fidler, Kripke, Nicolson, Folkman and others). In the 1980s and 1990s, several of the molecules involved were identified and their network interactions elucidated. These three volumes of *Current Topics in Microbiology and Immunology* compile the most recent developments on these metastasis-related molecules; their interactions, regulation, and ways to interfere with their action. It became evident that metastasis-related molecules are confined to distinct cellular compartments, such as the extracellular space, the cell membrane, the cytoplasmic signalling network, and the nuclear regulatory system.

For the complex metastatic cascade, proteolysis and alterations in adhesive functions are the most obvious and thus one of the most thoroughly investigated processes. Various proteases and precursors (metalloproteinases and serine proteases) and their inhibitors (tissue inhibitors of metalloproteases, plasminogen activator inhibitors and serpins) exhibit a sensitive complex of interplay – we are particularly fascinated by their highly regulated nature. Not only the proteases and their inhibitors are important in all the different stages of metastasis formation, but also to the same extent adhesive and "de-adhesive" interactions: metastatic cells must constantly detach themselves from their old partners and reattach to new ones, as mainly outlined in the first volume and partly in the second volume. Among the widespread members of the adhesion molecule families, certain immunoglobulins, integrins, cadherins, selectins, and hyaluronic acid receptors as well as their ligands are implicated in the spread of metastatic cells. The control of the metastatic cascade by these extracellularly acting molecules is delicately balanced, and slight changes could affect the establishment of the normal cellular organization and consequently promote metastasis formation. Strikingly, some genes of adhesion molecules have recently been identified as tumor suppressor genes in model organisms (e.g. Drosophila) and are in fact mutated in metastasizing human tumors.

Growth of primary tumors and metastases is strictly dependent on angiogenesis, the formation of new blood vessels. How this process is regulated by cytokines is another topic of the second volume. Cytokines are not only important in angiogenesis but are essential for the direct migration of metastatic cells. Cytokines act through specific receptors which mediate signals by different means, e.g., tyrosine phosphorylation. A recent discovery is that cytoplasmic signal transduction components, transcription factors, and cell cycle regulators are also metastasis-related. Many of the presently described genes in metastasis were known as activated oncogenes for several years, but apparently the encoded gene products have a broader spectrum of action than was originally assumed.

We have recently learned that the spread of metastatic cells, especially of micrometastases, is far more extensive than previously expected. A successful antimetastatic therapy therefore requires new strategies: for this reason the third volume comprises novel approaches such as immunotherapy, transfer of tumor-inhibiting genes and anti-sense constructs, as well as interference with signal transduction pathways. Promising new therapeutic approaches also involve the use of anti-angiogenic factors or of recombinant soluble metastasis-related molecules which interfere with ligand interactions.

As the process of metastatic spread is presently regarded as a multifactor event which is yet to be sufficiently understood in the multitude of its aspects, approaches to clinical treatment have to be polypragmatic. Methods of treatment are based on chemotherapy and radiotherapy, refined and adapted to the type of tumor pertaining and the pattern of metastatic spread. Increasingly, therapies which incorporate new insights from immunology and molecular biology are adopted for clinical use. To present a rounded scope of the topic, these current strategies are covered by the third volume in particular. Surgical treatment options are indicated in cases where a curative intervention is feasible e.g. in solitary metastases of colorectal carcinoma, soft tissue, and kidney tumors.

We hope that the reader of these volumes is impressed by the quality of the contents. Metastasis has obviously emerged as a serious discipline of natural sciences due to the fact that the molecular biology of various metastasis-related molecules and their complex interplay became transparent. We are, nevertheless, still in the beginning phase and await further progress from which patients will finally benefit.

Most, if not all of the metastasis-specific processes described are also known to be involved in embryonic development and pattern formation, as well as in leukocyte biology. The disciplines of metastasis research, developmental biology, and immunology can, therefore, profit from and stimulate each other. The genetic analysis of candidate molecules and their interplays in transgenic mice will certainly further broaden our understanding of the molecular basis of metastasis formation.

We would like to thank the authors who have spent their valuable time in writing a chapter for this series. Without their expertise and cooperation, this compilation of newest developments in metastasis research would not have been attainable. Leslie Nicklin (Basel) assisted the edition of this series with her competent skills; we are most grateful for her contribution.

Basel — Ursula Günthert
Berlin — Walter Birchmeier
Berlin — Peter M. Schlag

List of Contents

List of Contents of Companion Volume 213/II

List of Contents of Companion Volume 213/III

List of Contributors

(Their addresses can be found at the beginning of their respective chapters.)

Boyer, B. 179
Bracke, M.E. 123
Braun, S. 163
Dall, P. 249
Dunon, D. 195
Elble, R.C. 107
Gisler, R.H. 195
Günthert, U. 271
Hekele, A. 249
Hendrix, M.J.C. 51
Herrlich, P. 249
Imhof, B.A. 195
Johnson, J.P. 95
Mareel, M.M. 123
Matrisian, L.M. 1
Mignatti, P. 33
Moll, J. 249
Montuori, N. 205
Mueller, B.M. 65
Pantel, K. 163
Pauli, B.U. 107
Pemberton, P. 51
Piali, L. 195
Ponta, H. 249
Powell, W.C. 1
Reed, M.J. 81
Rifkin, D.B. 33
Rothbächer, U. 95
Rummel, M.M. 95
Sage, E.H 81
Sager, R. 51
Seiki, M. 23
Sers, C. 95
Sheng, S. 51
Sherman, L. 249
Sleeman, J. 249
Sobel, M.E. 205
Thiery, J.P. 179
Vallés, A.M. 179
Van Roy, F.M. 123
Zöller, M. 215

Complex Roles of Matrix Metalloproteinases in Tumor Progression

W.C. Powell and L.M. Matrisian

1 Introduction

The local invasion and metastatic spread of tumor cells throughout the body is one of the primary concerns of the oncologist. If all tumors were confined within a defined space, most neoplasms could be removed and subsequently cured with the surgeon's scalpel. However, this is not the case within the scope of cancer biology. Some neoplasms are highly metastatic (melanoma), while others are rarely metastatic (brain tumors), but given enough time and/or negligence most tumors will spread to distant sites. The intent of this series is to understand the mechanisms that tumor cells use to invade, disseminate and form viable metastatic colonies as well as discuss potential therapies for metastatic disease. This chapter will address the role of the matrix metalloproteinases (MMPs) and their inhibitors in the process of tumor progression, invasion, and metastasis.

1.1 The Matrix Metalloproteinase Multigene Family

The metalloproteinase multigene family is a continually growing group of enzymes that have links to both normal cellular processes, for example the menstrual cycle (Rodgers et al. 1994), and neoplastic invasion and metastasis.

Department of Cell Biology, Vanderbilt University School of Medicine, Nashville, TN 37232, USA

The proteinases that comprise the MMP family have several distinguishing characteristics:

1. The proteins have a characteristic pattern of conserved domains.
2. The substrates for these enzymes are the proteins that make up the extracellular matrix (ECM) and basal lamina (BL).
3. Proteolytic activity is inhibited by tissue inhibitors of metalloproteinases (TIMPs).
4. The enzymes are either secreted or transmembrane pro-enzymes that require activation to exert their matrix degrading activity.
5. The active site contains a zinc ion and requires a second metal cofactor such as calcium.
6. Enzymatic activity is optimal in the physiological pH range.

In the past, the MMP family has been frequently divided into three classes based on their substrate specificity: (1) the collagenases degrade fibrillar collagens: (2) the gelatinases degrade denatured and basement membrane collagens, and (3) the stromelysins degrade proteoglycans and glycoproteins. This classification scheme becomes problematic when new MMPs are cloned or new substrates are identified for an enzyme and has resulted in the need to place some enzymes in an "other" category until sufficient information is available. We propose a methodology for classifying MMPs based primarily on the protein domain structure. This approach has been used to subdivide other protein families, for example the growth factor receptor/kinase family of proteins. This classification reduces the variability associated with substrate specificity being the determining factor for classification, since not all enzymes have been tested on all ECM and BL proteins and allows for a more rapid and unbiased classification of new MMP family members.

1.2 Matrix Metalloproteinase Protein Structure

The MMP family members share several conserved and easily recognizable protein domains. Under the proposed classification system, the enzymes are categorized based on the number and characteristics of specific protein domains (Fig. 1). The minimal domain enzyme, matrilysin, contains three domains that are present in all MMPs and make up the structural basis for the entire family. The signal sequence or "pre" domain contains a series of hydrophobic amino acids that direct the protein for export but are rapidly removed prior to secretion. The propeptide domain contain a highly conserved segment of eight amino acids, PRCGVPDV, with an unpaired cysteine that interacts with the zinc ion in the active site. This model, termed the "cysteine switch" mechanism of MMP activation (Park et al. 1991; Springman et al. 1990; Van Wart and Birkedal-Hansen 1990), was based on the ability of various compounds that activate MMPs to disrupt the interaction between the zinc and the conserved cysteine either directly or by altering the protein conformation of the

ENZYME	DOMAIN STRUCTURE	SUBSTRATES	ACTIVATION OF OTHER MMPs
Minimal Domain MMP			
Matrilysin[1] (Pump-1, MMP-7)	Pre · Pro · Catalytic Zn++	Proteoglycans[2], Laminin[3], Fibronectin[4], Gelatins[4], Collagen IV[3], Elastin[5], Entactin[6], Tenascin[7]	Interstitial Collagenase[4], Gelatinase A[8]
Hemopexin Domain MMPs			
Interstitial Collagenase[9] (MMP-1)	Pre · Pro · Catalytic Zn++ · H · Hemopexin	Collagens I[10], II[10], III[10], VII[11], X[12], Gelatins[10]	Unknown
Neutrophil Collagenase[13] (MMP-8)	Pre · Pro · Catalytic Zn++ · H · Hemopexin	Collagens I[14], II[14], III[14]	Unknown
Collagenase-3[15] (MMP-13)	Pre · Pro · Catalytic Zn++ · H · Hemopexin	Collagen I[15]	Unknown
Metalloelastase[16] (MMP-12)	Pre · Pro · Catalytic Zn++ · H · Hemopexin	Elastin[16]	Unknown
Stromelysin-1[17] (Transin, MMP-3)	Pre · Pro · Catalytic Zn++ · H · Hemopexin	Proteoglycans[18], Laminin[18], Fibronectin[18], Collagen III[19], IV[18], V[19], IX[20], Gelatins[18]	Interstitial Collagenase[19], Neutrophil Collagenase[21], Gelatinase B[22]
Stromelysin-2[1] (Transin-2, MMP-10)	Pre · Pro · Catalytic Zn++ · H · Hemopexin	Proteoglycans[23], Fibronectin[19], Collagen III[19], IV[19], V[19], Gelatins[19]	Neutrophil Collagenase[24]
Stromelysin-3[25] (MMP-11)	Pre · Pro · F · Catalytic Zn++ · H · Hemopexin	Laminin[26], Fibronectin[26] (Very weakly)	Unknown
Transmembrane Domain MMP			
MT-MMP-1[27] (MMP-14)	Pre · Pro · F · Catalytic Zn++ · H · Hemopexin · TMD	Unknown	Gelatinase A[27]
MT-MMP-2[28]	Pre · Pro · F · Catalytic Zn++ · H · Hemopexin · TMD	Unknown	Unknown
MT-MMP-3[29]	Pre · Pro · F · Catalytic Zn++ · H · Hemopexin · TMD	Unknown	Gelatinase A[29]
Fibronectin Domain MMPs			
Gelatinase A[30] (72kD gelatinase, MMP-2)	Pre · Pro · Catalytic · FN · Zn++ · H · Hemopexin	Gelatins[30], Collagens IV[30], V[30], VII[30], X[31], Elastin[32], Fibronectin[30]	Unknown
Gelatinase B[33] (92kD gelatinase, MMP-9)	Pre · Pro · Catalytic · FN · Zn++ · Col · H · Hemopexin	Gelatins[34], Collagens IV[33], V[33], Elastin[32]	Unknown

Fig. 1. Classification of matrix metalloprotienases (*MMPs*). *FN*, fibronectin-like domain; *F*, furin-recognition domain; *H*, hinge domain; *TMD*, transmembrane domain; *Col*, collagen domain. *References*: *1*, MULLER et al. 1988; *2*, MURPHY et al. 1991; *3*, MIYAZAKI et al. 1990; *4*, QUANTIN et al. 1989; *5*, MURPHY et al. 1991; *6*, SIRES et al. 1993; *7*, SIRI et al. 1995; *8*, CRABBE et al. 1994; *9*, GOLDBERG et al. 1986; *10*, MILLER et al. 1976; *11*, SELTZER et al. 1989; *12*, SCHMID et al. 1986; *13*, HASTY et al. 1990; *14*, HASTY et al. 1987; *15*, FREIJE et al. 1994; *16*, SHAPIRO et al. 1992; *17*, MATRISIAN et al. 1985; *18*, CHIN et al. 1985; *19*, NICHOLSON et al. 1989; *20*, OKADA et al. 1989; *21*, KNAUPER et al. 1993; *22*, OGATA et al. 1992; *23*, FOSANG et al. 1991; *24*, KNAUPER et al. 1994; *25*, BASSET et al. 1990; *26*, MURPHY et al. 1993; *27*, SATO et al. 1994; *28*, WILL et al. 1995; *29*, TAKINO et al. 1995; *30*, COLLIER et al. 1988; *31*, WELGUS et al. 1990; *32*, SENIOR et al. 1991; *33*, WILHELM et al. 1989; *34*, MOLL et al. 1990

prodomain. This event allows a water molecule to be the fourth zinc ligand, thus displacing the cysteine (Birkedal-Hansen et al. 1993). The release of the prodomain from the active site zinc leads to the removal of the prodomain via an autocatalytic mechanism to produce the mature enzymatic form. The catalytic domain contains three conserved histidines in the peptide sequence HEXGHXXGXXHS. These histidines are required for ligation of the zinc ion and the activity of the MMP (Windsor et al. 1994 and references therein). The catalytic domain of human collagenase has been crystallized and the structure solved to a 2.0Å resolution (Bode et al. 1994; Lovejoy et al. 1994). This work verified that the zinc contained in the catalytic site is coordinated by the three histidines of the conserved catalytic domain and that the protein contained a second "structural" zinc. The authors indicated that the structure of the catalytic domain should hold true for the other members of the MMP family with the exception of gelatinases A and B, which contain a 182 amino acid insertion (the fibronectin domain, see below) in the catalytic domain.

Matrilysin is the only family member to date that does not contain a COOH- terminal domain that has homology to a heme binding protein (hemopexin) and the ECM component vitronectin. In this classification scheme, the addition of this domain distinguishes the "hemopexin domain" MMPs from matrilysin (Fig. 1). Several functions have been assigned to this domain in different MMP family members. The hemopexin domains in progelatinase A and B have been associated with interactions with the inhibitors TIMP-2 and TIMP-1, respectively (Baragi et al. 1994; Howard and Banda 1991; Howard et al. 1991a; Murphy et al. 1992). In interstitial and neutrophil collagenase, the hemopexin domain is associated with substrate and inhibitor binding (Bigg et al. 1994; Sanchez-Lopez et al. 1993; Schnierer et al. 1993). The hemopexin domain of gelatinase A is required for cell surface activation (Murphy et al. 1992; Ward et al. 1994).

The hemopexin domain and the catalytic domain are linked by a "hinge region" that is of variable length and composition (Birkedal-Hansen et al. 1993). Interestingly, this hinge region has a conserved size and peptide sequence in MMPs that are able to degrade fibrillar collagen, and this sequence has been shown to be important in conferring the ability of collagenases to degrade fibrillar collagen (Hirose et al. 1993). This structural difference in the hinge region allows for a subclassification within the hemopexin domain enzymes that differentiates the collagenases from the stromelysins and other MMPs that have similar structures but different substrates.

The most recently described members of the MMP family contain a domain that is common to many proteins but is newly characterized in the MMPs. Membrane-type MMP-1 (MT-MMP-1) has the domain structure found in the hemopexin MMPs with the addition of a membrane spanning domain. The transmembrane domain of MT-MMP-1 has been verified by deletion and translocation analysis (Cao et al. 1995). Membrane localization is required for its only known activity, activating gelatinase A on cell membrane surfaces (Cao et al. 1995). Recently, MT-MMP-2 and MT-MMP-3 have been cloned and their domain structure is very similar to MT-MMP-1 (WILL et al. 1995; TAKINO et al. 1995).The recent cloning of MT-MMP-1 has indicated a second potential new domain, one they share with stromelysin-3. This domain is a ten amino acid insertion following the PRCGVPDV sequence that contains the consensus sequence RXKR, which is a recognition site for furin-like enzymes (ROEBROEK et al. 1994). The furin-like enzymes function in protein processing and have been shown to function in a number of protein maturation pathways (BRESNAHAN et al. 1990). Recently, stromelysin-3 has been shown to be activated intracellularly by furin in COS cells and, furthermore, the transfer of the ten amino acid sequence to interstitial collagenase causes intracellular activation (PEI and WEISS 1995). MT-MMP-1 and stromelysin-3 share another characteristic in that neither enzyme appears to efficiently degrade ECM proteins. Stromelysin-3 requires a COOH-terminal cleavage as a second activation step and still has very low activity against ECM substrates (MURPHY et al. 1993), and the only described substrate for MT-MMP-1 is gelatinase A (SATO et al. 1994). These observations suggest the possibility that the mechanism of activation may affect substrate recognition, or that there maybe other functions associated with MMPs other than degrading ECM and BL proteins directly.

The last subclass of MMPs is based on the presence of the fibronectin domain. Both gelatinase A and B contain a domain that is homologous to the collagen binding region of fibronectin (COLLIER et al. 1988; WILHELM et al. 1989). This region has been shown to be required for gelatinase activity as well as collagen binding of gelatinase A (MURPHY et al. 1994). A separate study indicates that the fibronectin domain, when expressed by itself, had a high affinity for gelatin and could displace native fibronectin from gelatin (BANYAI et al. 1994). Gelatinase B is the only MMP to contain a sequence similar to the α_2 chain of type V collagen in the hinge region but the functional significance of this insertion is unclear.

We have proposed a new classification system for the MMP family based on protein domain structure. This system allows for faster classification of newly cloned genes and eliminates many of the discrepancies that arose from the substrate specificity-based system. Like their ECM substrates, it is now recognized that the MMPs are a large family of highly modular proteins and that the combination of specific domains relates to the functional characteristics of the individual enzyme.

1.3 Matrix Metalloproteinase Activation

It is generally accepted that degradation of the BL or ECM is the end result of a proteolytic cascade involving members of both the serine and metalloproteinase families (LIOTTA et al. 1991; MATRISIAN 1992; NAGASE et al. 1990; SUZUKI et al. 1990; WOESSNER 1991). These models are based on data showing that in vitro, both MMPs and serine proteinases can act on the proforms of one another, and on inhibitor studies that demonstrate that both MMPs and serine proteinases are necessary for invasion of tumor and endothelial cells through BL components (MIGNATTI et al. 1986). Plasmin has been shown to activate most of the MMPs by making one cleavage in the prodomain which initiates the autoactivation of the MMP (EECKHOUT and VAES 1977; SANTIBAÑEZ and MARTÍNEZ 1993). The MMP matrilysin can act on urokinase to liberate the NH_2-terminal fragment containing the receptor binding domain from high molecular weight prourokinase (MARCOTTE et al. 1992), which could affect subsequent cell surface activation of urokinase. These activations have been shown to occur in tissue culture (HE et al. 1989; MARCOTTE et al. 1992), but there is little direct evidence that these types of activations occur in vivo. One of the recently discovered MMPs, MT-MMP, has been shown to be responsible for the membrane localized activation of gelatinase A in tumor cells (SATO et al. 1994). The concept of a proteolytic cascade formed co-operatively by different cell types is a subject of considerable interest that requires further investigation.

1.4 Tissue Inhibitors of Metalloproteinases

The activated form of these MMPs can be inhibited by a family of secreted proteins known as tissue inhibitors of metalloproteinases. Currently there are three known TIMPs; TIMP-1, TIMP-2 and TIMP-3. The TIMPs are highly structured proteins that contain 12 conserved cysteines that form six disulfide bridges between protein segments (WILLIAMSON et al. 1990). The TIMPs are expressed in most tissues and can be coregulated or differentially regulated with the MMPs depending on the tissue or cell type. TIMP-1 inhibits all of the MMPs but preferentially binds progelatinase A and inhibits other members of the MMP family (HOWARD et al. 1991 a,b). The most recently discovered TIMP, TIMP-3, has been shown to be expressed in a broad range of tissues, as was found with TIMPs 1 and 2 (APTE et al. 1994; LECO et al. 1994; SILBIGER et al. 1994). The biochemistry of TIMP-3 interactions with the MMP family members has yet to be addressed. Currently TIMP-2 is being tested for clinical efficacy in treating human cancers that exhibit high rates of metastatic spread (PARKINS 1994). Thus TIMPs represent a growing family of broad spectrum MMP inhibitors that play a critical role in the regulation of ECM degradation/remodeling.

2 Matrix Metalloproteinases and Cancer

2.1 Matrix Metalloproteinase Expression in Neoplasia

The MMPs have long been associated with malignant transformation. Stromelysin-1 and gelatinase A were originally cloned from viral or oncogene transformed cells (COLLIER et al. 1988; MATRISIAN et al. 1985). Many oncogenes have been shown to regulate MMP levels, and to date all but one MMP family member has been shown to have an AP-1 transcription element in their promoter (GAIRE et al. 1994 and references within) which has been shown in several cases to be involved in mediating this response (GUTMANN and WASYLYK 1990 for review). MMPs can also be induced by many growth factors and cytokines in cultured cells, including epidermal growth factor, (EGF) platelet-derived growth factor, (PDGF) tumor necrosis factor-α (TNFα) and transforming growth factor-α (TGF)α (for review see MATRISIAN and HOGAN 1990). Although MMPs are induced by oncogenes in several cell types, in situ hybridization studies have demonstrated that most MMPs are produced by normal stromal cells surrounding the tumor (BASSET et al. 1990; NEWELL et al. 1994, for example), suggesting that tumor produced growth factors may be involved in MMP induction in vivo. Matrilysin is unusual among the MMPs in that it is primarily expressed by epithelial cells and tumor cells of epithelial origin (MCDONNELL et al. 1991; PAJOUH et al. 1991; RODGERS et al. 1993). In early neoplastic events this segregation of MMP production appears to remain intact, i.e. glandular epithelial derived carcinoma cells express matrilysin while several other MMPs are expressed in the surrounding connective tissue cells (NEWELL et al. 1994, for example). However, in some late stage squamous cell carcinomas stromelysin-1 and collagenase can be expressed by epithelial derived tumor cells (WRIGHT et al. 1994 and unpublished data), potentially reflecting the epithelial/stromal conversion that occurs frequently in late-stage carcinogenesis.

2.2 Tumor Growth, Invasion and Metastasis

The process by which neoplasms metastasize is composed of a complex series of events (FIDLER 1991; KHOKHA et al. 1989; LIOTTA et al. 1991, NICOLSON 1989). Liotta has proposed a three step model of tumor cell invasion: (1) tumor cell attachment to the BL, (2) localized proteolysis of the BL, and (3) migration through the BL and stroma (LIOTTA 1986). In the case of glandular epithelia, many tissues contain a basal cell layer next to the BL. As carcinoma in situ develops, these basal cells are displaced and the carcinoma cells come in contact with and attach to the BL. The degradation of the BL through the secretion of proteinases can be mediated by not only metalloproteinases and serine proteinases, as described previously, but lysosomal cathepsins of both the cysteine and aspartyl class can also be secreted by tumor cells and degrade components of the BL (KANE and GOTTESMAN 1990; ROCHEFORT et al.

1990). Finally the tumor cells migrate through the digested BL and into the ECM with subsequent access to the microvasculature and lymph systems. This three step process of tumor cell invasion is one critical portion of the larger process of metastasis which includes cell motility, intravasation, immune system evasion, extravasation and tumor colony formation (LIOTTA et al. 1991; NICOLSON 1991).

The evidence of MMP involvement in invasion and metastasis comes from three main sources: correlative evidence demonstrating the expression of MMPs in advanced-stage tumors (discussed in detail below), in vitro models of invasion, and in vivo models of invasion and metastasis. The best evidence of the role MMPs perform in invasion and metastasis have come from studies using the TIMPs as a method of inhibiting all MMP family members. Using in vitro models of invasion with modified Boyden chambers and a membranous barrier of human amnion, reconstituted BL (matrigel), or smooth muscle cells, TIMP-1 (DECLERCK et al. 1991, for example) has been shown to inhibit tumor cells from entering or crossing the membrane. When TIMP-1 levels in a nontumorigenic fibroblast cell line were reduced by antisense RNA, the resulting cells were tumorigenic and were able to form distant metastases when injected subcutaneously into mice (KHOKHA et al. 1989). Mice that were treated i.p. with TIMP-1 every 12 h and injected i.v. with B16-F10 melanoma cells had fewer experimental metastases (SCHULTZ et al. 1988). TIMP-2 overexpression in metastatic H-*ras* transformed rat embryo cells inhibited their ability to invade the surrounding tissue and partially inhibited experimental metastasis (DECLERCK et al. 1992).

Although the effect of MMPs and TIMPs on tumor metastasis has been presumed to be related to the ability of the cell to degrade microvascular basement membranes and extravasate, recent evidence points to a role in tumor growth. Using intravital microscopy, B16-F10 cells transfected with TIM-1 have been shown to extravasate as efficiently as controls; however, they have a decreased growth potential at the new site and form fewer metastatic tumors in a chick embryo assay (KHOKHA et al. 1992; KOOP et al. 1994). Another highly metastatic melanoma cell line, M24met, that expresses collagenase and gelatinases A and B has been transfected with TIMP-2 (MONTGOMERY et al. 1994). These cells have reduced growth potential in vivo and in in vitro collagen gels, but their metastatic ability remains the same. Transfection based expression of matrilysin in SW480 colon cancer cells increased the cells tumorgenicity when implanted into the cecum of nude mice, whereas in SW620 cells, antisense ablation of endogenous matrilysin mRNA expression decreases tumorgenicity compared to parental cell line controls (WITTY et al. 1994). The mechanism underlying the growth effects of MMPs is unclear; it may be an indirect effect through alterations in cellular morphology, or degradation of ECM and BL proteins by MMPs may release growth factors that have been sequestered in the matrix. A number of groups have shown that fragments of matrix proteins can alter gene transcription via interaction with the integrin family of cell surface receptors (DAMSKY and WERB

1992; TREMBLE et al. 1992; YAMAMOTO et al. 1994) and similar signal transduction mechanisms could stimulate proliferation. Thus MMP expression may play a number of roles in the production, proliferation and metastatic spread of neoplastic cells.

The role of MMPs in metastasis has been more difficult to verify when the system is reduced to altering the expression of a single MMP. The redundancy in number and overlapping substrate specificity of the MMPs has been problematic in the determination of the function of a specific MMP during the metastatic process. Transgenic mice with stromelysin-1 under the control of either the mouse mammary tumor virus (MMTV) promoter or the whey acidic protein promoter show an altered branching phenotype during mouse mammary gland development (SYMPSON et al. 1994; WITTY et al. 1995a), but mammary tumors induced in these mice do not appear to be significantly more aggressive (WITTY et al. 1995b). However, the expression of matrilysin in tumor cells lacking this MMP has been shown to effect BL penetration (POWELL et al. 1993) and tumorgenicity (WITTY et al. 1994). Although studies with TIMP suggest that the MMP family functions in invasion and metastasis, its individual members alone may not be sufficient to alter the entire metastatic phenotype in vivo. In tumors that express an MMP there tend to be several MMPs that are up-regulated (BASSET et al. 1990; NEWELL et al. 1994; PAJOUH et al. 1991 for example), so it may be the interaction of several MMPs that function in increasing the metastatic ability of tumor cells.

2.3 Other Effects of Matrix Metalloproteinases Influencing Tumorigenesis

Metalloproteinases also contribute to tumor progression by their involvement in the process of angiogenesis. Most studies focusing on angiogenesis have used TIMPs and/or synthetic MMP inhibitors to block angiogenesis in vitro and in vivo. An in vitro angiogenesis model using umbilical vein endothelial cells and type I collagen gels has shown that TIMP-1 and BB-94 (a synthetic MMP inhibitor) can inhibit the phorbol ester induced formation of vascular-like structures in these gels (FISHER et al. 1994). A recent study investigated the ability of TIMP-2 to inhibit Kaposi's sarcoma (KS) induced neovascularization (ALBINI et al. 1994). The study showed that TIMP-2 could inhibit vasculariation of injected Matrigel combined with either conditioned media from KS cells or basic-fibroblast growth factor (bFGF). The interaction of matrilysin and prourokinase is of potential interest in relationship to angiogenesis. Matrilysin can cleave prourokinase such that the NH_2-terminal fragment (ATF) which binds the urokinase cell surface receptor is released from high molecular weight prourokinase (MARCOTTE et al. 1992). The ATF has had a number of activities linked to it, including increasing cell motility and mitogenic stimulation. The ability to increase chemotactic cell motility has been shown for epithelial cells (DEL ROSSO et al. 1993), endothelial cells (ODEKON et al. 1992) and fibroblasts

(ANICHINI et al. 1994). Thus the production of ATF by tumor cells could effect tumor cell migration as well as neovascularization of the tumor mass.

MMPs have been linked to the process of apoptosis (programmed cell death), which is an important developmental process that can be exploited during the treatment of hormone dependent cancer. Prostate and mammary glands undergo apoptotic cell death during castration induced involution and postlactation involution, respectively (TENNISWOOD et al. 1992 for review). Matrilysin is expressed in the epithelial cells of prostatic ducts during the involution of the rat ventral prostate (POWELL et al. 1995). In situ hybridization indicates that matrilysin is not produced by the secretory epithelial cells undergoing apoptosis, but rather is produced in the cuboidal epithelium that lines the primary and secondary ducts leading to the urethra (unpublished data). There are preliminary indications that the androgen responsive prostate cell line LNCaP produces matrilysin in response to phorbol ester (TPA) induced apoptosis (BOWDEN, personal communication). In the involuting mammary gland, stromelysin-3 is expressed by the fibroblasts surrounding the degenerating ducts (LEFEBVRE et al. 1992). Two independent transgenic mouse models of stromelysin-1 overexpression in mammary epithelial cells indicate that degradation of the BL leads to an increase in epithelial cell apoptosis (BOUDREAU et al. 1995; WITTY et al. 1995b). These studies suggest the possibility that, during apoptosis, the MMPs degrade the BL and disrupt the interaction between the epithelial cells and their normal substratum, potentially leading to the degeneration of ductal structures and wide spread tissue remodelling.

2.4 Scope of Matrix Metalloproteinase Expression and Cancer

The expression of MMPs has been examined in numerous tumors by three primary methods; northern blotting, in situ hybridization and zymographic analys sis. Both northern analysis and in situ measure mRNA content; however, northern analysis is more quantitative and in situ hybridization is used to localize the mRNA to a particular cell type. Zymography uses a substrate impregnated SDS-PAGE gel to assess protein level. Table 1 is a list of MMPs and the neo-plasms that have been shown to express the indicated MMP by one of the previously described methods. Breast, colon and prostate cancers are among the most completely described systems and will be discussed in more detail below.

Breast cancer has been receiving more attention as of late due to a rising incidence and increased public awareness (BORING et al. 1993). Metastatic spread to the axillary lymph nodes is an important prognostic tool as the number of lymph nodes that are positive for metastatic colonies is directly related to the patient's survival (HARRIS et al. 1993). A number of MMPs have been associated with metastatic breast cancer; however, only stromelysin-3 has thus far been associated with grade/stage of the tumor. This MMP was originally cloned from a mammary carcinoma and found to be expressed by the stromal cells surrounding the tumor (BASSET et al. 1993). To date, stromelysin-3 has been found in virtually all metastatic beast cancers studied

Table 1. Metalloproteinase expression in human tumors

Metalloproteinase	Neoplasia	Localization	References
Matrilysin	Prostate	Tumor	PAJOUH et al. 1991; POWELL et al. 1993
	Colon	Tumor	MCDONNELL et al. 1991; NEWELL et al. 1994; WITTY et al. 1994; YAMAMOTO et al. 1994
	Head and neck	ND	MULLER et al. 1991
	Breast	Tumor	BASSET et al. 1990
	Gastric	Tumor	MCDONNELL et al. 1991
	Basal cell carcinoma	Tumor	KARELINA et al. 1994
Interstitial collagenase	Gastrointestinal	Stroma	MCDONNELL et al. 1991; OKADA et al. 1995
	Head and neck	Stroma/tumor	OKADA et al. 1995; POLETTE et al. 1991
	Breast	Stroma	OKADA et al. 1995
Collagenase-3	Breast	Tumor	FREIJE et al. 1994
Stromelysin-1	Colon	Stroma	NEWELL et al. 1994
	SCC esophagus	Tumor	SHIMA et al. 1992
	Basal cell carcinoma	Stroma	MAJMUDAR et al. 1994a
	Head and neck	ND	MULLER et al. 1991
	Brain	ND	NAKANO et al. 1993
Stromelysin-2	Head and neck	Stroma/tumor	MULLER et al. 1993; POLETTE et al. 1991
Stromelysin-3	Breast	Stroma	BASSET et al. 1993; ENGEL et al. 1994; HÄHNEL et al. 1993
	Basal cell carcinoma	Stroma	MAJMUDAR et al. 1994b; WAGNER et al. 1992; WOLF et al. 1992
	Head and neck	Stroma	MULLER et al. 1993; OKADA et al. 1995
	Colon	Stroma	NEWELL et al. 1994; OKADA et al. 1995
	Lung	ND	URBANSKI et al. 1992
MT-MMP-1	Breast	Stroma	OKADA et al. 1995
	Colon	Stroma	OKADA et al. 1995
	Head and neck	Stroma	OKADA et al. 1995
Gelatinase A	Cervix	Stromal/tumor	NUOVO et al. 1995
	Colon	Stroma	NEWELL et al. 1994
	SCC esophagus	Tumor	SHIMA et al. 1992
	Lung	Stroma/tumor	BROWN et al. 1993; NAKAGAWA and YAGIHASHI 1994
	Prostate	ND	PAJOUH et al. 1991
	Thyroid	Tumor	CAMPO et al. 1992
	Breast	Stroma	OKADA et al. 1995
Gelatinase B	Cervix	Stroma/tumor	NUOVO et al. 1995
	Lymphoma	Tumor/macrophage	KOSSAKOWSKA et al. 1993
	Prostate	ND	HAMDY et al. 1994
	Brain	ND	RAO et al. 1993
	Lung	Stroma/tumor	URBANSKI et al. 1992

ND, not determined; SCC, squamous cell carcinoma; MMP, matrix metalloproteinase

(95%–97%) (Basset et al. 1994; Kawami et al. 1993). When stromelysin-3 expression by in situ hybridization was compared to patient survival a significant positive correlation was found with fatal metastatic disease (Engel et al. 1994). The question arises, if ECM proteins are poor substrates for stromelysin-3 then what effect does stromelysin-3 have on tumor cell invasion and metastasis? The answer may not be in the ECM substrates but in other proteins that function in the metastatic process. Stromelysin-3 has been shown to proteolytically inactivate α1-proteinase inhibitor, a serine proteinase inhibitor (Pei et al. 1994). This activity fits with the previously discussed role for both stromelysin-3 and MT-MMP in not degrading the matrix, but to aid the invasive phenotype by activating other proteinases or inactivating inhibitors of proteinases involved in invasion. Further evidence that activation of MMPs is an important step in breast cancer is provided by a number of in vitro and in vivo studies showing activation of gelatinase A in breast cancer (Azzam et al. 1993; Brown et al. 1993a,b; Noël et al. 1994; Thompson et al. 1994). It is known that normal mammary epithelial cells are highly dependent on interactions with the BL and myoepithelial cells for maintaining normal mammary functions and morphology (Sympson et al. 1993; Talhouk et al. 1992). Consequently, neoplastic transformation of mammary epithelial cells and expression of MMPs by both the tumor and stromal cells may aid in the further phenotypic changes associated with mammary cancer.

Colon cancer has been one of the most studied neoplasms, and as a consequence a very defined set of genetic alterations have been shown during the progression from normal mucosa to malignant colon cancer (Fearon and Vogelstein 1990). McDonnell et al. (1991) examined the expression of MMPs by northern analysis and found significant levels of matrilysin mRNA. A through study of MMP expression by in situ hybridization in colon cancer was presented by Newell et al. (1994). This study indicated that matrilysin was expressed exclusively in tumor epithelium and was expressed focally in eight of 17 early adenomas (< 1.0 cm in diameter) and at higher levels in nine of ten late-stage carcinomas. Stromelysins 1 and 3 and gelatinase A are all seen exclusively in the stroma of carcinomas, but not all MMPs are expressed within the same tumor (Newell et al. 1994). Witty et al. (1994) have provided evidence that matrilysin may have an effect on colon cancer cell growth in an orthotopic model of colon cancer. SW480 human colon cancer cells were transfected with the matrilysin cDNA under a constitutive promoter and when these cells were implanted into the cecum of nude mice the tumors that arose were faster growing than nontransfected SW480 cells. In the matched cell line SW620 (SW480 from the primary tumor and SW620 from a lymph node metastasis) which constitutively secretes matrilysin, antisense ablation of the matrilysin mRNA reduced the number of tumors (Witty et al. 1994). Thus the expression patterns during colon tumor progression suggest a role for matrilysin in early stages of colon cancer, i.e., tumor cell growth, whereas the MMPs expressed by stromal cells may play a role in later stage events such as tumor cell invasion and metastasis.

Prostate cancer is a slow growing neoplasm believed to arise from the epithelial cells of the gland. Given its deep anatomical location and slow growth rate, its symptomatic onset often allows for metastatic spread prior to diagnosis. Matrilysin and gelatinase A were found to be over expressed in human prostate adenocarcinomas compared to normal prostates from young accident victims (PAJOUH et al. 1991). This study found that matrilysin was localized to the epithelial component of invasive adenocarcinomas in 72% of cases and 27% of normal controls. Gelatinase A expression by northern analysis was seen in 60% of adenocarcinomas and not detected in normal prostate tissue. Gelatinase B has also been shown to be expressed in invasive malignant prostate cancer by both zymography and western blotting (HAMDY et al. 1994). In an attempt to understand the functional role for MMP expression in prostate adenocarcinomas, the matrilysin cDNA was transfected into the nonmetastatic, weakly invasive prostate tumor cell line DU-145 (POWELL et al. 1993). When these cells and control transfected cells were injected i.p. into SCID mice, tumor colonies formed on the diaphragm. The diaphragm was removed and cross-sectioned and the extent of invasion examined. The cells transfected with the matrilysin gene invaded past the BL in 66.7% of the diaphragms compared to 11.1% for the control transfected cells. In this model no increase in metastatic ability was seen. These data indicated that matrilysin plays a role in early events in invasion, primarily in degrading the BL as discussed previously. Recent data from our laboratory show that matrilysin is expressed in the postcastration involuting rat ventral prostate (POWELL et al. 1995). This is important to a discussion of prostate cancer because chemical or physical castration is one of the primary modes of treatment for metastatic disease (HANKS et al. 1993). The potential implication is that castration of the patient may cause a proteolytic burst of not only matrilysin but also urokinase over a period of days from the tumor cells (ANDREASEN et al. 1990). This consequence could further spread the already metastatic prostate cancer cells and potentially hasten the patient's disease.

Breast, colon and prostate cancer are among the most common cancers diagnosed in the Western world and as such have been extensively studied with regards to clinical progression. These three neoplasms, whose progression to the malignant phenotype may be linked to the expression of MMPs, are just a sample of the wide range of neoplasias that have been investigated. As indicated in Table 1, many other neoplasms have been shown to express MMPs during some point in tumor development.

3 Conclusions

The MMP family continues to expand with the cloning of two new MMPs in 1994 and the recent cloning of TIMP-3. These new additions add more complexity to the field, while at the same time opening new avenues of research. The cloning of MT-MMPs and analysis of their cell surface activity will undoubtedly lead to further investigations into cell membrane proteolysis mediated by MMPs. The identification of the furin recognition site in stromelysin-3 and the MT-MMPs and the activation of stromelysin-3 by furin has indicated that intracellular trafficking and processing of MMPs is an important area of research that needs more attention. The solution of the MMP active site structure will aid in the study of MMP biochemistry and allow for the design of specific inhibitors of each MMP. The increasing complexity of the structure and functions of the MMPs prompted the new MMP classification system. It is becoming apparent that one must think of each MMP member not only in terms of its substrate specificity, but also what other effects it may have on the phenotype of a cell. The classification of the MMPs by protein structure rather than substrate specificity may free the MMPs from the view that they are enzymes that only degrade ECM and BL.

The expression of MMPs by both tumor and stromal cells can be linked to a number of inducing agents including growth factors, oncogenes and integrin signaling. The initiation of constitutive MMP expression by tumor and stromal cells may function to destabilize the ECM and BL surrounding the invading tumor cells, thus allowing access to the circulatory and lymphatic systems and subsequent distant metastasis. MMPs appear to have other effects on tumorigenesis such as increased angiogenesis and proliferative potential of the tumor cells. Thus the metalloproteinases are not limited to simply degrading structural proteins that surround the cell, but appear to have a more generalized role in the interactions of cells with their matrix environment, affecting basic cellular processes such as differentiation, proliferation and apoptosis. Such important processes regulated by a multigene family of enzymes with overlapping activities implies that the MMPs play a critical role in maintaining normal cellular homeostasis and enhances the observations that misregulation of these proteinases can have important consequences in the neoplastic process.

Acknowledgments. Research in the authors' laboratory is supported by grants from the NIH (CA46843, CA60867) and the DOD (DAMD17-94-J-4226). W.C.P. is supported by a postdoctoral fellowship from the DOD (DAMD17-94-J-4442).

References

Albini A, Fontanini G, Masiello L, Tacchetti C, Bigini D, Luzzi P, Noonan DM, StetlerStevenson WG (1994) Angiogenic potential in vivo by Kaposi's sarcoma cell-free supernatants and HIV-1 tat product: inhibition of KS-like lesions by tissue inhibitor of metalloproteinase-2. AIDS 8: 1237–1244

Andreasen PA, Kristensen P, Lund LR, Danø K (1990) Urokinase-type plasminogen activator is increased in the involuting ventral prostate of castrated rats. Endocrinology 126: 2567–2576

Anichini E, Fibbi G, Pucci M, Caldini R, Chevanne M, Del Rosso M (1994) Production of second messengers following chemotactic and mitogenic urokinase-receptor interaction in human fibroblasts and mouse fibroblasts transfected with human urokinase receptor. Exp Cell Res 231: 438–448

Apte SS, Hayashi K, Seldin MF, Mattei MG, Hayashi M, Olsen BR (1994) Gene encoding a novel murine tissue inhibitor of metalloproteinases (TIMP), TIMP-3, is expressed in developing mouse epithelia, cartilage, and muscle, and is located on mouse chromosome 10. Dev Dyn 200: 177–197

Azzam HS, Arand G, Lippman ME, Thompson EW (1993) Association of MMP-2 activation potential with metastatic progression in human breast cancer cell lines independent of MMP-2 production. J Natl Cancer Inst 85: 1758–1764

Banyai L, Tordai H, Patthy L (1994) The gelatin-binding site of human 72 kDa type IV collagenase (gelatinase A). Biochem J 298: 403–407

Baragi VM, Fliszar CJ, Conroy MC, Ye QZ, Shipley JM, Welgus HG (1994) Contribution of the C-terminal domain of metalloproteinases to binding by tissue inhibitor of metalloproteinases. C-terminal truncated stromelysin and matrilysin exhibit equally compromised binding affinities as compared to full-length stromelysin. J Biol Chem 269: 12692–12697

Basset P, Bellocq JP, Wolf C, Stoll I, Hutin P, Limacher JM, Podhajcer OL, Chenard MP, Rio MC, Chambon P (1990) A novel metalloproteinase gene specifically expressed in stromal cells of breast carcinomas. Nature 348: 699–704

Basset P, Wolf C, Chambon P (1993) Expression of the stromelysin-3 gene in fibroblastic cells of invasive carcinomas of the breast and other human tissues: a review. Breast Cancer Res Treat 24: 185–193

Basset P, Wolf C, Rouyer N, Bellocq J-P, Rio M-C, Chambon P (1994) Stromelysin-3 in stromal tissue as a control factor in breast cancer behavior. Cancer Suppl 74: 1045–1049

Bigg HF, Clark IM, Cawston TE (1994) Fragments of human fibroblast collagenase: Interaction with metalloproteinase inhibitors and substrates. Biochim Biophys Acta 1208: 157–165

Birkedal-Hansen H, Moore WGI, Bodden MK, Windsor LJ, Birkedal-Hansen B, DeCarlo A, Engler JA (1993) Matrix metalloproteinases: a review. Crit Rev Oral Biol Med 4: 197–250

Bode W, Reinemer P, Huber R, Kleine T, Schnierer S, Tschesche H (1994) The X-ray crystal structure of the catalytic domain of human neutrophil collagenase inhibited by a substrate analogue reveals the essentials for catalysis and specificity. EMBO J 13: 1263–1269

Boring CC, Squires TS, Tong T (1993) Cancer Statistics, 1993. CA Cancer J Clin 43: 7–26

Boudreau N, Sympson CJ, Werb Z, Bissel MJ (1995) Suppression of ICE and apoptosis in mammary epithelial cells by extracellular matrix. Science 267: 891–893

Bresnahan PA, Leduc R, Thomas L, Thorner J, Gibson HL, Brake AJ, Barr PJ, Thomas G (1990) Human fur gene encodes a yeast KEX2-like endoprotease that cleaves pro-beta-NGF in vivo. J Cell Biol 111: 2851–2859

Brown PD, Bloxidge RE, Anderson E, Howell A (1993a) Expression of activated gelatinase in human invasive breast carcinoma. Clin Exp Metastasis 11: 183–189

Brown PD, Bloxidge RE, Stuart NS, Gatter KC, Carmichael J (1993b) Association between expression of activated 72-kilodalton gelatinase and tumor spread in nonsmall-cell lung carcinoma. J Natl Cancer Inst 85: 574–578

Campo E, Merino MJ, Liotta L, Neumann R, Stetler-Stevenson W (1992) Distribution of the 72-kd type IV collagenase in nonneoplastic and neoplastic thyroid tissue. Hum Pathol 23: 1395–1401

Cao J, Sato H, Takino T, Seiki M (1995) The C-terminal region of membrane type matrix metalloproteinase is a functional transmembrane domain required for progelatinase A activation. J Biol Chem 270: 801–805

Chin JR, Murphy G, Werb Z (1985) Stromelysin, a connective tissue-degrading metalloendopeptidase secreted by stimulated rabbit synovial fibroblasts in parallel with collagenase. Biosynthesis, isolation, characterization, and substrates. J Biol Chem 260: 12367–12376

Collier IE, Smith J, Kronberger A, Bauer EA, Wilhelm SM, Eisen AZ, Goldberg GI (1988) The structure of the human skin fibroblast collagenase gene. J Biol Chem 263: 10711–10713

Crabbe T, Smith B, O'Connell J, Docherty A (1994) Human progelatinase A can be activated by matrilysin. FEBS Lett 345: 14–16

Damsky CH, Werb Z (1992) Signal transduction by integrin receptors for extracellular matrix: cooperative processing of extracellular information. Curr Opin Cell Biol 4: 772–781

DeClerck YA, Yean TD, Chan D, Shimada H, Langley KE (1991) Inhibition of tumor invasion of smooth muscle cell layers by recombinant human metalloproteinase inhibitor. Cancer Res 51: 2151–2157

DeClerck YA, Perez N, Shimada H, Boone TC, Langley KE, Taylor SM (1992) Inhibition of invasion and metastasis in cells transfected with an inhibitor of metalloproteinases. Cancer Res 52: 701–708

Del Rosso M, Anichini E, Pedersen N, Blasi F, Fibbi G, Pucci M, Ruggiero M (1993) Urokinase-urokinase receptor interaction: non-mitogenic signal transduction in human epidermal cells. Biochem Biophys Res Commun 190: 347–352

Eeckhout Y, Vaes G (1977) Further studies on the activation of procollagenase, the latent precursor of bone collagenase: effects of lysosomal cathepsin B, plasmin and kallikrein, and spontaneous activation. Biochem J 166: 21–29

Engel G, Heselmeyer K, Auer G, Bäckdahl M, Eriksson E, Linder S (1994) Correlation between stromelysin-3 mRNA level and outcome of human breast cancer. Int J Cancer 58: 830–835

Fearon E, Vogelstein B (1990) A genetic model for colorectal tumorigenesis. Cell 61: 759–767

Fidler IJ (1991) Cancer metastasis. Br Med Bull 47: 157–177

Fisher C, Gilbertson-Beadling S, Powers EA, Petzold G, Poorman R, Mitchell MA (1994) Interstitial collagenase is required for angiogenesis in vitro. Dev Biol 162: 499–510

Fosang AJ, Neame PJ, Hardingham TE, Murphy G, Hamilton JA (1991) Cleavage of cartilage proteoglycan between G1 and G2 domains by stromelysins. J Biol Chem 266: 15579–15582

Freije JM, Diez-Itza I, Balbin M, Sanchez LM, Blasco R, Tolivia J, Lopez-Otin C (1994) Molecular cloning and expression of collagenase-3, a novel human matrix metalloproteinase produced by breast carcinomas. J Biol Chem 269: 16766–16773

Gaire M, Magbanua Z, McDonnell S, McNeil L, Lovett DH, Matrisian LM (1994) Structure and expression of the human gene for the matrix metalloproteinase matrilysin. J Biol Chem 269: 2032–2040

Goldberg G, Wilhelm S, Kronberger A, Bauer E, Grant G, Eisen A (1986) Human fibroblast collagenase; complete primary structure and homology to an oncogene transformation-induced rat protein. J Biol Chem 261: 6600–6605

Gutmann A, Wasylyk B (1990) The collagenase gene promoter contains a TPA and oncogene-responsive unit encompassing the PEA3 and AP-1 binding sites. EMBO J 9: 2241–2246

Hähnel E, Harvey JM, Joyce R, Robbins PD, Sterrett GF, Hähnel R (1993) Stromelysin-3 expression in breast cancer biopsies: clinico-pathological correlations. Int J Cancer 55: 771–774

Hamdy FC, Fadlon EJ, Cottam D, Lawry J, Thurrell W, Silcocks PB, Anderson JB, Williams JL, Rees RC (1994) Matrix metalloproteinase 9 expression in primary human prostatic adenocarcinoma and benign prostatic hyperplasia. Br J Cancer 69: 177–182

Hanks GE, Myers CE, Scardino PT (1993) Cancer of the prostate. In: DeVita VTJ, Hellman S, Rosenberg SA (eds) Cancer: principles and practice, 4th edn. Lippincott, Philadelphia, pp 1073–1113

Harris JR, Morrow M, Bonadonna G (1993) Cancer of the breast. In: DeVita VTJ, Hellman S, Rosenberg SA (eds) Cancer: principles and practice, 4th edn. Lippincott, Philadelphia, pp 1264–1332

Hasty KA, Jeffery JJ, Hibbs MS, Welgus HG (1987) The collagen substrate specificity of human neutrophil collagenase. J Biol Chem 262: 10048–10052

Hasty KA, Pourmotabbed TF, Goldberg GI, Thompson JP, Spinella DG, Stevens RM, Mainardi CL (1990) Human neutrophil collagenase: a distinct gene product with homology to other matrix metalloproteinases. J Biol Chem 265: 11421–11424

He CS, Wilhelm SM, Pentland AP, Marmer BL, Grant GA, Eisen AZ, Goldberg GI (1989) Tissue cooperation in a proteolytic cascade activating human interstitial collagenase. Proc Natl Acad Sci USA 86: 2632–2636

Hirose T, Patterson C, Pourmotabbed T, Mainardi CL, Hasty KA (1993) Structure-function relationship of human neutrophil collagenase: identification of regions responsible for substrate specificity and general proteinase activity. Proc Natl Acad Sci USA 90: 2569–2573

Howard EW, Banda MJ (1991) Binding of tissue inhibitor of metalloproteinases 2 to two distinct sites on human 72-kDa gelatinase. Identification of a stabilization site. J Biol Chem 266: 17972–17977

Howard EW, Bullen EC, Banda MJ (1991a) Preferential inhibition of 72- and 92-kDa gelatinases by tissue inhibitor of metalloproteinases-2. J Biol Chem 266: 13070–13075

Howard EW, Bullen EC, Banda MJ (1991b) Regulation of the autoactivation of human 72-kDa progelatinase by tissue inhibitor of metalloproteinases-2. J Biol Chem 266: 13064–13069

Kane SE, Gottesman MM (1990) The role of cathepsin L in malignant transformation. Semin Cancer Biol 1: 127–136

Karelina TV, Goldberg GI, Eisen AZ (1994) Matrilysin (PUMP) correlates with dermal invasion during appendageal development and cutaneous neoplasia. J Invest Dermatol 103: 482–487

Kawami H, Yoshida K, Ohsaki A, Kuroi K, Nishiyama M, Toge T (1993) Stromelysin-3 mRNA expression and malignancy: Comparison with clinicopathological features and type IV collagenase mRNA expression in breast tumors. Anticancer Res 13: 2319–2324

Khokha R, Waterhouse P, Yagel S, Lala P, Overall C, Norton G, Denhardt D (1989) Antisense RNA-induced reduction in murine TIMP levels confers oncogenicity on Swiss 3T3 cells. Science 244: 947–950

Khokha R, Zimmer MJ, Wilson SM, Chambers AF (1992) Up-regulation of TIMP-1 expression in B16-F10 melanoma cells suppresses their metastatic ability in chick embryo. Clin Exp Metastasis 10: 365–370

Knduper V, Murphy G, Tschesche H (1994) Neutrophil procollagenase can be activated by stromelysin-2. Ann NY Acad Sci 732: 367–368

Knduper V, Wilhelm SM, Seperack PK, DeClerck YA, Langley KE, Osthues A, Tschesche H (1993) Direct activation of human neutrophil procollagenase by recombinant stromelysin. Biochem J 295: 581–586

Koop S, Khokha R, Schmidt EE, MacDonald IC, Morris VL, Chambers AF, Groom AC (1994) Overexpression of metalloproteinase inhibitor in B16F10 cells does not affect extravasation but reduces tumor growth. Cancer Res 54: 4791–4797

Kossakowska AE, Urbanski SJ, Watson A, Hayden LJ, Edwards DR (1993) Patterns of expression of metalloproteinases and their inhibitors in human malignant lymphomas. Oncol Res 5: 19–28

Leco KJ, Khokha R, Pavloff N, Hawkes SP, Edwards DR (1994) Tissue inhibitor of metalloproteinases-3 (TIMP-3) is an extracellular matrix-associated protein with a distinctive pattern of expression in mouse cells and tissues. J Biol Chem 269: 9352–9360

Lefebvre O, Wolf C, Limacher J-M, Hutin P, Wendling C, LeMeur M, Basset P, Rio M-C (1992) The breast cancer-associated stromelysin-3 gene is expressed during mouse mammary gland apoptosis. J Cell Biol 119: 997–1002

Liotta L (1986) Tumor invasion and metastasis – role of the extracellular matrix: Rhoads Memorial Lecture. Cancer Res 46: 1–7

Liotta LA, Steeg PS, Stetler-Stevenson WG (1991) Cancer metastasis and angiogenesis: an imbalance of positive and negative regulation. Cell 64: 327–336

Lovejoy B, Cleasby A, Hassell AM, Longley K, Luther MA, Weigl D, McGeehan G, McElroy AB, Drewry D, Lambert MH, Jordan SR (1994) Structure of the catalytic domain of fibroblast collagenase complexed with an inhibitor. Science 263: 375–377

Majmudar G, Nelson BR, Jensen TC, Johnson TM (1994a) Increased expression of matrix metalloproteinase-3 (stromelysin-1) in cultured fibroblasts and basal cell carcinomas of nevoid basal cell carcinoma syndrome. Mol Carcinog 11: 29–33

Majmudar G, Nelson BR, Jensen TC, Voorhees JJ, Johnson TM (1994b) Increased expression of stromelysin-3 in basal cell carcinomas. Mol Carcinog 9: 17–23

Marcotte PA, Kozan IM, Dorwin SA, Ryan JM (1992) The matrix metalloproteinase pump-1 catalyzes formation of low molecular weight (pro)urokinase in cultures of normal human kidney cells. J Biol Chem 267: 13803–13806

Matrisian L, Glaichenhaus N, Gesnel M, Breathnach R (1985) Epidermal growth factor and oncogenes induce transcription of the same cellular mRNA in rat fibroblasts. EMBO J 4: 1435–1440

Matrisian LM (1992) The matrix-degrading metalloproteinases. Bioessays 14: 455–463

Matrisian LM, Hogan BLM (1990) Growth factor-regulated proteases and extracellular matrix remodeling during mammalian development. Curr Top Dev Biol 24: 219–259

McDonnell S, Navre M Jr, Coffey RJ, Matrisian LM (1991) Expression and localization of the matrix metalloproteinase PUMP-1 (MMP-7) in human gastric and colon carcinomas. Mol Carcinog 4: 527–533

Mignatti P, Robbins E, Rifkin DB (1986) Tumor invasion through the human amniotic membrane: requirement for a proteinase cascade. Cell 47: 487–498

Miller EJ, Harris ED Jr, Chung E, Finch JE Jr, McCroskery PA, Butler WT (1976) Cleavage of type II and III collagens with mammalian collagenase: site of cleavage and primary structure at the NH2-terminal portion of the smaller fragment released from both collagens. Biochemistry 15: 787–792

Miyazaki K, Hattori Y, Umenishi F, Yasumitsu H, Umeda M (1990) Purification and characterization of extracellular matrix-degrading metalloproteinase, matrin (pump-1), secreted from human rectal carcinoma cell line. Cancer Res 50: 7758–7764

Moll UM, Lane B, Zucker S, Suzuki K, Nagase H (1990) Localization of collagenase at the basal plasma membrane of a human pancreatic carcinoma cell line. Cancer Res 50: 6995–7002

Montgomery AMP, Mueller BM, Reisfeld RA, Taylor SM, DeClerck YA (1994) Effect of tissue inhibitor of the matrix metalloproteinases- 2 expression on the growth and spontaneous metastasis of a human melanoma cell line. Cancer Res 54: 5467–5473

Muller D, Quantin B, Gesnel M, Millon-Collard R, Abecassis J, Breathnach R (1988) The collagenase gene family in humans consists of at least four members. Biochem J 253: 187–192

Muller D, Breathnach R, Engelmann A, Millon R, Bronner G, Flesch H, Dumont P, Eber M, Abecassis J (1991) Expression of collagenase-related metalloproteinase genes in human lung or head and neck tumours. Int J Cancer 48: 550–556

Muller D, Wolf C, Abecassis J, Millon R, Engelmann A, Bronner G, Rouyer N, Rio MC, Eber M, Methlin G (1993) Increased stromelysin 3 gene expression is associated with increased local invasiveness in head and neck squamous cell carcinomas. Cancer Res 53: 165–169

Murphy G, Cockett MI, Ward RV, Docherty AJ (1991) Matric metalloproteinase degradation of elastin, type IV collagen and proteoglycan. A quantitative comparison of the activities of 95 kDa and 72 kDa gelatinases, stromelysins-1 and -2 and punctuated metalloproteinase (PUMP). Biochem J 277: 277–279

Murphy G, Willenbrock F, Ward RV, Cockett MI, Eaton D, Docherty AJP (1992) The C-terminal domain of 72 kDa gelatinase A is not required for catalysis, but is essential for membrane activation and modulates interactions with tissue inhibitors of metalloproteinases. Biochem J 283: 637–641

Murphy G, Segain JP, O'Shea M, Cockett M, Ioannou C, Lefebvre O, Chambon P, Basset P (1993) The 28-kDa N-terminal domain of mouse stromelysin-3 has the general properties of a weak metalloproteinase. J Biol Chem 268: 15435–15441

Murphy G, Nguyen Q, Cockett MI, Atkinson SJ, Allan JA, Knight CG, Willenbrock F, Docherty AJP (1994) Assessment of the role of the fibronectin-like domain of gelatinase A by analysis of a deletion mutant. J Biol Chem 269: 6632–6636

Nagase H, Enghild JJ, Suzuki K, Salvesen G (1990) Stepwise activation mechanisms of the precursor of matrix metalloproteinase 3 (stromelysin) by proteinases and (4-aminophenyl)mercuric acetate. Biochemistry 29: 5783–5789

Nakagawa H, Yagihashi S (1994) Expression of type IV collagen and its degrading enzymes in squamous cell carcinoma of lung. Jpn J Cancer Res 85: 934–938

Nakano A, Tani E, Miyazaki K, Furuyama J, Matsumoto T (1993) Expression of matrilysin and stromelysin in human glioma cells. Biochem Biophys Res Commun 192: 999–1003

Newell KJ, Witty JP, Rodgers WH, Matrisian LM (1994) Expression and localization of matrix-degrading metalloproteinases during colorectal tumorigenesis. Mol Carcinog 10: 199–206

Nicholson R, Murphy G, Breathnach R (1989) Human and rat malignant-tumor-associated mRNAs encode stromelysin-like metalloproteinases. Biochemistry 28: 5195–5203

Nicolson G (1991) Molecular mechanisms of cancer metastasis: tumor and host properties and the role of oncogenes and suppressor genes. Curr Opin Oncol 3: 75–92

Nicolson GL (1989) Metastatic tumor cell interactions with endothelium, basement membrane and tissue. Curr Opin Cell Biol 1: 1009–1019

Noël AC, Polette M, Lewalle J-M, Munaut C, Emonard HP, Birembaut P, Foidart J-M (1994) Coordinate enhancement of gelatinase A mRNA and activity levels in human fibroblasts in response to breast-adenocarcinoma cells. Int J Cancer 56: 331–336

Nuovo GJ, MacConnell PB, Simsir A, Valea F, French DL (1995) Correlation of the in situ detection of polymerase chain reaction-amplified metalloproteinase complementary DNAs and their inhibitors with prognosis in cervical carcinoma. Cancer Res 55: 267–275

Odekon L, Sato Y, Rifkin DB (1992) Urokinase-type plasminogen activator mediates basic fibroblast growth factor-induced bovine endothelial cell migration independent of its proteolytic activity. J Cell Physiol 150: 258–263

Ogata Y, Enghild JJ, Nagase H (1992) Matrix metalloproteinase 3 (stromelysin) activates the precursor for the human matrix metalloproteinase 9. J Biol Chem 267: 3581–3584

Okada Y, Konomi H, Yada T, Kimata K, Nagase H (1989) Degradation of type IX collagen by matrix metalloproteinase 3 (stromelysin) from human rheumatoid synovial cells. FEBS Lett 244: 473–476

Okada A, Bellocq J-P, Rouyer N, Chenard M-P, Rio M-C, Chambon P, Basset P (1995) Membrane-type matrix metalloproteinase (MT-MMP) gene is expressed in stromal cells of human colon, breast and head and neck carcinomas. Proc Natl Acad Sci USA 92: 2730–2734

Pajouh M, Nagle R, Breathnach R, Finch J, Brawer M, Bowden G (1991) Expression of metalloproteinase genes in human prostate cancer. J Cancer Res Clin Oncol 117: 144–150

Park AJ, Matrisian LM, Kells AF, Pearson R, Yuan Z, Navre M (1991) Mutational analysis of the transin (rat stromelysin) autoinhibitor region demonstrates a role for residues surrounding the "cysteine switch". J Biol Chem 266: 1584–1590

Parkins T (1994) Researchers test TIMP-2 as potential tumor terminator. J Natl Cancer Inst 86: 174–175

Pei D, Weiss SJ (1995) Furin-dependent intracellular activation of the human stromelysin-3 zymogen. Nature 375: 244–247

Pei D, Majmudar G, Weiss SJ (1994) Hydrolytic inactivation of a breast carcinoma cell-derived serpin by human stromelysin-3. J Biol Chem 269: 25849–25855

Polette M, Clavel C, Muller D, Abecassis J, Binninger I, Birembaut P (1991) Detection of mRNAs encoding collagenase I and stromelysin 2 in carcinomas of the head and neck by in situ hybridization. Invasion Metastasis 11: 76–83

Powell WC, Knox JD, Navre M, Grogan TM, Kittelson J, Nagle RB, Bowden GT (1993) Expression of the metalloproteinase matrilysin in DU-145 cells increases their invasive potential in severe combined immunodeficient mice. Cancer Res 53: 417–422

Powell WC, Domann R, Nagle RB, Matrisian LM, Bowden GT (1995) Castration induced expression of matrilysin in the rat ventral prostate. Prostate (submitted)

Quantin B, Murphy G, Breathnach R (1989) Pump-1 cDNA codes for a protein with characteristics similar to those of classical collagenase family members. Biochem J 28: 5327–5333

Rao JS, Steck PA, Mohanam S, Stetler-Stevenson WG, Liotta LA, Sawaya R (1993) Elevated levels of Mr 92,000 type IV collagenase in human brain tumors. Cancer Res 53: 2208–2211

Rochefort H, Capony F, Garcia M (1990) Cathepsin D: a protease involved in breast cancer metastasis. Cancer Metastasis Rev 9: 321–331

Rodgers WH, Osteen KG, Matrisian LM, Navre M, Giudice LC, Gorstein F (1993) Expression and localization of matrilysin, a matrix metalloproteinase, in human endometrium during the reproductive cycle. Am J Obstet Gynecol 168: 253–260

Rodgers WH, Matrisian LM, Giudice LC, Dsupin B, Cannon P, Svitek C, Gorstein F, Osteen KG (1994) Patterns of matrix metalloproteinase expression of cycling endometrium imply differential functions and regulation by steroid hormones. J Clin Invest 94: 946–953

Roebroek AJ, Creemers JW, Ayoubi TA, van de Ven WJ (1994) Furin-mediated proprotein processing activity: involvement of negatively charged amino acid residues in the substrate binding region. Biochimie 76: 210–216

Sanchez-Lopez R, Alexander CM, Behrendtsen O, Breathnach R, Werb Z (1993) Role of zinc-binding- and hemopexin domain-encoded sequences in the substrate specificity of collagenase and stromelysin-2 as revealed by chimeric proteins. J Biol Chem 268: 7238–7247

Santibañez JF, Martínez J (1993) Membrane-associated procollagenase of leukemic cells is activated by urokinase-type plasminogen activator. Leuk Res 17: 1057–1062

Sato H, Takino T, Okada Y, Cao J, Shinagawa A, Yamamoto E, Seiki M (1994) A matrix metalloproteinase expressed on the surface of invasive tumour cells. Nature 370: 61–65

Schmid TM, Mayne R, Jeffery JJ, Linsenmayer TF (1986) Type X collagen contains two cleavage sites for a vertabrate collagenase. J Biol Chem 261: 4184–4189

Schnierer S, Kleine T, Gote T, Hillemann A, Knäuper V, Tschesche H (1993) The recombinant catalytic domain of human neutrophil collagenase lacks type 1 collagen substrate specificity. Biochem Biophys Res Commun 191: 319–326

Schultz RM, Silberman S, Persky B, Bajkowski AS, Carmichael DF (1988) Inhibition by human recombinant tissue inhibitor of metalloproteinases of human amnion invasion and lung colonization by murine B16-F10 melanoma cells. Cancer Res 48: 5539–5545

Seltzer JL, Eisen AZ, Bauer EA, Morris NP, Glanville RW, Burgeson RE (1989) Cleavage of type VII collagen by interstitial collagenase and type IV collagenase (gelatinase) derived from human skin. J Biol Chem 264: 3822–3826

Senior RM, Griffin GL, Fliszar CJ, Shapiro SD, Goldberg GI, Welgus HG (1991) Human 92- and 72-kilodalton type IV collagenases are elastases. J Biol Chem 266: 7870–7875

Shapiro SD, Griffin GL, Gilbert DJ, Jenkins NA, Copeland NG, Welgus HG, Senior RM, Ley TJ (1992) Molecular cloning, chromosomal localization, and bacterial expression of a murine macrophage metalloelastase. J Biol Chem 267: 4664–4671

Shima I, Sasaguri Y, Kusukawa J, Yamana H, Fujita H, Kakegawa T, Morimatsu M (1992) Production of matrix metalloproteinase-2 and metalloproteinase-3 related to malignant behavior of esophageal carcinoma. A clinicopathologic study. Cancer 70: 2747–2753

Silbiger SM, Jacobsen VL, Cupples RL, Koski RA (1994) Cloning of cDNAs encoding human TIMP-3, a novel member of the tissue inhibitor of metalloproteinase family. Gene 141: 293–297

Sires UI, Griffin GL, Broekelmann TJ, Mecham RP, Murphy G, Chung AE, Welgus HG, Senior RM (1993) Degradation of entactin by matrix metalloproteinases. Susceptibility to matrilysin and identification of cleavage sites. J Biol Chem 268: 2069–2074

Siri A, Knauper V, Veirana N, Caocci F, Murphy G, Zardi L (1995) Different susceptability of small and large human tenascin-C isoforms to degradation by matrix metalloproteinases. J Biol Chem 270: 8650–8654

Springman E, Angleton E, Birkedal-Hansen H, Van Wart H (1990) Multiple modes of activation of latent human fibroblast collagenase: evidence for the role of a Cys 73 active-site zinc complex in latency and a 'cysteine switch' mechanism for activation. Proc Natl Acad Sci USA 87: 364–368

Suzuki K, Enghild JJ, Morodomi T, Salvesen G, Nagase H (1990) Mechanisms of activation of tissue procollagenase by matrix metalloproteinase 3 (stromelysin) Stepwise activation mechanisms of the precursor of matrix metalloproteinase 3 (stromelysin) by proteinases and (4-aminophenyl) mercuric acetate. Biochemistry 29: 5783–5789

Sympson CJ, Alexander CM, Chin JR, Werb Z, Bissell MJ (1993) Transgenic expression of stromelysin from the WAP promoter alters branching morphogenesis during mammary gland development and results in precocious expression of milk genes. Mol Biol Cell 4: 188a

Sympson CJ, Talhouk RS, Alexander CM, Chin JR, Clift SM, Bissell MJ, Werb Z (1994) Targeted expression of stromelysin-1 in mammary gland provides evidence for a role of proteinases in branching morphogenesis and the requirement for an intact basement membrane for tissue-specific gene expression. J Cell Biol 125: 681–693

Takino T, Sato H, Shinagawa A, Seiki M (1995) Identification of the second membrane-type matrix metalloproteinase (MT-MMP-2) gene from a human placenta cDNA library. J Biol Chem 270: 23013–23020

Talhouk RS, Bissell MJ, Werb Z (1992) Coordinated expression of extracellular matrix-degrading proteinases and their inhibitors regulates mammary epithelial function during involution. J Cell Biol 118: 1271–1282

Tenniswood MP, Guenette RS, Lakins J, Mooibroek M, Wong P, Welsh J-E (1992) Active cell death in hormone-dependent tissues. Cancer Metastasis Rev 11: 197–220

Thompson EW, Yu M, Bueno J, Jin L, Maiti SN, Palao-Marco FL, Pulyaeva H, Tamborlane JW, Tirgari R, Wapnir I, Azzam H (1994) Collagen induced MMP-2 activation in human breast cancer. Breast Cancer Res Treat 31: 357–370

Tremble PM, Damsky CH, Werb Z (1992) Fibronectin fragments, but not intact fibronectin, signalling through the fibronectin receptor induce metalloproteinase gene expression in fibroblasts. Matrix [Suppl] 1: 212–214

Urbanski SJ, Edwards DR, Maitland A, Leco KJ, Watson A, Kossakowska AE (1992) Expression of metalloproteinases and their inhibitors in primary pulmonary carcinomas. Br J Cancer 66: 1188–1194

Van Wart HE, Birkedal-Hansen H (1990) The cysteine switch: a principle of regulation of metalloproteinase activity with potential applicability to the entire matrix metalloproteinase gene family. Proc Natl Acad Sci USA 87: 5578–5582

Wagner SN, Ruhri C, Kunth K, Holecek BU, Goos M, Hofler H, Atkinson MJ (1992) Expression of stromelysin 3 in the stromal elements of human basal cell carcinoma. Diagn Mol Pathol 1: 200–205

Ward RV, Atkinson SJ, Reynolds JJ, Murphy G (1994) Cell surface-mediated activation of progelatinase A: demonstration of the involvement of the C-terminal domain of progelatinase A in cell surface binding and activation of progelatinase A by primary fibroblasts. Biochem J 304: 263–269

Welgus HG, Fliszar CJ, Seltzer JL, Schmid TM, Jeffrey JJ (1990) Differential susceptibility of type X collagen to cleavage by two mammalian interstitial collagenases and 72-kDa type IV collagenase. J Biol Chem 265: 13521–13527

Wilhelm SM, Collier IE, Marmer BL, Eisen AZ, Grant GA, Goldberg GI (1989) SV40-transformed human lung fibroblasts secrete a 92-kDa type IV collagenase which is identical to that secreted by normal human macrophages [published erratum appears in J Biol Chem 1990 Dec 25; 265(36): 22570]. J Biol Chem 264: 17213–17221

Will H, Hinzmann B (1995) cDNA sequence and mRNA tissue distribution of novel human matrix metalloproteinase with a potential transmembrane segment. Eur J Biochem 231: 602–608

Williamson RA, Marston FA, Angal S, Koklitis P, Panico M, Morris HR, Carne AF, Smith BJ, Harris TJ, Freedman RB (1990) Disulphide bond assignment in human tissue inhibitor of metalloproteinases (TIMP). Biochem J 268: 267–274

Windsor LJ, Bodden MK, Birkedal-Hansen B, Engler JA, Birkedal-Hansen H (1994) Mutational analysis of residues in and around the active site of human fibroblast-type collagenase. J Biol Chem 269: 26201–26207

Witty JP, McDonnell S, Newell KJ, Cannon P, Navre M, Tressler RJ, Matrisian LM (1994) Modulation of matrilysin levels in colon carcinoma cell lines affects tumorigenicity in vivo. Cancer Res 54: 4805–4812

Witty JP, Wright J, Matrisian LM (1995a) Mammary epithelial cell proliferation and premature alveolar development in virgin transgenic mice expressing the metalloproteinase stromelysin. Mol Biol Cell (Submitted)

Witty JP, Lempka T, Coffey RJ Jr, Matrisian LM (1995b) Decreased tumor formation in 7,12-dimethylbenzanthracene-treated stromelysin-1 transgenic mice is associated with alterations in mammary epithelial cell apoptosis. Cancer Res 55: 1401–1406

Woessner JF Jr (1991) Matrix metalloproteinases and their inhibitors in connective tissue remodeling. FASEB J 5: 2145–2154

Wolf C, Chenard MP, Durand de Grossouvre P, Bellocq JP, Chambon P, Basset P (1992) Breast-cancer-associated stromelysin-3 gene is expressed in basal cell carcinoma and during cutaneous wound healing. J Invest Dermatol 99: 870–872

Wright JH, McDonnell S, Portella G, Bowden GT, Balmain A, Matrisian LM (1994) A switch from stromal to tumor cell expression of stromelysin-1 mRNA associated with the conversion of squamous to spindle carcinomas during mouse skin tumor progression. Mol Carcinog 10: 207–215

Yamamoto H, Itoh F, Hinoda Y, Senota A, Yoshimoto M, Nakamura H, Imai K, Yachi A (1994) Expression of matrilysin mRNA in colorectal adenomas and its induction by truncated fibronectin. Biochem Biophys Res Commun 201: 657–664

Membrane Type-Matrix Metalloproteinase and Tumor Invasion

M. SEIKI

1 Introduction

Matrix metalloproteinases (MMPs) are frequently expressed in a variety of tumor tissues at higher levels than in normal tissues. However, overexpression itself does not mean immediate use of the enzymes for tumor invasion or for the remodeling of the extracellular matrix (ECM), processes associated with tumor growth. MMPs are produced and secreted as zymogens (pro-MMPs) that need proteolytic activation for the enzymes to function (WOESSNER 1991; BIRKEDAL et al. 1993). Thus, invasive tumor cells should have devices to activate proMMPs to degrade the ECM at the periphery.

Serine proteases, such as plasmin, elastase and trypsin, can cleave propeptide domains of proMMPs at basic amino acid residues and consequently induce autocatalytic activation of proMMP-1, proMMP-3 and pro-MMP9 (HE et al. 1989; NAGASE et al. 1990). Some of the activated MMPs can further activate other proMMPs. For example, MMP-3 activates proMMP-1

Department of Molecular Virology and Oncology, Cancer Research Institute, Kanazawa University, 13-1 Takara-machi, Kanazawa 920, Japan

and proMMP-9 (He et al. 1989; Suzuki et al. 1990). Thus, the serine proteases are expected to be the initiator for a complex array of activation cascades of proMMPs in tissue.

Activation of proMMP-2, which degrades type IV collagen in the basement membrane and therefore is believed to play an essential role in tumor invasion and ultimately in metastasis (Liotta et al. 1991), is distinct from other pro-MMPs (Kleiner and Stetler 1993). It is resistant to serine proteases (Nagase et al. 1992; Okada et al. 1990) but activated in a cell-mediated manner (Ward et al. 1991; Azzam and Thompson 1992; Brown et al. 1993a; Strongin et al. 1993). Activation of proMMP-2 can be induced by cells (such as fibroblasts, human fibrosarcoma HT1080 and breast carcinoma cell lines) treated with phorbol ester (TPA) or concanavalin A (Brown et al. 1990; Ward et al. 1991; Azzam et al. 1993; Strongin et al. 1993) or cultivated in collagen gel (Azzam and Thompson 1992). Since exogenously added proMMP-2 can be activated by cells and isolated plasma membrane retained this activity, the proMMP activator was thought to be on the cell surface. The activator could be extracted by detergent from the plasma membrane fraction of cells (Azzam et al. 1993; Brown et al. 1993a; Strongin et al. 1993), but was resistant to further purification. Cell-mediated activation was not inhibited by serine protease inhibitors (Atkinson et al. 1992), but was fully abrogated by tissue inhibitor of metalloproteinase (TIMP)-2 and poorly by TIMP-1 (Ward et al. 1991; Strongin et al. 1993; Willenbrock et al. 1993). Strongin et al. (1993) have demonstrated that cell-mediated activation of proMMP-2 occurs in two steps. Cleavage in the propeptide domain at Asn^{37} -Leu generated a 64 kDa intermediate form and the NH_2-terminal of the activated 62 kDa form was Tyr^{81}. Thus the cleavage between the Asn-Leu is thought to be mediated by the cell surface activator followed by autocatalytic activation.

Activated MMP-2 can be detected frequently in surgical specimens from the lung (Brown et al. 1993b), breast (Brown et al. 1993C), and gastric and colon carcinomas (Yamagata et al. 1991) but not in normal tissue counterparts. The activation rate of MMP-2 in the tissue correlated well to local invasiveness and lymph node metastasis (Brown et al. 1993b,c). Thus, the activator of proMMP-2 is thought to be expressed in malignant tumor tissues. The human breast carcinoma cell line MDA-MB231, which activates proMMP-2, is more metastatic in nude mice than MCF-7, which cannot activate the zymogen (Azzam et al. 1993).

The COOH-terminal domain of MMP-2, which is important for complex formation with TIMP-2, is not required for its enzyme activity (Murphy et al. 1992). However deletion of this domain abolished cell-mediated activation of proMMP-2 (Murphy et al. 1992; Ward et al. 1994). Overexpression of the COOH-terminal fragment inhibited proMMP-2 activation (Strongin et al. 1993; Ward et al. 1994). Thus, the COOH-terminal domain of MMP-2 has an essential function that is activated by the cell surface activator.

Cell-mediated activation of proMMP-2 is thought to be facilitated by specific association of the enzyme with the cell surface (Ward et al. 1994). A high-affinity cell surface receptor for proMMP-2 has been demonstrated on some human breast carcinoma cell lines (EMONARD et al. 1992), on HT1080 cells stimulated with TPA (STRONGIN et al. 1995) and fibroblasts treated with concanavalin A (WARD et al. 1994). In tumor tissue, MMP-2 was frequently immunolocalized onto carcinoma cells (LIOTTA et al. 1991; STETLER et al. 1993), although it is mainly expressed in fibroblasts in the tumor stroma, as determined by in situ hybridization (POLETTE et al. 1993; POULSOM et al. 1993; PYKE et al. 1993; WOLF et al. 1993). This observation suggests that proMMP-2 produced by fibroblasts diffuses to adjacent tumor cells, binds them and is activated there. MONSKEY et al. (1993) have reported that transformed chicken fibroblasts activate proMMP-2 and that the activated enzyme was found on the invadopodia, where ECM degradation actively takes place. Thus, activated MMP-2 on the tumor cell surface is expected to play a crucial role in invasion by degrading type IV collagen in the basement membrane beneath the tumor cells. The activator might be a new member of the MMP family, since the activation process is sensitive to chelating agents and TIMP-2.

2 Identification of a New Member of the Matrix Metalloproteinase Family Having a Transmenbrane Domain

A cDNA fragment that shows sequence homology to MMP genes was amplified from human placenta RNA by reverse transcriptase-polymerase chain reaction (RT-PCR) using degenerate oligonucleotide primers corresponding to the conserved amino acid sequences among MMPs (TAKINO et al. 1995). Amino acid sequence homology of the putative MMP fragment to other MMP family members was as follows: interstitial collagenase/MMP-1 (50%), 72 kDa-type IV collagenase/gelatinase A/MMP-2 (48%), stromelysin-1/MMP-3 (52%), matrilysin/MMP-7 (44%), 92 kDa type IV collagenase/gelatinase B/MMP-9 (48%) and stromelysin-3/MMP-11 (43%), human macrophage metalloelastase (HME)/MMP-12 (52%). Isolated cDNA obtained from a placental cDNA library using the putative MMP fragment contained a long open reading frame coding for a 582 amino acid protein. The amino acid sequence aligned well with the reported MMPs, keeping the typical domain structure of members of the MMP family. However, the new member has three unique insertions that are not found in other family members (Fig. 1). These are: (1) 11 amino acids between the propeptide and catalytic domains (Ins-1); (2) eight amino acids in the catalytic domain (Ins-2); and (3) 66 amino acids at the COOH-terminal of the hemopexin-like domain (Ins-3). In the third insertion at the COOH-terminal, there is a transmembrane (TM)-like structure with a stretch of 24 hydrophobic amino acid residues.

Monoclonal antibodies against synthetic peptides reacted with a 63 kDa protein (zymogen form) in transfected cells by immunoprecipitation and the product was localized onto the cell surface by immunostaining and

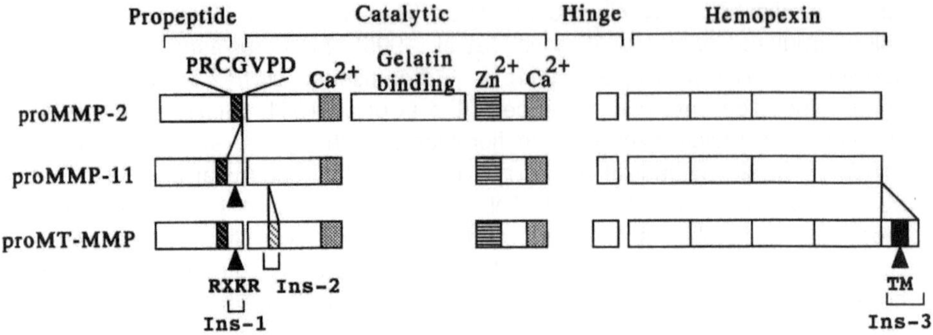

Fig. 1. Domain structure of membrane-type metalloproteinase (MT-MMP). The domain structure of MT-MMP was illustrated in relation to that of proMMP-2 and proMMP-11 (Stromelysin 3). The domains are as follows from the N-terminus: *propeptide*, propeptide domain that contains the PRCGVPD sequence that is essential to keep the enzyme in an inactive state; *Catalytic*, catalytic domain contains calcium and zinc-binding sites (Ca^{2+} and zn^{2+}, respectively); *hinge*, Hinge domain is rich in prolins; *Hemopexin*, hemopexin-like domain at the C-terminus. *Ins-1 to -3* are the insertions specific to MT-MMP compared with other reported MMPs. However, Ins-1 exists in proMMP-11 at the same position with the conserved *RXKR* sequence. In the Ins-3, there is a transmembrane domain (*TM*) composed of hydrophobic amino acids

antibody-binding assay (CAO et al. 1995; SATO et al. 1994). The COOH-terminal portion of the protein was indeed a membrane linker, as demonstrated by making a fusion protein with TIMP-1 (CAO et al. 1995). TIMP-1 is a naturally soluble protein and not retained on the cell surface. By fusing TIMP-1 with the COOH-terminal portion (TM domain) of the 63 kDa protein, however, the localization of TIMP-1 was changed onto the cell surface without being released into the media. Therefore, the 63 kDa MMP having a TM domain was referred to as membrane type-MMP (MT-MMP). MT-MMP expressed in transiently transfected cells was not detectable in the culture media, although the released protein may be degraded rapidly.

3 Membrane Type-Matrix Metalloproteinase Is an Activator of proMMP-2 on the Cell Surface

Cell surface localization of MT-MMP fits the profile of a hypothetical activator of proMMP-2. This possibility was examined by transfecting the MT-MMP plasmid into human fibrosarcoma HT1080 and mouse fibroblast NIH3T3 cell lines, both of which secrete proMMP-2 and proMMP-9. Expression of MT-MMP in the transfected cells induced processing of proMMP-2, similar to cells treated with concanavalin A, via a 64 kDa intermediate form (SATO et al. 1994). The processing was specific to proMMP-2 and proMMP-9 was not affected at all. Since MT-MMP increases the gelatinolytic activity in the media, processing of proMMP-2 is thought to generate active MMP-2. Activation was inhibited efficiently by TIMP-2 but only partially by TIMP-1. The activity causing processing of proMMP-2 was colocalized with MT-MMP in the plasma membrane fraction of transfected cells (CAO et al. 1995; SATO et al. 1994).

Membrane anchoring of MT-MMP through the TM domain was essential for activity but did not show sequence specificity. Deletion of the TM domain (COOH-terminal 47 amino acids) abrogated activation function of proMMP-2, but a fusion of the mutant with a heterologous TM domain of interleukin-2 receptor (α-chain) restored the activity (CAO et al. 1995). The deletion mutant was still detected on the cell surface while a substantial amount of the product was secreted into the media. Thus, in spite of localization on the cell surface, the mutant could not induce proMMP-2 activation without membrane-anchoring through the TM domain. This domain of MT-MMP is thought to play an important role as a determinant of the appropriate conformation, orientation, or localization of the enzyme on the cell surface.

TIMP-2 has dual effects on activation of proMMP-2 (STRONGIN et al. 1993). It enhances activation of proMMP-2 at a low concentration range and, by contrast, inhibits this same process at higher concentrations. This result indicates that TIMP-2 plays an important regulatory role in the activation of proMMP-2 on the cell surface (KLEINER and STETLER 1993; STRONGIN et al. 1993). STRONGIN et al. (1995) purified TIMP-2 and an activated form of MT-MMP as a trimolecular complex from a lubrol extract of HT1080 cell membranes using an affinity column with the COOH-terminal fragment of MMP-2. TIMP-2 may act as a bridge between proMMP-2 and cell surface MT-MMP, since the COOH-terminal domain of MMP-2 can bind to the COOH-terminal portion of TIMP-2 and TIMP-2 binds to the cell surface activator. The COOH-terminal domain of proMMP-2 was shown to be essential for cell-mediated activation (MURPHY et al. 1992; STRONGIN et al. 1993). Thus, the TIMP-2/MT-MMP complex on the cell surface may act as both a receptor and an activator for proMMP-2 (STRONGIN et al. 1995).

However, there have been conflicting reports concerning the activation of proMMP-2 complexed with TIMP-2. STRONGIN et al. (1993) demonstrated that the proMMP-2/TIMP-2 compled was not activated by the membrane fraction of HT1080, but others observed that the complex was activated along with free proMMP-2 (BROWN et al. 1993a; KLEINER and STETLER 1993). The exact mechanism and cascade of the interplay between proMMP-2, TIMP-2 and MT-MMP on the cell surface remain to be elucidated.

4 Membrane Type-Matrix Metalloproteinase Enhances the Invasive Potential of Tumor Cells

Activated MMP-2 degrades extracellular macromolecules in the basement membrane, such as type IV collagen, laminin and fibronectin. Thus, the expression of MT-MMP in cells producing proMMP-2 is expected to enhance their invasive potential in reconstituted basement membrane (Matrigel). Indeed, this was the case when MT-MMP was expressed in HT1080 and NIH3T3 cells by transient transfection of the cDNA and the cells were seeded onto Matrigel-coated filters in modified Boyden chambers (SATO et al. 1994). MT-MMP expression increased the number of invaded cells by two- to three fold without affecting cell motility. The invasion was inhibited by recombinant TIMP-2.

5 Membrane Type-Matrix Metalloproteinase Expression in Tumors

The appearance of the activated form of gelatinase A in tumor tissue has been reported to be associated with advanced stages of breast (BROWN et al. 1993c; DAVIES et al. 1993) and lung carcinomas (BROWN et al. 1993b). MT-MMP was immunolocalized in/on carcinoma cells in lung (SATO et al. 1994), colon, head and neck, and gastric tumors (NOMURA et al. 1995). Fibroblasts adjacent to the carcinoma cells and endothelial cells in the tissue were occasionally positive in the advanced stages of tumors. In situ hybridization of colon and gastric carcinoma tissues also supported the finding that MT-MMP is expressed in carcinoma cells and adjacent fibroblasts while gelatinase A was expressed exclusively in stroma cells (OHTANI et al., manuscript in preparation). However, OKADA et al. (1995) reported that MT-MMP is expressed in stroma cells but not in carcinoma cells of breast carcinoma tissues, as determined by in situ hybridization. Thus, there is a discrepancy concerning MT-MMP expression in carcinoma cells. Although there is no definite explanation for the difference, a possibility is as follows: since RNA is labile and had different stabilities depending on cell types in tissue, RNA signals in carcinoma cells may be more sensitive to differences in sample preparations.

MT-MMP mRNA expression correlated well with the activation rate of gelatinase A in lung carcinoma tissues analyzed by gelatin zymography (SATO et al. 1994), and the activation rate showed a positive correlation to vascular invasion and lymph node metastasis (BROWN et al. 1993b; TOKURAKU et al. 1995). Thus, MT-MMP in tumor tissue is thought to be responsible for activation of proMMP-2 at the invasive carcinoma cell nest followed by degradation of the basement membrane beneath the cells (Fig. 2).

6 Regulation of Membrane Type-Matrix Metalloproteinase Expression

Expression of MT-MMP in tumor tissue is induced at the RNA level and is not found in normal epithelial cells and fibroblasts. Although details of the regulation of MT-MMP gene expression remain unknown, treatment of fibroblasts and human breast carcinoma (HBC) cell lines with TPA or concanavalin A is

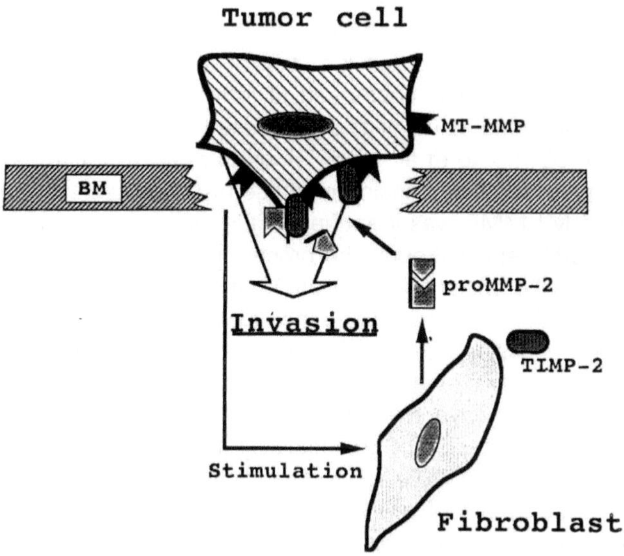

Fig. 2. Schematic illustration of the basement membrane (*BM*) invasion by the tumor cells expressing membrane-type metalloproteinase (*MT-MMP*). The role of MT-MMP and MMP-2 in the basement membrane invasion by tumor cells was illustrated hypothetically. A substantial amount of proMMP-2 is thought to be embedded already into the ECM and the basement membrane during formation. MT-MMP expressed on the tumor cells initially utilizes the pre-existing proMMP-2 at the periphery of the cells, and the degradation of the basement membrane beneath the tumor cells starts to occur. The degradation products of the basement membrane (such as the fragments from laminin and fibronectin, and the released TGF-β) stimulate the vicinity of the fibroblasts to produce more proMMP-2. Thus, this is an example of tumor-stroma cooperation for the basement membrane invasion

known to cause proMMP-2 activation (AZZAM et al. 1993; BROWN et al. 1993a; STRONGIN et al. 1993; WARD et al. 1991). Activation was found to be associated with increased MT-MMP mRNA in cells (YU et al. 1995). Thus, MT-MMP is a device for proMMP-2 activation in these cells, and gene expression is thought to be regulated by extracelluar signals mediated by cell surface receptors. Mutations of oncogenes and tumor suppresser genes in carcinoma cells may affect the signaling pathway to the MT-MMP gene and consequently up-regulate the expression. Activation of proMMP-2 also can be seen in fibroblasts cultivated in collagen gel; treatment of the cells with anti-β1 integrin antibody abolished the activation (SELTZER et al. 1994). Although MT-MMP expression in cells has not been studied yet, it may be modulated by cell-ECM interaction through integrins.

7 Posttranslational Regulation

Membrane type-matrix metalloproteinase has a domain structure typical of members of the MMP family, i.e., propetide domain that can be cleaved resulting in enzyme activation. However, MT-MMP also has a unique insertion of 11 amino acids between the propeptide and catalytic domains (Fig. 1). Only stromelysin-3 in the MMP family has such an insertion at the same position. Between the two, RXKR is interestingly conserved at immediately upstream of the reported NH_2-terminal amino acids of the processed enzymes (Basset et al. 1990; Murphy et al. 1993; Takino et al. 1995; Strongin et al. 1995). Thus, RXKR is thought to be the signal for processing of these pro-enzymes.

The processed form of MT-MMP was identified in the plasma membrane extract of TPA-treated HT1080 cells, although the processing enzyme has not been identified yet. Subtilisin-like enzymes (Furin etc.) may be responsible for the processing since RXKR is the consensus sequence for these enzymes and the sequence exists in the processing sites of many growth factors, hormones and viral envelope proteins (Hosaka et al. 1991). Serine proteases such as urokinase and plasmin may also cleave this site (Strongin et al. 1995). Keski et al. (1992) reported that overexpression of urokinase by transfection induced activation of proMMP-2. Saiki et al. (1993) found that an inhibitor of aminopeptidase N (CD13) blocks activation of proMMP-2 by cells and suppresses Matrigel invasion. These proteases may affect the processing of proMT-MMP directly or indirectly.

8 Physiological Roles

Sufficient information is not available to discuss the physiological roles of MT-MMP. Like other MMPs, MT-MMP is thought to be required for ECM remodeling, as suggested by expression in fibroblasts in the vicinity of tumor cells or cells treated with TPA or concanavalin A. However, MT-MMP may be more important in degrading pericellular ECM components, coupled with some particular cell functions such as cell locomotion to traverse the tissue or proliferation. Studies on these possibilities are now underway in may laboratories including ours.

References

Atkinson SJ, Ward RV, Reynolds JJ, Murphy G (1992) Cell-mediated degradation of type IV collagen and gelatin films is dependent on the activation of matrix metalloproteinases. Biochem J 288: 605–611

Azzam HS, Thompson EW (1992) Collagen-induced activation of the M(r) 72,000 types IV collagenase in normal and malignant human fibroblastoid cells. Cancer Res 52: 4540–4504

Azzam HS, Arand G, Lippman ME, Thompson EW (1993) Association of MMP-2 activation potential with metastatic progression in human breast cancer cell lines independent of MMP-2 activation. J Natl Cancer Inst 85: 1758–1764

Basset P, Bellocq JP, Wolf C, Stoll I, Hutin P, Limacher JM, Podhajcer OL, Chenard MP, Rio MC, chambon P (1990) A novel metalloproteinase gene specifically expressed in stromal cells of breast carcinomas. Nature 348: 699–704

Birkedal HH, Moore WG, Bodden MK, Windsor LJ, Birkedal HB, DeCarlo A, Engler JA (1993) Matrix metalloproteinases: a review. Crit Rev Oral Biol Med 4: 197–250

Brown PD, Levy AT, Margulies IM, Liotta LA, Stetler SW (1990) Independent expression and cellular processing of Mr 72,000 type IV collagenase and interstitial collagenase in human tumorigenic cell lines. Cancer Res 50: 6184–6191

Brown PD, Bloxidge RE, Stuart NSA, Gatter KC, Carmichael J (1993a) Association between expression of activated 72-Kilodaloton gelatinase and tumor spread in non-small-cell lung carcinoma. J Natl Cancer Inst 85: 574–578

Brown PD, Bloxidge RE, Anderson E, Howell A (1993b) Expression of activated gelatinase in human invasive breast carcinoma. Clin Exp Metastasis 11: 183–189

Brown PD, Kleiner DE, Unsworth EJ, Stetler SW (1993c) Cellular activation of the 72 kDa type IV procollagenase/TIMP-2 complex. Kidney Int 43: 163–170

Cao J, Sato H, Takino T, Seiki M (1995) The C-terminal region of membrane type matrix metalloproteinase is a functional transmembrane domain required for pro-gelatinase A activation. J Biol Chem 270: 801–805

Davies B, Miles DW, Happerfield LC, Naylor MS, Bobrow LG, Rubens RD, Balkwill FR (1993) Activity of type IV collagenases in benign and malignant breast disease. Br J Cancer 67: 1126–1131

Emonard HP, Remacle AG, Noel AC, Grimaud JA, Stetler SW, Foidart JM (1992) Tumor cell surface-associated binding site for the M(r) 72,000 type IV collagenase. Cancer Res 52: 5845–5848

He CS, Wilhelm SM, Pentland AP, Marmer BL, Grant GA, Eisen AZ, Goldberg GI (1989) Tissue cooperation in a proteolytic cascade activating human interstitial collagenase. Proc Natl Acad Sci USA 86: 2632–2636

Hosaka M, Nagahama M, Kim WS, Watnabe T, Hatsuzawa K, Ikemizu J, Murakami K, Nakayama K (1991) Arg-X-Lys/Arg-Arg motif as a signal for precursor cleavage catalyzed by furin within the constitutive secretory pathway. J Biol Chem 266: 12127–12130

Kesei OJ, Lohi J, Tuuttila A, Tryggvason K, Vartio T (1992) Proteolytic processing of the 72,000-Da type IV collagenase by urokinase plasminogen activator. Exp Cell Res 202: 471–476

Kleiner DJ, Stetler SW (1993) Structural biochemistry and activation of matrix metalloproteases. Curr Opin Cell Biol 5: 891–897

Liotta LA, Steeg PS, Stetler-Stevenson WG (1991) Cancer metastasis and angiogenesis: an imbalance of positive and negative regulation. Cell 64: 327–336

Monsky WL, Kelly T, Lin CY, Yeh Y, Stetler-Stevenson WG, Mueller SC, Chen WT (1993) Binding and localization of Mr 72,000 matrix metalloproteinase at cell surface invadopondia. Cancer Res 53: 3159–3164

Murphy G, Willenbrock F, Ward RV, Cockett MI, Eaton D, Docherty AJ (1992) The C-terminal domain of 72 kDa gelatinase A is not required for catalysis, but is essential for membrane activation and modulates interactions with tissue inhibitors of metalloproteinases. Biochem J 283: 637–641

Murphy G, Segain JP, O'Shea M, Cockett M, Ioannou C, Lefebvre O, Chambon P, Basset P (1993) The 28-kDa N-terminal domain of mouse stromelysin-3 has the general properties of a weak metalloproteinase. J Biol Chem 268: 15435–15441

Nagase H, Enghild JJ, Suzuki K, Salvesan G (1990) Stepwise activation mechanisms of the precursor of matrix metalloproteinase 3 (stromelysin) by proteinases and (4-aminophenyl) mercuric acetate. Biochemistry 29: 5783–5789

Nagase H, Suzuki K, Morodomi T, Enghild JJ, Salvesan G (1992) Activation mechanisms of the precursors of matrix metalloproteinases 1,2 and 3. Matrix [suppl] 1: 237–244

Nomura H, Sato H, Seiki M, Mai M, Okada Y (1995) Expression of membrane-type matrix metalloproteinase in human gastric carcinomas. Cancer Res 55: 3263–3266

Okada A, Bellocq KP, Rouyer N, Chenard MP, Rio MC, Chambon P, Basset P (1995) Membrane-type matrix metalloproteinase (MT-MMP) gene is expressed in stromal cells of human colon, breast, and head and neck carcinomas. Proc Natl Acad Sci USA (in press)

Okada Y, Morodomi T, Enghild JJ, Suzuki K, Yasui A, Nakanishi I, Salvesen G, Nagase H (1990) Matrix metalloproteinase 2 from human rheumatoid synovial fibroblasts. Purification and activation of the precursor and enzymic properties. Eur J Biochem 194: 721–730

Polette M, Clavel C, Cockett M, Girod dBS, Murphy G, Birembaut P (1993) Detection and localization of mRNAs encoding matrix metalloproteinases and their tissue inhibitor in human breast pathology. Invasion Metastasis 13: 31–37

Poulsom R, Hanby AM, Pignatelli M, Jeffery RE, Longcroft JM, Rogers L, Stamp GW (1993) Expression of gelatinase A and TIMP-2 mRNAs in desmoplastic fibroblasts in both mammary carcinomas and basal cell carcinomas of the skin. J Clin Pathol 46: 429–436

Pyke C, Ralfkiaer E, Tryggvason K, Dano K (1993) Messenger RNA for two type IV collagenases is located in stromal cells in human colon cancer. Am J Pathol 142: 359–365
Saiki I, Fujii H, Yoneda J, Abe F, Nakajima M, Tsuruo T, Azuma I (1993) Role of aminopeptidase N (CD13) in tumor-cell invasion and extracellular matrix degradation. Int J Cancer 54: 137–143
Sato H, Takino T, Okada Y, Cao J, Shinagawa A, Yamamoto E, Seiki M (1994) A matrix metalloproteinase expressed on the surface of invasive tumor cells. Nature 370: 61–65
Seltzer JL, Lee AY, Akers KT, Sudbeck B, Southon EA, Wayner EA, Eisen AZ (1994) Activation of 72-kDa type IV collagenase/gelatinase by normal fibroblasts in collagen lattices is mediated by integrin receptors but is not related to lattice contraction. Exp Cell Res 213: 356–374
Stetler SW, Aznavoorian S, Liotta LA (1993) Tumor cell interactions with the extracellular matrix during invasion and metastasis. Annu Rev Cell Biol 9: 541–573
Strongin AY, Marmer BL, Grant GA, Goldberg GI (1993) Plasma membrane-dependent activation of the 72-kDa type IV collagenase is prevented by complex formation with TIMP-2. J Biol Chem 268: 14033–14039
Strongin AY, Collier I, Bannikov G, Marmer BL, Grant GA, Goldberg GI (1995) Mechanism of cell surface activation of 72 kDa type IV collagenase: isolation of the activated form of the membrane metalloproteinase. J Biol Chem (in press)
Suzuki K, Enghild JJ, Morodomi T, Salvesan G, Nagase H (1990) Mechanisms of activation of tissue procollagenase by matrix metalloproteinase 3 (stromelysin). Biochemistry 29: 10261–10270
Takino T, Sato H, Yamamoto E, Seiki M (1995) Cloning of a gene potentially encoding a novel matrix metalloproteinase having a trans-membrane domain at the C-terminus. Gene 155: 293–298
Tokuraku M, Sato H, Murakami S, Okada Y, Watanabe Y, Seiki M (1995) Activation of the precursor of gelatinase A/72-kDa type IV collagenase/MMP-2 in lung carcinomas correlates with the expression of membrane-type matrix metalloproteinase (MT-MMP) and with lymph node metastasis. Int J Cancer (Pred Oncol) 64: 355–359
Ward RV, Atkinson SJ, Slocombe PM, Docherty AJ, Reynolds JJ, Murphy G (1991) Tissue inhibitor of metalloproteinases-2 inhibits the activation of 72 kDa progelatinase by fibroblast membranes. Biochim Biophys Acta 1079: 242–246
Ward RV, Atkinson SJ, Reynolds JJ, Murphy G (1994) Cell surface-mediated activation of progelatinase A: demonstration of the involvement of the C-terminal domain of progelatinase A in cell surface binding and activation of progelatinase A by primary fibroblasts. Biochem J 304: 263–269
Willenbrock F, Crabbe T, Slocombe PM, Sutton CW, Docherty AJ, Cockett MI, O'Shea M, Brocklehurst K, Phillips IR, Murphy G (1993) The activity of the tissue inhibitors of metalloproteinases is regulated by C-terminal domain interactions: a kinetic analysis of the inhibition of gelatinase A. Biochemistry 32: 4330–4337
Woessner JFJ (1991) Matrix metalloproteinases and their inhibitors in connective tissue remodeling. FASEB J 5: 2145–2154
Wolf C, Rouyer N, Lutz Y, Adida C, Loriot M, Bellocq JP, Chambon P, Basset P (1993) Stromelysin 3 belongs to a subgroup of proteinases expressed in breast carcinoma fibroblastic cells and possibly implicated in tumor progression. Proc Natl Acad Sci USA 90: 1843–1847
Yamagata S, Yoshii Y, Suh JG, Tanaka R, Shimizu S (1991) Occurrence of an active form of gelatinase in human gastric and colorectal carcinoma tissues. Cancer Lett 59: 51–55
Yu M, Sato H, Seiki M, Thompson EW (1995) Complex regulation of membrane-type metalloproteinase expression and matrix metalloproteinase-2 activation by concanavalin A in MDA-MB-231 human breast cancer cells. Cancer Res 55: 3272–3277

Plasminogen Activators and Angiogenesis

P. Mignatti[1] and D.B. Rifkin[2]

1 Introduction

The formation of capillaries from preexisting blood vessels (angiogenesis) occurs in a variety of normal and pathological conditions, including organ development, would healing and tumor growth. Angiogenesis requires a strict temporal modulation of opposing cell functions: cell proliferation and migration, and extracellular matrix (ECM) degradation in the initial steps; arrest of cell proliferation and migration, ECM deposition and morphogenesis in the final steps. In the initial steps of angiogenesis, microvascular endothelial cells behave similarly to invasive tumor cells, as they cross basement membranes and interstitial stroma and invade adjacent tissues. However, unlike tumor invasion, the invasive process that occurs during angiogenesis is spatially and temporally restricted. As for tumor cell invasion, proteinases are of fundamental importance for the degradation of the perivascular ECM and endothelial cell invasion into the tissue to be vascularized. A number of studies have shown

[1]Dipartimento di Genetica e Microbiologia, Università di Pavia, Via Abbiategrasso 207, 27100 Pavia, Italy
[2]Department of Cell Biology and Kaplan Cancer Center, and the Raymond and Beverly Sackler Foundation Laboratory, New York University Medical Center, 550 First Avenue, New York, NY 10016, USA

that, among the proteinases involved in angiogenesis, components of the plasminogen activator (PA)-plasmin system play a central role in endothelial cell migration and invasion.

2 General Features of the Plasminogen Activator-Plasmin System

The urokinase-type (uPA) and tissue-type (tPA) plasminogen activators are serine proteinases that convert the zymogen plasminogen, a ubiquitous plasma protein, to plasmin, a broad-spectrum enzyme. Plasmin degrades several ECM components, including fibronectin, laminin and the protein core of proteoglycans (WERB et al. 1980). It does not degrade elastin and collagens but can degrade gelatins, the partially degraded or denatured forms of the collagens.

The most relevant feature of the PA-plasmin system is the amplification achieved by the conversion of plasminogen to plasmin. Because of the high concentration of plasminogen in virtually all tissues, the production of small amounts of PA can result in high local concentrations of plasmin. Besides degrading a variety of ECM components, plasmin also activates several matrix-degrading metalloproteinases (MMPs), including interstitial collagenase, stromelysin and the type IV collagenases (WERB et al. 1977; MOSCATELLI and RIFKIN 1988; MATRISIAN 1990; MURPHY et al. 1992; KLEINER and STETLER-STEVENSON 1993; MIGNATTI and RIFKIN 1993), as well as latent elastase (CHAPMAN and STONE 1984). Thus, the production of even small amounts of PA can result in the generation of high local concentrations of active serine- and metalloproteinases with differing substrate specificity. This cascade can be blocked at different levels by specific inhibitors. The blockade of plasminogen activation by specific inhibitors of PAs (see below) will inhibit all subsequent events; the blockade of plasmin formation will result in the repression of MMP and elastase activation. In contrast, inhibition of MMPs by their specific tissue inhibitors (TIMPs) will block these enzymes but leave plasmin unaffected.

3 Endothelial Cell Expression of Components of the Plasminogen Activator-Plasmin System

Vascular endothelial cells produce PAs both in vitro and in vivo. Immunohistochemical surveys of tissue sections have shown that human endothelial cells produce tPA but no detectable uPA (KRISTENSEN et al. 1984). Primary cultures of human endothelial cells secrete tPA and at later passages both tPA and uPA, or only uPA (LAUG et al. 1980; BYKOWSKA et al. 1982; LEVIN and LOSKUTOFF 1982; PHILIPS et al. 1984). Long-term or immortalized cultures of endothelial cells express primarily uPA (TSUBOI et al. 1990; PEVERALI et al. 1994). Although the PA produced by vascular endothelial cells in vivo is tPA, the endothelial cells of growing capillaries appear to express uPA (BACHARACH et al. 1992).

Endothelial cell PA expression is modulated at the transcriptional level by a number of agents, including cytokines and growth factors that also induce angiogenesis (see below). In addition, uPA activity is regulated extracellularly by several mechanisms, including control of proenzyme activation, interaction with binding sites on the cell membrane and ECM, and inhibition by specific tissue inhibitors.

Endothelial cells express high levels of PA inhibitors both in vivo and in vitro. The major inhibitor produced by endothelial cells is the type 1 PA inhibitor (PAI-1), a 45 kDa protein also present in platelets and plasma (LOSKUTOFF and EDGINGTON 1977; HEKMAN and LOSKUTOFF 1985). Other inhibitors that may contribute to modulating endothelial cell PA activity include the type 2 PA inhibitor (PAI-2), a 46.6 kDa protein expressed most notably by cells of the monocyte-macrophage lineage (KAWANO et al. 1970; ASTEDT et al. 1985; KRUITHOF et al. 1986), and protease nexin I (PN I), a 45 kDa protein originally purified from cultured fibroblasts but also produced by several other cell types (BAKER et al. 1980; EATON et al. 1984). A fourth, less characterized inhibitor, called PAI-3, isolated from human urine, is identical to the protein C inactivator but is considerably less efficient in inhibiting PAs than the other inhibitors (HEEB et al. 1987). Both PAI-1 and PAI-2 bind the active two-chain (tc) forms of uPA and tPA, rapidly forming 1:1 molar complexes, but have a poor affinity for the single-chain (sc) form of either PA (see below). The association rate constant of PAI-1 for tc-uPA and tc-tPA ($k_{ass}=10^7$–10^8 M^{-1} s^{-1}) is higher than that of PAI-2 ($k_{ass}=10^5$–10^6 M^{-1} s^{-1}). PN I is less specific for PA than PAI-1 and PAI-2. It inhibits tc-uPA effectively ($k_{ass}=10^5$ M^{-1} s^{-1}) but has virtually no effect on sc-uPA, sc- and tc-tPA. In contrast, PN I is an extremely rapid inhibitor of thrombin and also inactivates trypsin and plasmin (SAKSELA and RIFKIN 1988). As is the case for PAs, endothelial cell expression of PAI-1 is also modulated by a variety of cytokines, some of which have angiogenic activity.

As is the case for all extracellular serine- and metalloproteinases, PAs are secreted as single-chain proenzymes (pro-uPA or sc-uPA; pro-tPA or sc-tPA) that are converted to active two-chain forms by limited proteolysis. Trace amounts of plasmin activate pro-uPA, generating a self-maintained feedback mechanism of pro-uPA and plasminogen activation. This amplification loop is further modulated by the high-affinity interaction (k_D = 50 –150 pM) of uPA with its cell membrane receptor or binding site. The uPA receptor (uPAR) of endothelial cells is a highly glycosylated, 35–46 kDa protein linked to the plasma membrane by a glycosyl-phosphatidyl inositol (GPI) anchor (BARNATHAN et al. 1990a,b; BEHRENDT et al. 1990; ROLDAN et al. 1990; MIGNATTI et al. 1991a; PLOUG et al. 1991; MØLLER et al. 1993; MANDRIOTA et al. 1995). Following secretion, pro-uPA binds to uPAR through a specific NH_2-terminal sequence (APPELLA et al. 1987). The bound zymogen is then activated, and remains active on the cell surface for several hours (VASSALLI et al. 1985). The interaction of uPA with uPAR on the plasma membrane has three important consequences:

(1) enzyme activity is localized to focal contact sites (Hébert and Baker 1988; Pöllanen et al. 1988); (2) the k_M for plasminogen activation is dramatically lowered (40-fold) (Ellis et al. 1991; Lee et al. 1994); (3) after complex formation with PAIs, uPAR-bound uPA is internalized and rapidly degraded (Cubellis et al. 1990). As is the case for uPA and PAIs, uPAR expression is also up-regulated in vascular endothelial cells by several angiogenic factors, including basic fibroblast growth factor (bFGF) and vascular endothelial growth factor (VEGF) (Mignatti et al. 1991a; Mandriota et al. 1995).

Binding sites for tPA and/or tPA-PAI-1 complexes have also been described on the membrane of endothelial cells and other cell types (Hajjar and Hamel 1990; Felez et al. 1993). A high-affinity binding site for tPA is associated with a 40 kDa protein that also binds plasminogen on the membrane of human vascular endothelial cells (Hajjar and Hamel 1990; Hajjar 1991, 1993; Felez et al. 1993). This receptor has complete sequence identity with annexin II (Ann-II), a calcium- and phospholipid-binding protein (Hajjar et al. 1994). Purified native Ann-II binds tPA, plasminogen and plasmin with high affinity (k_D=25 n*M*, 161 n*M*, and 75 n*M*, respectively) and confers an approximately 60-fold increase in catalytic efficiency upon tPA-dependent activation of plasminogen. Thus, Ann-II-mediated assembly of plasminogen and tPA may promote and localize constitutive plasmin generation on the surface of the vessel wall (Cesarman et al. 1994). However, PAI-1 associated with the surface of endothelial cells also appears to be a major binding site for tPA (Ramakrishnan et al. 1990; Russell et al. 1990; Wittwer and Sanzo 1990). Unlike uPA-uPAR interactions, binding of tPA-PAI-1 complexes requires elements of the PAI-1 moiety and/or regions of the proteinase domain of tPA (Morton et al. 1990). Binding of plasmin and tPA on the surface of endothelial cells protects these enzymes from their physiologic inhibitors, α_2-antiplasmin and PAI-1 (Shih and Hajjar 1993).

Although high-affinity plasma membrane binding sites for plasminogen have not yet been characterized, an α-enolase-related molecule and the Hyman nephritis autoantigen (gp330) have been implicated as candidate receptors (Kanalas and Makker 1991; Miles et al. 1991). In general, cell surface proteins with COOH-terminal lysyl residues appear to function as plasminogen binding sites; α-enolase is a prominent representative of this class of receptors (Miles et al. 1991). This molecule, as well as the endothelial cell tPA receptor (Ann-II), also interacts with tPA (Miles et al. 1991; Felez et al. 1993). Interestingly, binding of plasminogen to Ann-II, and Ann-II-mediated enhancement of t-PA-dependent plasminogen activation are inhibited by ε-aminocaproic acid or by cleavage of Ann-II by carboxypeptidase B, indicating a COOH-terminal lysine-dependent interaction. These findings suggest a novel mechanism whereby a plasmin-like serine proteinase may cleave Ann-II at Lys_{307}-Arg_{308}, and expose a new COOH-terminal lysine residue (Lys_{307}) for binding and efficient activation of plasminogen (Hajjar et al. 1994). Thus, several high-affinity binding sites for tPA may be shared with plasminogen. In addition, low-affinity, high-capacity binding sites for plasminogen are present in the chondroitin sulfate proteoglycans of the ECM and of the cell surface (Miles and

PLOW 1985; HAJJAR et al. 1986; PLOW et al. 1986; PLOW and MILES 1990). Interestingly, uPA has a significant affinity for heparin and heparan sulfate proteoglycans (ANDRADE-GORDON and STRICKLAND 1986). Binding sites for plasminogen and tPA are also present on fibronectin and laminin (MOSER et al. 1993). Fibronectin binds both plasminogen and tPA via a 55 kDa NH_2-terminal fragment (MOSER et al. 1993). Unlike fibrin, intact fibronectin does not enhance the rate of tPA-catalyzed plasminogen activation; however, a mixture of proteolytically degraded fibronectin fragments stimulates the activation reaction, resulting in an 11-fold increase in the k_{cat}/K_M (STACK and PIZZO 1993). Therefore, both plasminogen and PAs are colocalized either on the cell surface and/or in the ECM (PLOW et al. 1986).

4 Involvement of the Plasminogen Activator-Plasmin System in Angiogenesis

The earliest indication for an involvement of PAs in angiogenesis was provided in the early 1980s by findings that angiogenic preparations, including bovine retinal extract, mouse adipocyte-conditioned medium, human hepatoma cell lysate, or the tumor promoter 12-O-tetradecanoylphorbol 13-acetate (TPA), stimulate endothelial cell PA and collagenase activities in a dose-dependent manner (MOSCATELLI et al. 1980; GROSS et al. 1983). The subsequent development of in vitro assays for angiogenesis, and the purification of a variety of angiogenic factors added further evidence that several components of the PA-plasmin system play multiple roles in capillary formation. Bovine microvascular endothelial cells treated with the tumor promoter phorbol 12-myristate 13-acetate (PMA), vanadate (an inhibitor of phosphotyrosine-specific phosphatases) or with bFGF, all of which markedly increase production of collagenase and PA, invade three-dimensional collagen or fibrin gels where they form an extensive network of capillary-like tubular structures (in vitro angiogenesis) (MONTESANO and ORCI 1985; MONTESANO et al. 1986, 1987, 1988). Cultured aortic explants express uPA in capillary sprouts but not in underlying endothelial cell sheets, suggesting that uPA expression depends on the histological context of the endothelial cell (BACHARACH et al. 1992). However, PMA-treated endothelial cells invade fibrin gels only in the presence of fibrinolytic inhibitors. In the absence of inhibitors, treatment with PMA results in progressive lysis without invasion of the fibrin matrix, suggesting that uncontrolled matrix degradation is not permissive for angiogenesis (MONTESANO and ORCI 1985; MONTESANO et al. 1987). This hypothesis is also supported by the finding that, in coculture systems in which microvascular endothelial cells are induced by glial cells to form capillary-like structures, endothelial cell uPA activity decreases several-fold when capillary-like structures are formed. Incubation of cocultures with PMA, which up-regulates uPA activity, or with proteolytically active low-M_r uPA inhibits endothelial differentiation. By contrast, addition of serine proteinase inhibitors increases glia-induced capillary formation (LATERRA et al. 1994). These findings indicate the critical importance of an appropriate balance between proteinases and proteinase inhibitors in angiogenesis. Whereas ECM invasion and capillary lumen formation are inhibited in the presence of excess proteinase inhibitors, excessive proteolysis is incompatible with normal capillary morphogenesis. Thus, a precisely regulated proteolytic balance is necessary for normal capillary morphogenesis.

A variety of cytokines and growth factors with angiogenic activity have different effects on the expression of components of the PA-plasmin system by endothelial cells. These include bFGF, transforming growth factor β-1 (TGFβ1), hepatocyte growth factor/scatter factor (HGF/SF), and VEGF (ROBERTS et al. 1986; YANG and MOSES 1990; FERRARA et al. 1991; PEPPER et al. 1991; BASILICO and MOSCATELLI 1992; GRANT et al. 1993). Basic FGF and TGFβ1 have opposing effects on the PA activity of endothelial cells. The former is a potent inducer of uPA and uPAR expression and has a relatively modest effect on PAI-1 synthesis. The latter strongly down-regulates uPA and up-regulates PAI-1 but has no effect on uPAR expression (SAKSELA et al. 1987; PEPPER et al. 1990; MIGNATTI et al. 1991a). In contrast, bFGF and VEGF have a potent synergistic effect on endothelial cell PA production and angiogenesis (PEPPER et al. 1991, 1992a). HGF/SF, which has structural homology to plasminogen and is activated by uPA (NALDINI et al. 1992; MIZUNO and NAKAMURA 1993), also up-regulates uPA expression in endothelial and epithelial cells (PEPPER et al. 1991; BUSSOLINO et al. 1992), indicating the existence of an amplification mechanism of uPA production and HGF/SF activation. In human umbilical vein endothelial cells, retinoic acid up-regulates t-PA expression without affecting PAI-1 synthesis (THOMPSON et al. 1991). In contrast, tumor necrosis factor, lymphotoxin, and interleukin-1, but not interleukin-6, stimulate human endothelial cell production of PAI-1 but have no effect on PA expression (VAN HINSBERGH et al. 1990).

Some of these angiogenic factors have relevant effects on endothelial cell invasiveness in vitro. Addition of exogenous bFGF promotes basement membrane invasion by cultured capillary endothelial cells. Invasion is abolished by TGFβ1, which inhibits endothelial cell uPA activity, and by plasmin inhibitors or antibodies to PA (MIGNATTI et al. 1989). PA inhibitors, including PAI-1 and PAI-2, also block basement membrane invasion by endothelial cells. In contrast, PN I, which inhibits thrombin more efficiently than PAs (SAKSELA and RIFKIN 1988), has no effect on endothelial cell invasion (authors' unpublished data). Endothelial cell PA activity appears to be independent of the endogenous bFGF expressed by endothelial cells, whereas type IV collagenase expression and in vitro invasiveness correlate with the level of endogenous bFGF. Thus, other factors, including metalloproteinases, are important for endothelial cell invasion of natural basement membranes (TSUBOI et al. 1990).

The concerted action of several angiogenic factors on endothelial cell PA activity indicates that plasmin formation must be finely modulated during angiogenesis. This view is also supported by the abnormal behavior of endothelial cells that express high levels of uPA. Endothelioma cells expressing the polyoma virus middle T (mT) oncogene (End cells) produce high amounts of uPA and low levels of PA inhibitors. When End cells are grown within fibrin gels, they invade the substrate and form large, hemangioma-like, cystic structures. Neutralization of excess proteolytic activity by exogenous serine proteinase inhibitors results in the formation of capillary-like tubular structures (MONTESANO et al. 1990). Similar results have recently been obtained with transformed murine endothelial cells that express high levels of uPA and form cysts in fibrin gels. Inhibition of plasmin prevents cyst formation. However, nontransformed murine endothelial cells organize into capillary-like structures regardless of perturbations in their fibrinolytic activity. Thus, plasmin is essential for hemangioma formation but appears not to be critical for capillary formation by normal endothelial cells (DUBOIS-STRINGFELLOW et al. 1994). This view is also supported by the recent finding that PA-deficient transgenic mice develop normally and appear to have normal physiologic angiogenesis (CARMELIET et al. 1994). In the light of the fundamental importance of angiogenesis in organ development, it appears most likely that complex organisms must have developed compensatory mechanisms for angiogenesis.

Studies of the neovascularization of ovarian follicles, the corpus luteum, and the maternal decidua have shown interesting features of endothelial cell uPA and PAI-1 expression in vivo. uPA mRNA is expressed in the ovary along the route of capillary extension, in capillary sprouts within the developing corpus luteum, and in endothelial cell cords traversing the maternal decidua in the direction of the newly implanted embryo. Interestingly, uPA mRNA expression in endothelial cells is down-modulated upon completion of neovascularization, suggesting that uPA expression is part of the angiogenic response. During corpus luteum development and decidual neovascularization, as well as in aortic explants, PAI-1 expression is preferentially activated in cells in the vicinity of uPA-expressing capillary-like structures. These findings suggest a functional interplay between uPA- and PAI-1-expressing cells and support the hypothesis that during angiogenesis PAI-1 protects neovascularized tissues from excessive proteolysis (BACHARACH et al. 1992).

The involvement of PAs in angiogenesis is also indicated by the finding that certain steroids, including glucocorticoids, and a novel class of "angiostatic" steroids that block endothelial cell PA activity also block angiogenesis in several systems. This effect appears to be mediated by a significant increase in endothelial cell production of PAI-1 (BLEI et al. 1993; WOLFF et al. 1993).

5 The Urokinase-Type Plasminogen Activator-Plasmin System in Tumor Angiogenesis

The observation that tumor growth and metastasis are dependent upon the process of vascularization (FOLKMAN 1986, 1990; FOLKMAN et al. 1989; WEIDNER et al. 1991) has generated considerable interest in the mechanisms that modulate tumor angiogenesis. A variety of tumor cells in vitro and in vivo express angiogenic factors, including bFGF, VEGF, TGFβ1 and HGF/SF. As these cytokines modulate PA, uPAR and PAI expression in endothelial cells, these components of the PA/plasmin system may also play relevant roles in tumor angiogenesis.

The finding of high levels of bFGF in the capillary endothelial cells of some tumors (SCHULZE-OSTHOFF et al. 1990) has generated the hypothesis that tumor angiogenesis may be triggered by an increase in endothelial cell bFGF, rather than by a paracrine effect of bFGF derived from other cells or the ECM. Because bFGF can modulate a variety of cell functions, including migration and uPA expression, with an autocrine mechanism (SATO and RIFKIN 1988; MIGNATTI et al. 1991b, 1992; BIKFALVI et al. 1995) this effect can result in increased endothelial cell motility and uPA activity. Recent studies have shown that the conditioned medium of certain tumor cells rapidly up-regulates endothelial cell expression of bFGF and in vitro angiogenesis. Incubation of endothelial cells with tumor cell-conditioned medium also results in increased expression of uPA (PEVERALI et al. 1994). Although tumor cell-conditioned media contain no bFGF (MOSCATELLI et al. 1986a), addition of anti-recombinant bFGF IgG abolishes the up-regulation of uPA and blocks in vitro angiogenesis, indicating that both the increase in uPA production and formation of capillary-like structures are mediated by endogenous bFGF expressed by the endothelial cells. Both the bFGF/uPA-inducing activity and the angiogenic activity of SK-Hep1 cell-conditioned medium copurify with a relatively acid-resistant peptide that has moderate affinity for heparin and M_r<18 kDa>3.5 kDa. This peptide may represent a novel tumor-derived angiogenic factor that modulates the concerted expression by endothelial cells of cytokines and proteolytic enzymes required for capillary formation (PEVERALI et al. 1994). Other peptide factor(s) in tumor cell-conditioned medium also up-regulate endothelial cell expression of uPAR (authors' unpublished results). In addition, recent unpublished observations indicate that the endogenous bFGF of endothelial cells up-regulates VEGF expression in the same cells with an autocrine mechanism (authors' unpublished results). As bFGF and VEGF have synergistic effects (PEPPER et al. 1992a), up-regulation of endogenous bFGF by tumor-derived factors may have relevant effects on endothelial cell production of uPA.

A recent intriguing finding also implicates plasmin degradation in tumor angiogenesis. The surgical removal of a solitary neoplasm is often followed by the appearance of remote metastases. In mice bearing Lewis lung carcinoma the neovascularization and growth of distant metastases is inhibited by a circulating inhibitor. Serum and urine from tumor-bearing animals, but not from control mice, specifically inhibit endothelial cell proliferation. This activity copurifies with a 38 kDa plasminogen fragment that has been named angiostatin. A corresponding fragment from human plasminogen has similar activity. Systemic administration of angiostatin, but not of intact plasminogen, potently blocks the growth of both the primary tumor and distant metastases. Thus, inhibition of metastasis by a primary tumor is mediated, at least in part, by angiostatin (FIDLER and ELLIS 1994; O'REILLY et al. 1994). Although the mechanism of action of angiostatin is not understood, it may inhibit plasmin formation by competition with intact plasminogen for binding to endothelial cells and thus deprive vascular endothelial cells of the proteolytic activity required for angiogenesis.

6 Urokinase-Type Plasminogen Activator and Endothelial Cell Migration

The role of uPA in angiogenesis appears to be mediated not only by its catalytic activity but also by a paracrine/autocrine effect on endothelial cell migration independent of plasmin generation. In two-dimensional in vitro models for wound repair, endothelial cells migrating from the edges of an experimental wound display an increase in uPA, uPAR and PAI-1 that reverts to background levels upon wound closure and cessation of cell movement (PEPPER et al. 1987, 1992b, 1993). This result demonstrates a direct temporal relationship between endothelial cell migration and uPA activity, and suggests that induction of uPA activity is a component of the migratory process. Nanomolar concentrations of native uPA or of its noncatalytic A chain stimulate endothelial cell motility in Boyden chamber assays. This effect is abolished by a monoclonal antibody that inhibits uPA binding to uPAR, suggesting that endothelial cell motility is mediated by this specific interaction and independent of proteolytic activity (FIBBI et al. 1988).

Although these findings implicate a role for PAs and PAIs in endothelial cell migration and angiogenesis, the mechanism(s) of action of PA in cell movement is/are not yet understood. Plasmin has been shown to disrupt actin cables in rat embryo cells (POLLACK and RIFKIN 1975), suggesting that PA or plasmin may be involved in the continuous rearrangement of cytoskeletal components that occurs during cell migration. However, this hypothesis contrasts with several findings that the uPA-mediated stimulation of cell motility requires interaction of the uPA NH_2-terminal peptide with uPAR and is independent of the enzyme's catalytic activity (FIBBI et al. 1988; ODEKON et al. 1992). Addition of uPA or its noncatalytic NH_2-terminal peptide to fibroblasts or epidermal cells that express uPAR results in increased migration and/or cell proliferation. These effects correlate with an increase in de novo synthesis of intracellular diacylglycerol, indicating that uPA-uPAR interactions can trigger intracellular signaling through a mechanism independent of uPA activity (DEL ROSSO et al. 1990, 1993; ANICHINI et al. 1994). More recently, it has been shown that addition of pro-uPA to human epithelial cells that express uPAR but no uPA results in a dose- and time-dependent enhancement of migration (BUSSO et al. 1994). Urokinase binding to uPAR elicits a time- and protein kinase C ε-dependent increase in the phosphorylation of cytokeratins 8 and 18 on serine residues, a cell shape change and redistribution of the cytokeratin filaments.

Although these effects appear to be restricted to certain cell types, they strongly suggest that uPAR serves not only as an anchor for uPA but also participates in a signal transduction pathway resulting in a pronounced biological response (Busso et al. 1994). However, because uPAR is a GPI-anchored protein and has no cytoplasmic domain (Ploug et al. 1991), the mechanism(s) by which intracellular signaling can be generated remain unclear.

The importance of PAs in angiogenesis is also supported by other findings in vitro and in vivo. It has long been known that plasma-derived fibrinolytic agents interact with ECM components and play important roles in maintaining vascular integrity. Treatment of ECM with plasmin or trypsin stimulates in vitro endothelial cell organization into capillary-like structures (Maciag 1984). This finding has generated the hypothesis that ECM degradation initiates a cascade of events that modulate angiogenesis. It must be considered that in vivo PAs and other ECM-degrading proteinases originate from a variety of cell types, including granulocytes, macrophages and fibroblasts. A scenario of cell cooperation may occur that is mediated by proteolytic enzymes. Fibrin, fibronectin and collagens, and their respective degradation products, are chemotactic for endothelial cells and potent angiogenesis inducers in vivo (Bowersox and Sorgente 1982; Alessandri et al. 1983). ECM degradation products are produced early after the initiation of tissue injury, inflammation and tumor growth, conditions that also involve fibrin deposition. As soon as clot lysis is initiated by PAs, fibrin- and fibronectin-derived peptides are released at the site of the lesion. This may promote angiogenesis by attracting endothelial cells from adjacent capillaries and/or venules, and by favoring their organization into new capillaries.

7 Plasminogen Activator-Cytokine Interactions

Recent findings have indicated that PAs, besides mediating endothelial cell migration and invasion, also modulate tumor angiogenesis through two parallel mechanisms: (1) mobilization of bFGF and VEGF from the cell surface or ECM, and (2) activation of latent TGFβ1 (Vlodavsky et al. 1990).

An important feature of bFGF is its high affinity for the heparan sulfate proteoglycans of the ECM (Saksela et al. 1988; Bashkin et al. 1989; Saksela and Rifkin 1990). The growth factor is found associated with the ECM in vitro and with basement membranes in vivo (Bashkin et al. 1989; DiMario et al. 1989) where it retains its biological activity and is protected from proteolytic degradation (Saksela et al. 1988; Rogelj et al. 1989). Moreover, binding to the ECM and cell surface-associated proteoglycans is a prerequisite for bFGF interaction with its high-affinity plasma membrane receptors (Yayon et al. 1991). Also uPA and plasminogen have significant affinities for heparan sulfate and chondroitin sulfate (Miles and Plow 1985; Andrade-Gordon and Strickland 1986; Hajjar et al. 1986; Plow et al. 1986). Thus, both growth factors and components of the PA-plasmin system are located in the ECM in an insoluble state.

Several lines of experimental evidence indicate that bFGF release from the cell surface and ECM can be mediated by plasmin. When plasmin activity is increased, the release of bFGF-proteoglycan complexes is heightened. By contrast, when plasmin formation is inhibited, the release of bFGF-proteoglycan complexes is suppressed (SAKSELA and RIFKIN 1990). This process of growth factor mobilization is of particular importance for the biological activity of bFGF because it permits bFGF to partition into the aqueous phase, rather than into the insoluble matrix, and to diffuse in the tissue and interact with its plasma membrane receptors (FLAUMENHAFT et al. 1990). The mobilization of bFGF from the ECM can be achieved not only through the action of plasmin but also by other degradative enzymes, including heparitinases (VLODAVSKI et al. 1983). However, the relative abundance of plasminogen in virtually all tissues and the efficient amplification mechanism achieved by PA-mediated plasminogen activation implicate plasmin as the most important proteinase in this process.

Plasmin also releases the 165-amino acid and 189-amino acid forms of VEGF from the cell surface or ECM. The released 34 kDa dimeric species are active as endothelial cell mitogens and as vascular permeability agents (HOUCK et al. 1992).

The mobilization of bFGF from the ECM contributes to the control of extracellular proteolysis mediated by vascular endothelial cells during angiogenesis. The bFGF released from the ECM stimulates PA, uPAR and collagenase expression by microvascular endothelial cells (MOSCATELLI et al. 1986b; MIGNATTI et al. 1991a). These proteinases are required to degrade the vascular BM and permit vascular endothelial cell invasion during blood vessel formation (MONTESANO and ORCI 1985; MONTESANO et al. 1986; MIGNATTI et al. 1986, 1989). This highly controlled invasive process may also be modulated by PAs through their activation of latent TGFβ1 in the ECM. TGFβ1, a potent inhibitor of cell proliferation, migration and proteinase production in vascular endothelial cells (HEIMARK et al. 1986; MULLER et al. 1987; SAKSELA et al. 1987), is secreted constitutively by a variety of cell types, including endothelial cells, as part of a high-M_r complex. Extracellularly, the growth factor (25 kDa) remains noncovalently linked to its 75 kDa latency-associated propeptide (LAP). LAP is disulfide-linked to the 125–190 kDa latent TGFβ1-binding protein (MIYAZONO et al. 1988). Mature TGFβ1 must be released from this complex to interact with its plasma membrane receptor and elicit a biological response. Plasmin is an activator of latent TGFβ1 both in vitro and in cell cultures (LYONS et al. 1988, 1990; SATO and RIFKIN 1989; SATO et al. 1990). Binding of uPA to uPAR on the cell surface considerably accelerates latent TGFβ1 activation in the presence of physiological concentrations of plasminogen (ODEKON et al. 1994). The active TGFβ1 formed in the ECM by the action of plasmin counteracts the stimulatory effects of bFGF on vascular endothelial cells by down-regulating uPA and collagenase gene expression and by stimulating PAI-1 and TIMP synthesis (SAKSELA et al. 1987; PEPPER et al. 1990). As a consequence, plasmin formation is blocked, active MMPs are inhibited, and pro-MMPs are no longer activated. However, this blockade of extracellular proteolysis also turns off the plasmin-mediated activation of TGFβ1. When no more active TGFβ1 is present, the effects of bFGF on endothelial cells again become prevalent and proteinase production increases (FLAUMENHAFT et al. 1992).

This self-regulatory mechanism has profound implications in angiogenesis. Electron microscopic analysis of capillary formation has shown that the proteolytic degradation of the vessels' basal lamina and endothelial cell migration are temporally followed by a stage in which the newly-forming capillaries synthesize and organize a new lamina propria. During this process, extracellular proteolysis must be locally inhibited to permit the deposition and assembly of ECM components. After a capillary loop is formed, degradation of the newly formed BM occurs at the tip of the loop. Endothelial cells invade from this location, and a new capillary sprout is formed (AUSPRUNK and FOLKMAN 1977). Thus, from a biochemical point of view the process of capillary formation during angiogenesis can be thought of as resulting from alternate cycles of activation and inhibition of extracellular proteolysis.

8 Conclusion and Perspectives

A consistent body of experimental evidence has shown multiple roles for PAs in angiogenesis: (a) PAs trigger a proteinase cascade required for endothelial cell degradation and invasion of the ECM; (b) they significantly contribute to the extracellular activation and mobilization of angiogenic factors; (c) uPA modulates endothelial cell functions, including migration and growth, through a mechanism independent of its proteolytic activity; (d) plasmin generated by PAs produced by endothelial cells or other cell types generates ECM degradation products that are chemotactic for endothelial cells. PA activity is modulated in endothelial cells by complex mechanisms that involve transcriptional regulation by angiogenic factors, and extracellular control by tissue inhibitors and interaction with binding sites on the cell surface or the ECM. A detailed comprehension of these control mechanisms for endothelial cell PA activity may afford the development of antiproteolytic strategies aimed at controlling angiogenesis-related diseases, including tumor growth and metastasis.

Acknowledgments. This work was supported by grants from the Italian Association for Cancer Research (AIRC) and the Italian Ministry of University, and Scientific and Technological Research (MURST 60% and 40%) to P.M., and by grants CA34282 and CA29753 from the National Institutes of Health, and 194169 from the Juvenile Diabetes Foundation to D.B.R.

References

Alessandri G, Rajn K, Gullino PM (1983) Mobilization of capillary endothelium in vitro induced by effectors of angiogenesis in vivo. Cancer Res 43: 1790–1797

Andrade-Gordon P, Strickland S (1986) Interaction of heparin with plasminogen activators and plasminogen: effects on the activation of plasminogen. Biochemistry 25: 4033–4040

Anichini E, Fibbi G, Pucci M, Caldini R, Chevanne M, Del Rosso M (1994) Production of second messengers following chemotactic and mitogenic urokinase-receptor interaction in human fibroblasts and mouse fibroblasts transfected with human urokinase receptor. Exp Cell Res 213: 438–448
Appella E, Robinson EA, Ullrich SJ, Stoppelli MP, Corti A, Cassani G, Blasi F (1987) The receptor-binding sequence of urokinase. A biological function for the growth factor module of proteases. J Biol Chem 262: 4437–4440
Astedt B, Lecander I, Brodin T, Ludblad A, Low K (1985) Purification of a specific placental plasminogen activator inhibitor by monoclonal antibody and its complex formation with plasminogen activator. Thromb Haemost 53: 122–125
Ausprunk DH, Folkman J (1977) Migration and proliferation of endothelial cells in preformed and newly formed blood vessels during tumor angiogenesis. Microvasc Res 14: 53–65
Bacharach E, Itin A, Keshet E (1992) In vivo patterns of expression of urokinase and its inhibitor PAI-1 suggest a concerted role in regulating physiological angiogenesis. Proc Natl Acad Sci USA 89: 10686–10690
Baker JB, Low DA, Simmer RL, Cunningham DD (1980) Protease-nexin: a cellular component that links thrombin and plasminogen activator and mediates their binding to cells. Cell 21: 37–45
Barnathan ES, Kuo A, Kariko K, Rosenfeld L, Murray SC, Behrendt N, Rønne E, Weiner D, Henkin J, Cines DB (1990a) Characterization of human endothelial cell urokinase-type plasminogen activator receptor protein and messenger RNA. Blood 76: 1795–1806
Barnathan ES, Kuo A, Rosenfeld L, Kariko K, Leski M, Robbiati F, Nolli ML, Henkin J, Cines DB (1990b) Interaction of single-chain urokinase-type plasminogen activator with human endothelial cells. J Biol Chem 265: 2865–2872
Bashkin P, Doctrow S, Klagsbrun M, Svahn CM, Folkman J, Vlodavski I (1989) Basic fibroblast growth factor binds to subendothelial extracellular matrix and is released by heparitinase and heparin-like molecules. Biochemistry 28: 1737–1743
Basilico C, Moscatelli D (1992) The FGF family of growth factors and oncogenes. Adv Cancer Res 59: 115–165
Behrendt N, Rønne E, Ploug M, Petri T, Løber D, Nielsen LS, Schleuning WD, Blasi F, Appella E, Danø K (1990) The human receptor for urokinase plasminogen activator. NH2-terminal sequence and glycosylation variants. J Biol Chem 265: 6453–6460
Bikfalvi A, Klein S, Pintucci G, Quarto N, Mignatti P, Rifkin DB (1995) Differential modulation of cell phenotype by different molecular weight forms of basic fibroblast growth factor: possible intracellular signaling by the high molecular weight forms. J Cell Biol 129: 233–243
Blei F, Wilson EL, Mignatti P, Rifkin DB (1993) Mechanism of action of angiostatic steroids: suppression of plasminogen activator activity via stimulation of plasminogen activator inhibitor synthesis. J Cell Physiol 155: 568–578
Bowersox JC, Sorgente N (1982) Chemotaxis of aortic endothelial cells in response to fibronectin. Cancer Res 42: 2547–2551
Busso N, Masur SK, Lazega D, Waxman S, Ossowski L (1994) Induction of cell migration by pro-urokinase binding to its receptor: possible mechanism for signal transduction in human epithelial cells. J Cell Biol 126: 259–270
Bussolino F, Di Renzo MF, Ziche M, Bocchietto E, Olivero M, Naldini L, Gaudino G, Tamagnone L, Coffer A, Comoglio PM (1992) Hepatocyte growth factor is a potent angiogenic factor which stimulates endothelial cell motility and growth. J Cell Biol 119: 629–641
Bykowska K, Levin EG, Rijken DC, Loskutoff DJ, Collen D (1982) Characterization of a plasminogen activator secreted by cultured bovine aortic endothelial cells. Biochim Biophys Acta 703: 113–115
Carmeliet P, Schoonjans L, Kieckens L, Ream B, Degen J, Bronson R, DeVos R, van den Oord JJ, Collen D, Mulligan RC (1994) Physiological consequences of loss of plasminogen activator gene function in mice. Nature 386: 419–425
Cesarman GM, Guevara CA, Hajjar KA (1994) An endothelial cell receptor for plasminogen/tissue plasminogen activator (t-PA). II. Annexin II-mediated enhancement of t-PA-dependent plasminogen activation. J Biol Chem 269: 21198–21203
Chapman HA, Stone OL Jr (1984) Cooperation between plasmin and elastase in elastin degradation by intact murine macrophages. Biochem J 222: 721–728
Cubellis MV, Wun TC, Blasi F (1990) Receptor-mediated internalization and degradation of urokinase is caused by its specific inhibitor PAI-1. EMBO J 9: 1079–1085
Del Rosso M, Fibbi G, Dini G, Grappone C, Pucci M, Caldini R, Magnelli L, Fimiani M, Lotti T, Panconesi E (1990) Role of specific membrane receptors in urokinase-dependent migration of human keratinocytes. J Invest Dermatol 94: 310–316

Del Rosso M, Anichini E, Pedersen N, Blasi F, Fibbi G, Pucci M, Ruggiero M (1993) Urokinase-urokinase receptor interaction: non-mitogenic signal transduction in human epidermal cells. Biochem Biophys Res Commun 190: 347–352
DiMario J, Buffinger N, Yamada S, Strohman RC (1989) Fibroblast growth factor in the extracellular matrix of dystrophic (mdx) mouse muscle. Science 244: 688–690
Dubois-Stringfellow N, Jonczyk A, Bautch VL (1994) Perturbations in the fibrinolytic pathway abolish cyst formation but not capillary-like organization of cultured murine endothelial cells. Blood 83: 3206–3217
Eaton DL, Scott RW, Baker JB (1984) Purification of human fibroblast urokinase proenzyme and analysis of its regulation by proteases and protease nexin. J Biol Chem 259: 6241–6247
Ellis V, Behrendt N, Danø K (1991) Plasminogen activation by receptor-bound urokinase. A kinetic study with both cell-associated and isolated receptor. J Biol Chem 266: 12752–12758
Felez J, Chanquia CJ, Fabregas P, PPlow EF, Miles LA (1993) Competition between plasminogen and tissue plasminogen activator for cellular binding sites. Blood 82: 2433–2441
Ferrara N, Houck KA, Jakeman LB, Winer J, Leung DW (1991) The vascular endothelial cell growth factor family of polypeptides. J Cell Biochem 47: 211–218
Fibbi G, Ziche M, Morbidelli L, Magnelli L, Del Rosso M (1988) Interaction of urokinase with specific receptors stimulates mobilization of bovine adrenal capillary endothelial cells. Exp Cell Res 179: 385–395 [published erratum in Exp Cell Res 186:196]
Fidler IJ, Ellis LM (1994) The implications of angiogenesis for the biology and therapy of cancer metastasis (comment). Cell 79: 185–188
Flaumenhaft R, Moscatelli D, Rifkin DB (1990) Heparin and heparan sulfate increase the radius of diffusion and action of basic fibroblast growth factor. J Cell Biol 111: 1651–1659
Flaumenhaft R, Abe M, Mignatti P, Rifkin DB (1992) Basic fibroblast growth factor-induced activation of latent transforming growth factor β in endothelial cells: regulation of plasminogen activator activity. J Cell Biol 118: 901-909
Folkman J (1986) How is blood vessel growth regulated in normal and neoplastic tissue? G.H.A. Clowes Memorial Award Lecture. Cancer Res 46: 467–473
Folkman J (1990) What is the evidence that tumors are angiogenesis dependent? J Natl Cancer Inst 82: 4–6 (editorial)
Folkman J, Watson K, Ingber D, Hanahan D (1989) Induction of angiogenesis during the transition from hyperplasia to neoplasia. Nature 339: 58–61
Grant DS, Kleinman HK, Goldberg ID, Bhargava MM, Nickoloff BJ, Kinsella JL, Polverini P, Rosen EM (1993) Scatter factor induces blood vessel formation in vivo. Proc Natl Acad Sci USA 90: 1937–1941
Gross JL,Moscatelli D, Rifkin DB (1983) Increased capillary endothelial cell protease activity in response to angiogenic stimuli in vitro. Proc Natl Acad Sci USA 80: 2623–2627
Hajjar KA (1991) The endothelial cell tissue plasminogen activator receptor. Specific interaction with plasminogen. J Biol Chem 266: 21962–21970
Hajjar KA (1993) Homocysteine-induced modulation of tissue plasminogen activator binding to its endothelial cell membrane receptor. J Clin Invest 91: 2873–2879
Hajjar KA, Hamel NM (1990) Identification and characterization of human endothelial cell membrane binding sites for tissue plasminogen activator and urokinase. J Biol Chem 265: 2908–2916
Hajjar KA, Harpel PC, Jaffe EA, Nachman RL (1986) Binding of plasminogen to cultured human endothelial cells. J Biol Chem 261: 11656–11662
Hajjar KA, Jacovina AT, Chacko J (1994) An endothelial cell receptor for plasminogen/tissue plasminogen activator. I. Identity with annexin II. J Biol Chem 269: 21191–21197
Hébert CA, Baker JB (1988) Linkage of extracellular plasminogen activator to the fibroblast cytoskeleton: colocalization of cell surface urokinase with vinculin. J Cell Biol 106: 1241–1248
Heeb MJ, Espana F, Geiger M, Collen D, Stump DC, Griffin JH (1987) Immunological identity of heparin-dependent plasma and urinary protein C inhibitor and plasminogen activator inhibitor-3. J Biol Chem 262: 15813–15816
Heimark RL, Twardzik DR, Schwartz SM (1986) Inhibition of endothelial cell regeneration by type-beta transforming growth factor from platelets. Science 233: 1078–1080
Hekman CM, Loskutoff DJ (1985) Endothelial cells produce a latent inhibitor of plasminogen activators that can be activated by denaturants. J Biol Chem 260: 11581–11587
Houck KA, Leung DW, Rowland AM, Winer J, Ferrara N (1992) Dual regulation of vascular endothelial growth factor bioavailability by genetic and proteolytic mechanisms. J Biol Chem 267: 26031–26037
Kanalas JJ, Makker SP (1991) Identification of the rat Heymann nephritis autoantigen (gp330) as a receptor site for plasminogen. J Biol Chem 266: 10825–10829

Kawano T, Morimoto K, Uemura Y (1970) Partial purification and properties of urokinase inhibitor from human placenta. J Biochem 67: 333–342
Kleiner DE, Stetler-Stevenson WG (1993) Structural biochemistry and activation of matrix metalloproteinases. Curr Opin Cell Biol 5: 891–897
Kristensen P, Larsson LI, Nielsen LS, Grøndahl-Hansen J, Andreasen PA, Danø K (1984) Human endothelial cells contain one type of plasminogen activator. FEBS Lett 168: 33–37
Kruithof EKO, Vassalli JD, Schleuning W-D, Mattaliano RJ, Bachman F (1986) Purification and characterization of a plasminogen activator inhibitor from the histiocytic lymphoma cell line U-937. J Biol Chem 261: 11207–11213
Laterra J, Indurti RR, Goldstein GW (1994) Regulation of in vitro glia-induced microvessel morphogenesis by urokinase. J Cell Physiol 158: 317–324
Laug WE, Tokes ZA, Benedict WF, Sorgente N (1980) Anchorage independent growth and plasminogen activator production by bovine endothelial cells. J Cell Biol 84: 281–293
Lee SW, Ellis V, Dichek DA (1994) Characterization of plasminogen activation by glycosylphosphatidylinositol-anchored urokinase. J Biol Chem 269: 2411–2418
Levin EG, Loskutoff DJ (1982) Cultured bovine endothelial cells produce both urokinase and tissue-type plasminogen activators. J Cell Biol 94: 631–636
Loskutoff DJ, Edgington TS (1977) Synthesis of a fibrinolytic activator and inhibitor by endothelial cells. Proc Natl Acad Sci USA 74: 3903–3907
Lyons RM, Keski Oja J, Moses HL (1988) Proteolytic activation of latent transforming growth factor-β from fibroblast conditioned medium. J Cell Biol 106: 1659–1665
Lyons RM, Gentry LE, Purchio AF, Moses HL (1990) Mechanism of activation of latent recombinant transforming growth factor β-1 by plasmin. J Cell Biol 110: 1361–1367
Maciag T (1984) Angiogenesis. Angiogenesis: the phenomenon. In: Spaet TH (ed) Thrombosis and haemostasis. Grune and Stratton, New York, pp 167–182
Mandriota S, Seghezzi G, Vassalli JD, Ferrara N, Wasi S, Mazzieri R, Mignatti P, Pepper MS (1995) Vascular endothelial growth factor increases urokinase receptor expression in vascular endothelial cells. J Biol Chem (in press)
Matrisian LM (1990) Metalloproteinases and their inhibitors in matrix remodeling. Trends Genet 6: 121–125
Mignatti P, Rifkin DB (1993) Biology and biochemistry of proteinases in tumor invasion. Physiol Rev 73: 161–195
Mignatti P, Robbins E, Rifkin DB (1986) Tumor invasion through the human amniotic membrane: requirement for a proteinase cascade. Cell 47: 487–498
Mignatti P, Tsuboi R, Robbins E, Rifkin DB (1989) In vitro angiogenesis on the human amniotic membrane: requirement for basic fibroblast growth factor-induced proteinases. J Cell Biol 108: 671–682
Mignatti P, Mazzieri R, Rifkin DB (1991a) Expression of the urokinase receptor in vascular endothelial cells is stimulated by basic fibroblast growth factor. J Cell Biol 113: 1193–1201
Mignatti P, Morimoto T, Rifkin DB (1991b) Basic fibroblast growth factor released by single, isolated cells stimulates their migration in an autocrine manner. Proc Natl Acad Sci USA 88: 11007–11011
Mignatti P, Morimoto T, Rifkin DB (1992) Basic fibroblast growth factor, a protein devoid of secretory signal sequence, is released by cells via a pathway independent of the endoplasmic reticulum-Golgi complex. J Cell Physiol 151: 81–93
Miles LA, Plow EF (1985) Binding and activation of plasminogen on the platelet surface. J Biol Chem 260: 4303–4311
Miles LA, Dahlberg CM, Plescia J, Felez J, Kato K, Plow EF (1991) Role of cell-surface lysines in plasminogen binding to cells: Identification of alpha-enolase as a candidate plasminogen receptor. Biochemistry 30: 1682–1691
Miyazono K, Hellman U, Westedt C, Heldin CH (1988) Latent high molecular weight complex of transforming growth factor β1. J Biol Chem 263: 6407–6415
Mizuno K, Nakamura T (1993) Molecular characteristics of HGF and the gene, and its biochemical aspects. EXS 65: 1–29
Møller LB, Pöllanen J, Rønne E, Pedersen N, Blasi F (1993) N-linked glycosylation of the ligand-binding domain of the human urokinase receptor contributes to the affinity for its ligand. J Biol Chem 268, 11152–11159
Montesano R, Orci L (1985) Tumor-promoting phorbol esters induce angiogenesis in vitro. Cell 42: 469–477
Montesano R, Vassalli JD, Baird A, Guillemin R, Orci L (1986) Basic fibroblast growth factor induces angiogenesis in vitro. Proc Natl Acad Sci USA 83: 7297–7301

Montesano R, Pepper MS, Vassalli JD, Orci L (1987) Phorbol ester induces cultured endothelial cells to invade a fibrin matrix in the presence of fibrinolytic inhibitors. J Cell Physiol 132: 509–516
Montesano R, Pepper MS, Belin D, Vassalli JD, Orci L (1988) Induction of angiogenesis in vitro by vanadate, an inhibitor of phosphotyrosine phosphatases. J Cell Physiol 134: 460–466
Montesano R, Pepper MS, Möhle-Steinlein U, Risau W, Wagner EF, Orci L (1990) Increased proteolytic activity is responsible for the aberrant morphogenetic behavior of endothelial cells expressing the middle T oncogene. Cell 62: 435–445
Morton PA, Owensby DA, Wun TC, Billadello J, Schwartz AL (1990) Identification of determinants involved in binding of tissue-type plasminogen activator-plasminogen activator inhibitor type 1 complexes to HepG2 cells. J Biol Chem 265: 14093–14099
Moscatelli D, Rifkin DB (1988) Membrane and matrix localization of proteinase: a common theme in tumor cell invasion and angiogenesis. Biochim Biophys Acta 948: 67–85
Moscatelli D, Jaffe E, Rifkin DB (1980) Tetradecanoyl phorbol acetate stimulates latent collagenase production by cultured human endothelial cells. Cell 20: 343–51
Moscatelli D, Presta M, Joseph-Silverstein J, Rifkin DB (1986a) Both normal and tumor cells produce basic fibroblast growth factor. J Cell Physiol 129: 273–276
Moscatelli D, Presta M, Rifkin DB (1986b) Purification of a factor from human placenta that stimulates capillary endothelial cell protease production, DNA synthesis and migration. Proc Natl Acad Sci USA 83: 2091–2095
Moser TL, Enghild JJ, Pizzo SV, Stack MS (1993) The extracellular matrix proteins laminin and fibronectin contain binding domains for human plasminogen and tissue plasminogen activator. J Biol Chem 268: 18917–18923
Muller G, Behrens J, Nussbaumer U, Bohlen P, Birchmeyer W (1987) Inhibitory action of transforming growth factor β on endothelial cells. Proc Natl Acad Sci USA 84: 5600–5604
Murphy G, Allan JA, Willenbrock F, Cockett MI, O'Connell JP, Docherty AJP (1992) The C-terminal domain in collagenase and stromelysin specificity. J Biol Chem 267: 9612–9618
Naldini L, Tamagnone L, Vigna E, Sachs M, Hartmann G, Birchmeier W, Daikuhara Y, Tsubouchi H, Blasi F, Comoglio PM (1992) Extracellular proteolytic cleavage by urokinase is required for activation of hepatocyte growth factor/scatter factor. EMBO J 11: 4825–4833
O'Reilly MS, Holmgren L, Shing Y, Chen C, Rosenthal RA, Moses M, Lane WS, Cao Y, Sage EH, Folkman J (1994) Angiostatin: a novel angiogenesis inhibitor that mediates the suppression of metastases by a Lewis lung carcinoma. Cell 79: 315–328
Odekon LE, Sato Y, Rifkin DB (1992) Urokinase-type plasminogen activator mediates basic fibroblast growth factor-induced bovine endothelial cell migration independent of its proteolytic activity. J Cell Physiol 150: 258–263
Odekon LE, Blasi F, Rifkin DB (1994) Requirement for receptor-bound urokinase in plasmin-dependent cellular conversion of latent TGF-beta to TGF-beta. J Cell Physiol 158: 398–407
Pepper MS, Vassalli JD, Montesano R, Orci L (1987) Urokinase-type plasminogen activator is induced in migrating capillary endothelial cells. J Cell Biol 105: 2535–2541
Pepper MS, Belin D, Montesano R, Orci L, Vassalli JD (1990) Transforming growth factor-beta 1 modulates basic fibroblast growth factor-induced proteolytic and angiogenic properties of endothelial cells in vitro. J Cell Biol 111: 743–755
Pepper MS, Ferrara N, Orci L, Montesano R (1991) Vascular endothelial cell growth factor (VEGF) induces plasminogen activators and plasminogen activator inhibitor-1 in microvascular endothelial cells. Biochem Biophys Res Commun 181: 902–906
Pepper MS, Ferrara N, Orci L, Montesano R (1992a) Potent synergism between vascular endothelial growth factor and basic fibroblast growth factor in the induction of angiogenesis in vitro. Biochem Biophys Res Commun 189: 824–831
Pepper MS, Sappino AP, Montesano R, Orci L, Vassalli JD (1992b) Plasminogen activator inhibitor-1 is induced in migrating endothelial cells. J Cell Physiol 153: 129–139
Pepper MS, Sappino AP, Stocklin R, Montesano R, Orci L, Vassalli JD (1993) Upregulation of urokinase receptor expression on migrating endothelial cells. J Cell Biol 122: 673–684
Peverali FA, Mandriota S, Ciana P, Marelli R, Quax P, Rifkin DB, Della Valle G, Mignatti P (1994) Tumor cells secrete an angiogenic factor that stimulates basic fibroblast growth factor and urokinase expression in vascular endothelial cells. J Cell Physiol 161: 1–14
Philips M, Juul AG, Thorsen S (1984) Human endothelial cells produce a plasminogen activator inhibitor and a tissue plasminogen activator-inhibitor complex. Biochim Biophys Acta 802: 99–110
Ploug M, Rønne E, Behrendt N, Jensen AL, Blasi F, Danø K (1991) Cellular receptor for urokinase plasminogen activator. Carboxyl-terminal processing and membrane anchoring by glycosyl-phosphatidylinositol. J Biol Chem 266: 1926–1933

Plow EF, Miles LA (1990) Plasminogen receptors in the mediation of pericellular proteolysis. Cell Differ Dev 32: 293–298

Plow EF, Freaney DE, Plescia J, Miles LA (1986) The plasminogen system and cell surfaces: evidence for plasminogen and urokinase receptors on the same cell type. J Cell Biol 103: 2411–2420

Pollack R, Rifkin DB (1975) Actin-containing cables within anchorage-dependent rat embryo cells are dissociated by plasmin and trypsin. Cell 6: 495–506

Pöllanen J, Hedman K, Nielsen LN, Danø K, Vaheri A (1988) Ultrastructural localization of plasma membrane-associated urokinase-type plasminogen activator at focal contacts. J Cell Biol 106: 87–95

Ramakrishnan V, Sinicropi DV, Dere R, Darbonne WC, Bechtol KB, Baker JB (1990) Interaction of wild-type and catalytically inactive mutant forms of tissue-type plasminogen activator with human umbilical vein endothelial cell monalayers. J Biol Chem 265: 2755–2762

Roberts AB, Sporn MB, Assoian RK, Smith JM, Roche NS, Wakefield LM, Heine UI, Liotta LA, Falanga V, Kehrl JH, Fauci AS (1986) Transforming growth factor type β: rapid induction of fibrosis and angiogenesis in vivo and stimulation of collagen formation in vitro. Proc Natl Acad Sci USA 83: 4167–4171

Rogelj S, Klagsbrun M, Atzmon R, Kurokawa M, Haimovitz A, Fuks Z, Vlodavski I (1989) Basic fibroblast growth factor is an extracellular matrix component required for supporting the proliferation of vascular endothelial cells and the diffferentiation of PC12 cells. J Cell Biol 109: 823–831

Roldan AL, Cubellis MV, Masucci MT, Behrendt N, Lund LR, Danø K, Appella E, Blasi F (1990) Cloning and expression of the receptor for human urokinase plasminogen activator, a central molecule in cell surface, plasmin dependent proteolysis. EMBO J 9: 467–474

Russell ME, Quertermous T, Declerck PJ, Collen D, Haber E, Homcy CJ (1990) Binding of tissue-type plasminogen activator with human endothelial cell monolayers. Characterization of the high affinity interaction with plasminogen activator inhibitor-1. J Biol Chem 265: 2569–2575

Saksela O, Rifkin DB (1988) Cell-associated plasminogen activation: regulation and physiological functions. Annu Rev Cell Biol 4: 93–126

Saksela O, Rifkin DB (1990) Release of basic fibroblast growth factor-heparan sulfate complexes from endothelial cells by plasminogen activator-mediated proteolytic activity. J Cell Biol 110: 767–775

Saksela O, Moscatelli D, Rifkin DB (1987) The opposing effects of basic fibroblast growth factor and transforming growth factor beta on the regulation of plasminogen activator activity in capillary endothelial cells. J Cell Biol 105: 957–963

Saksela O, Moscatelli D, Sommer A, Rifkin DB (1988) Endothelial cell-derived heparan sulfate binds basic fibroblast growth factor and protects it from proteolytic degradation. J Cell Biol 107: 743–751

Sato Y, Rifkin DB (1988) Autocrine activities of basic fibroblast growth factor: regulation of endothelial cell movement, plasminogen activator synthesis, and DNA synthesis. J Cell Biol 107: 1199–1205

Sato Y, Rifkin DB (1989) Inhibition of endothelial cell movement by pericytes and smooth muscle cells: activation of a latent transforming growth factor β-1-like molecule by plasmin. J Cell Biol 109: 309–315

Sato Y, Tsuboi R, Lyons RM, Moses HL, Rifkin DB (1990) Characterization of the activation of latent TGFβ by co-cultures of endothelial cells and pericytes or smooth muscle cells: a self-regulating system. J Cell Biol 111: 757–763

Schulze-Osthoff K, Risau W, Vollmer E, Sorg C (1990) In situ detection of basic fibroblast growth factor by highly specific antibodies. Am J Pathol 137: 85–92

Shih GC, Hajjar KA (1993) Plasminogen and plasminogen activator assembly on the human endothelial cell. Proc Soc Exp Biol Med 202: 258–64

Stack MS, Pizzo SV (1993) Modulation of tissue plasminogen activator-catalyzed plasminogen activation by synthetic peptides derived from the amino-terminal heparin binding domain of fibronectin. J Biol Chem 268: 18924–18928

Thompson EA, Nelles L, Collen D (1991) Effect of retinoic acid on the synthesis of tissue-type plasminogen activator and plasminogen activator inhibitor-1 in human endothelial cells. Eur J Biochem 201: 627–632

Tsuboi R, Sato Y, Rifkin DB (1990) Correlation of cell migration, cell invasion, receptor number, proteinase production, and basic fibroblast growth factor levels in endothelial cells. J Cell Biol 110: 511–517

van Hinsbergh VW, van den Berg EA, Fiers W, Dooijewaard G (1990) Tumor necrosis factor induces the production of urokinase-type plasminogen activator by human endothelial cells. Blood 75: 1991–1998

Vassalli J-D, Baccino D, Belin D (1985) A cellular binding site for the M_r 55,000 form of the human plasminogen activator, urokinase. J Cell Biol 100: 86–92

Vlodavsky I, Fuks Z, Bar-Ner M, Ariav Y, Schirrmacher V (1983) Lymphoma cell-mediated degradation of sulfated proteoglycans in the subendothelial extracellular matrix: relationship to tumor cell metastasis. Cancer Res 43: 2704–2711

Vlodavsky I, Korner G, Ishai-Michaeli R, Bashkin P, Bar-Shavit R, Fuks Z (1990) Extracellular matrix-resident growth factors and enzymes: possible involvement in tumor metastasis and angiogenesis. Cancer Metastasis Rev 9: 203–226

Weidner N, Semple JP, Welch WR, Folkman J (1991) Tumor angiogenesis and metastasis-correlation in invasive breast carcinoma. N Engl J Med 324: 1–8

Werb Z, Mainardi C, Vater CA, Harris ED (1977) Endogenous activation of latent collagenase by rheumatoid synovial cells. Evidence for a role of plasminogen activator. N Engl J Med 296: 1017–1023

Werb Z, Banda MJ, Jones PA (1980) Degradation of connective tissue matrices by macrophages. I Proteolysis of elastin, glycoproteins and collagen by proteinases isolated from macrophages. J Exp Med 152: 1340–1357

Wittwer AJ, Sanzo MA (1990) Effect of peptides on the inactivation of tissue plasminogen activator by plasminogen activator inhibitor-1 and on the binding of tissue plasminogen activator to endothelial cells. Thromb Haemost 64: 270–275

Wolff JE, Guerin C, Laterra J, Bressler J, Indurti RR, Brem H, Goldstein GW (1993) Dexamethasone reduces vascular density and plasminogen activator activity in 9L rat brain tumors. Brain Res 604: 79–85

Yang EY, Moses HL (1990) Transforming growth factor β1-induced changes in cell migration, proliferation, and angiogenesis in the chick chorioallantoic membrane. J Cell Biol 111: 731–741

Yayon A, Klagsbrun M, Esko JD, Leder P, Ornitz DM (1991) Cell surface, heparin-like molecules are required for binding of basic fibroblast growth factor to its high affinity receptor. Cell 64: 841–848

Maspin: A Tumor Suppressing Serpin

R. SAGER[1], S. SHENG[1], P. PEMBERTON[2], and M.J.C. HENDRIX[3]

1 Introduction

Maspin (mammary serpin) is a member of the serpin superfamily (POTEMPA et al. 1994). The gene was first identified by subtractive hybridization on the basis of its expression at the mRNA level in normal but not in tumor-derived mammary epithelial cells (ZOU et al. 1994). The cloned and sequenced cDNA consists of 2584 nucleotides encoding a 42 kDa peptide with the overall structure of a serpin. Maspin has been localized to chromosome 18q21.3–q23 (SAGER et al. 1994) closely linked to plasminogen activator inhibitor-2 (PAI-2), to BCL2, to the candidate tumor suppressor gene DCC (SAGER et al. 1994), and to the candidate tumor suppressor gene SCCA (SCHNEIDER et al. 1995).

When normal cells become malignant, a crucial development is the acquisition of elevated protease activity, opening the door to invasion across the basement membrane and subsequently to metastasis. Several lines of

[1]Department of Cancer Genetics, Dana-Farber Cancer Institute, 44 Binney Street, Boston, MA 02115, USA
[2]LXR Biotechnology, 1401 Marina Way South, Richmond, CA 94804, USA
[3]Pediatric Research Institute, Cardinal Glennon Children's Hospital, St. Louis University, School of Medicine, 3662 Park Avenue, St. Louis, MO 63110, USA

evidence support the role of maspin as a strong tumor suppressor gene, acting at the level of invasion and metastasis. These include: (1) sequence similarity to other serpins with protease inhibitor activity; (2) functional studies in nude mice showing that tumor transfectants expressing maspin are inhibited in tumor growth and metastasis; (3) invasion assays showing that endogenous maspin produced by tumor transfectants, or exogenous recombinant maspin, inhibit invasion through basement membrane matrix in culture, and that this inhibition is reversible by anti-maspin antibodies; and (4) motility assays using membrane filters or direct video tracking, showing that motility can be blocked by endogenous or exogenous maspin, and that this inhibition is reversible with anti-maspin antibodies.

These results, discussed below, support the tumor suppressor activity of maspin protein. Consistent with the functional evidence is the loss of expression of maspin in invasive mammary carcinomas, shown by immunostaining of well-characterized tissue specimens (Zou et al. 1994). The protein is expressed in normal tissue and in some carcinoma in situ specimens, but infrequently in invasive or metastatic tumors. These results suggest the potential use of maspin expression as a positive indicator of a favorable prognosis in primary breast cancer.

On the biochemical side, however, the molecular basis of maspin's biological activity is unknown. Cleavage of the reactive center is sufficient to destroy all biological activity (Sheng et al. 1994), but maspin does not act as an inhibitor of proteinases such as trypsin, chymotrypsin, elastase, plasmin, thrombin, or of plasminogen activators in vitro (Pemberton et al. 1995). The molecule is not secreted in cell culture, but rather it is found in the cytoplasm and in the membrane fraction (S. Sheng, unpublished). Thus its action is probably distinct from that of the classical serine protease inhibitors. Its cellular target is as yet unidentified.

2 Comparative Structural Properties of Maspin

Maspin cDNA was isolated from a normal human mammary epithelial cell library. The cDNA sequence contains 1125 nucleotides in the coding region with a polyadenylation signal located 16 nucleotides from the 3' terminal of the sequence shown in Fig. 1. The sequence as shown includes 75 nucleotides of the 5' untranslated region and 1381 nucleotides of 3' untranslated region. The initiation codon and surrounding nucleotides fit the Kozak consensus. The inferred protein consists of 375 amino acids with an NH_2-terminal methionine and COOH-terminal valine. Maspin contains eight cysteine residues and may utilize two or more disulfide bonds to stabilize its tertiary structure.

```
   1                                        GGCACGAGTTGTGCTCCTCGCTTGCCTGTTCCTTTTCCACGCATTTTCCAGGATAACTGTGACTCCAGGCCCGCA   75

  76 ATG GAT GCC CTG CAA CTA GCA AAT TCG GCT TTT GCC GTT GAT CTG TTC AAA CAA CTA TGT GAA AAG GAG CCA CTG GGC AAT GTC CTC TTC   165
   1 Met Asp Ala Leu Gln Leu Ala Asn Ser Ala Phe Ala Val Asp Leu Phe Lys Gln Leu Cys Glu Lys Glu Pro Leu Gly Asn Val Leu Phe   30

 166 TCT CCA ATC TGT CTC TCC ACC TCT CTG TCA CTT GCT CAA GTG GGT GCT AAA GGT GAÇ ACT GCA AAT GAA ATT GGA CAG GTT CTT CAT TTT   255
  31 Ser Pro Ile Cys Leu Ser Thr Ser Leu Ser Leu Ala Gln Val Gly Ala Lys Gly Asp Thr Ala Asn Glu Ile Gly Gln Val Leu His Phe   60

 256 GAA AAT GTC AAA GAT ATA CCC TTT GGA TTT CAA ACA GTA ACA TCG GAT GTA AAC AAA CTT AGT TCC TTT TAC TCA CTG AAA CTA ATC AAG   345
  61 Glu Asn Val Lys Asp Ile Pro Phe Gly Phe Gln Thr Val Thr Ser Asp Val Asn Lys Leu Ser Ser Phe Tyr Ser Leu Lys Leu Ile Lys   90

 346 CGG CTC TAC GTA GAC AAA TCT CTG AAT CTT TCT ACA GAG TTC ATC AGC TCT ACG AAG AGA CCC TAT GCA AAG GAA TTG GAA ACT GTT GAC   435
  91 Arg Leu Tyr Val Asp Lys Ser Leu Asn Leu Ser Thr Glu Phe Ile Ser Ser Thr Lys Arg Pro Tyr Ala Lys Glu Leu Glu Thr Val Asp   120

 436 TTC AAA GAT AAA TTG GAA GAA ACG AAA GGT CAG ATC AAC AAC TCA ATT AAG GAT CTC ACA GAT GGC CAC TTT GAG AAC ATT TTA GCT GAC   525
 121 Phe Lys Asp Lys Leu Glu Glu Thr Lys Gly Gln Ile Asn Asn Ser Ile Lys Asp Leu Thr Asp Gly His Phe Glu Asn Ile Leu Ala Asp   150

 526 AAC AGT GTG AAC GAC CAG ACC AAA ATC CTT GTG GTT AAT GCT GCC TAC TTT GTT GGC AAG TGG ATG AAG AAA TTT CCT GAA TCA GAA ACA   615
 151 Asn Ser Val Asn Asp Gln Thr Lys Ile Leu Val Val Asn Ala Ala Tyr Phe Val Gly Lys Trp Met Lys Lys Phe Pro Glu Ser Glu Thr   180

 616 AAA GAA TGT CCT TTC AGA CTC AAC AAG ACA GAC ACC AAA CCA GTG CAG ATG ATG AAC ATG GAG GCC ACG TTC TGT ATG GGA AAC ATT GAC   705
 181 Lys Glu Cys Pro Phe Arg Leu Asn Lys Thr Asp Thr Lys Pro Val Gln Met Met Asn Met Glu Ala Thr Phe Cys Met Gly Asn Ile Asp   210

 706 AGT ATC AAT TGT AAG ATC ATA GAG CTT CCT TTT CAA AAT AAG CAT CTC AGC ATG TTC ATC CTA CTA CCC AAG GAT GTG GAG GAT GAG TCC   795
 211 Ser Ile Asn Cys Lys Ile Ile Glu Leu Pro Phe Gln Asn Lys His Leu Ser Met Phe Ile Leu Leu Pro Lys Asp Val Glu Asp Glu Ser   240

 796 ACA GGC TTG GAG AAG ATT GAA AAA CAA CTC AAC TCA GAG TCA CTG TCA CAG TGG ACT AAT CCC AGC ACC ATG GCC AAT GCC AAG GTC AAA   885
 241 Thr Gly Leu Glu Lys Ile Glu Lys Gln Leu Asn Ser Glu Ser Leu Ser Gln Trp Thr Asn Pro Ser Thr Met Ala Asn Ala Lys Val Lys   270

 886 CTC TCC ATT CCA AAA TTT AAG GTG GAA AAG ATG ATT GAT CCC AAG GCT TGT CTG GAA AAT CTA GGG CTG AAA CAT ATC TTC AGT GAA GAC   975
 271 Leu Ser Ile Pro Lys Phe Lys Val Glu Lys Met Ile Asp Pro Lys Ala Cys Leu Glu Asn Leu Gly Leu Lys His Ile Phe Ser Glu Asp   300

 976 ACA TCT GAT TTC TCT GGA ATG TCA GAG ACC AAG GGA GTG GCC CTA TCA AAT GTT ATC CAC AAA GTG TGC TTA GAA ATA ACT GAA GAT GGT   1065
 301 Thr Ser Asp Phe Ser Gly Met Ser Glu Thr Lys Gly Val Ala Leu Ser Asn Val Ile His Lys Val Cys Leu Glu Ile Thr Glu Asp Gly   330

1066 GGG GAT TCC ATA GAG GTG CCA GGA GCA CGG ATC CTG CAG CAC AAG GAT GAA TTG AAT GCT GAC CAT CCC TTT ATT TAC ATC ATC AGG CAC   1155
 331 Gly Asp Ser Ile Glu Val Pro Gly Ala Arg Ile Leu Gln His Lys Asp Glu Leu Asn Ala Asp His Pro Phe Ile Tyr Ile Ile Arg His   360

1156 AAC AAA ACT CGA AAC ATC ATT TTC TTT GGC AAA TTC TGT TCT CCT TAA   1203
 361 Asn Lys Thr Arg Asn Ile Ile Phe Phe Gly Lys Phe Cys Ser Pro End   376

1204 GTGGCATAGCCCATGTTAAGTCCTCCCTGACTTTTCTGTGGATGCCGATTTCTGTAAACTCTGCATCCAGAGATTCATTTTCTAGATACAATAAATTGCTAATGTTGCTGGATCAGGAAG   1323
1324 CCGCCAGTACTTGTCATATGTAGCCTTCACACAGATAGACCTTTTTTTTTTTCCAATTCTATCTTTTGTTTCCTTTTTTCCCATAAGACAATGACATACGCTTTTAATGAAAAGGAATCA   1443
1444 CGTTAGAGGAAAAATATTTATTCATTATTTGTCAAATTGTCCGGGGTAGTTGGCAGAAATACAGTCTTCCACAAAGAAAATTCCTATAAGGAAGATTTGGAAGCTCTTCTTCCCAGCACT   1563
1564 ATGCTTTCCTTCTTTGGGATAGAGAATGTTCCAGACATTCTCGCTTCCCTGAAAGACTGAAGAAAGTGTAGTGCATGGGACCCACGAAACTGCCCTGGCTCCAGTGAAACTTGGGCACAT   1683
1684 GCTCAGGCTACTATAGGTCCAGAAGTCCTTATGTTAAGCCCTGGCAGGCAGGTGTTTATTAAAATTCTGAATTTTGGGGATTTTCAAAAGATAATATTTTACATACACTGTATGTTATAG   1803
1804 AACTTCATGGATCAGATCTGGGGCAGCAACCTATAAATCAACACCTTAATATGCTGCAACAAAATGTAGAATATTCAGACAAAATGGATACATAAAGACTAAGTAGCCCATAAGGGGTCA   1923
1924 AAATTTGCTGCCAAATGCGTATGCCACCAACTTACAAAAACACTTCGTTCGCAGAGCTTTTCAGATTGTGGAATGTTGGATAAGGAATTATAGACCTCTAGTAGCTGAAATGCAAGACCC   2043
2044 CAAGAGGAAGTTCAGATCTTAATATAAATTCACTTTCATTTTTGATAGCTGTCCCATCTGGTCATGTGGTTGGCACTAGACTGGTGGCAGGGGCTTCTAGCTGACTCGCACAGGGATTCT   2163
2164 CACAATAGCCGATATCAGAATTTGTGTTGAAGGAACTTGTCTCTTCATCTAATATGATAGCGGGAAAAGGAGAGGAAACTACTGCCTTTAGAAAATATAAGTAAAGTGATTAAAGTGCTC   2283
2284 ACGTTACCTTGACACATAGTTTTTCAGTCTATGGGTTTAGTTACTTTAGATGGCAAGCATGTAACTTATATTAATAGTAATTTGTAAAGTTGGGTGGATAAGCTATCCCTGTTGCCGGTT   2403
2404 CATGGATTACTTCTCTATAAAAAATATATATTTACCAAAAAATTTTGTGACATTCCTTCTCCCATCTCTTCCTTGACATGCATTGTAAATAGGTTCTTCTTGTTCTGAGATTCAATATTG   2523
2524 AATTTCTCCTATGCTATTGACAATAAAATATTATTGAACTACCAAAAAAAAAAAAAAAAAA   2584
```

Fig. 1. cDNA and predicted amino acid sequence of maspin. The polyadenylation signal is *underlined*

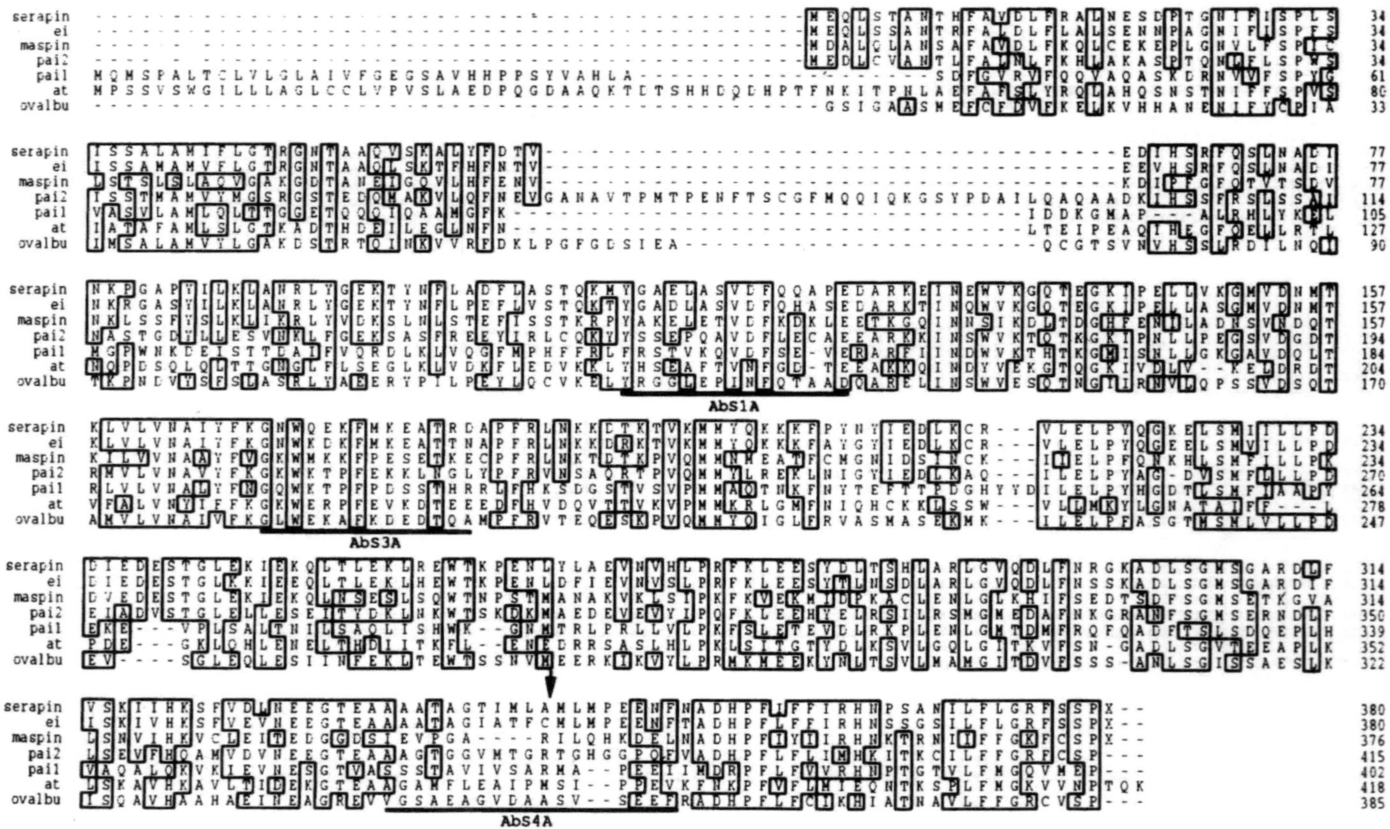

Fig. 2. Sequence comparison of maspin with related serpins. Identical residues are *boxed*. Three regions used for antibody production are *underlined*. *Arrow* denotes the proposed reactive center of maspin. *Serapin*, horse serapin; *ei*, human monocyte/neutrophil elastase inhibitor; *pai 2*, plasminogen activator inhibitor 2; *pai1*, plasminogen activator inhibitor 1; *at*, α1 protease inhibitor; *ovalb*, ovalbumin

Multiple alignment studies using BLAST analyzed by the GCG Pileup Program demonstrate close homology of maspin to the serpin superfamily (Fig. 2). Serpins are a diverse family of proteins related by primary sequence homology spanning the entire molecule and varying from 15% to 50% at the amino acid identity level. The first serpin to be analyzed crystallographically, α1 protease inhibitor (α1PI), is the prototype (HUBER and CARRELL 1989). Overall amino acid similarities compared to maspin are about 40% for plasminogen activator inhibitor (PAI)-2, neutroph il elastase inhibitor and ovalbumin and about 30% for α1PI and PAI-1. Recently identified serpins include PTI, (placental thrombin inhibitor, now renamed PI6) which is related to neutrophil elastase and interacts with trypsin, thrombin, and urokinase type plasminogen activator (uPA) (COUGHLIN et al. 1993) and SCCA which had previously been identified at the protein level as squamous cell carcinoma antigen. Two tandem linked genes, SCCA-1 and -2, have been described (SCHNEIDER et al. 1995).

Most of the amino acid substitutions in maspin do not alter the overall serpin conformation. However, significant differences are located in the hinge region, a peptide stretch located nine to 15 residues NH_2-terminal to the P1-P'1 peptide bond within the reactive site loop (RSL) (CARRELL and EVANS 1992). The reaction center is indicated in Fig. 2, with the putative reactive site marked by the arrow. In a typical reaction between a serine protease inhibitor and its protease target, the protease attempts to cleave the reactive site peptide bond but becomes trapped in a stable 1:1 stoichiometric complex in which the protease is inactive. Currently, all serpins of known inhibitory function (e.g. antitrypsin, antithrombin) undergo a transition from a stressed (S) to relaxed (R) stable molecule while serpins with no known inhibitory function (e.g., ovalbumin, angiotensinogen) do not undergo this transition. Thus the S-R transition is a useful predictor of serpin function.

Structurally, the stressed inhibitory (S) form was thought to result from partial insertion of the hinge region into β-pleated sheet A of the native molecule, following proteolysis with the RSL. The peptide sequence COOH-terminal to the hinge region then completely inserts into β-pleated sheet A to provide a more relaxed (R) stable structure. In maspin the amino acid substitutions in the hinge region are divergent from the consensus sequence usually observed in inhibitory serpins, and on this basis it has been suggested that maspin may not function as a protease inhibitor (HOPKINS et al. 1994). However, the recent crystallographic structure of α1 anti-chymotrypsinogen has revealed that partial insertion is not a prerequisite for inhibitory function (WEI et al. 1994). Thus, knowledge of serpin hinge region sequences is not a sufficient criterion on which to base a prediction of inhibitory function.

It was recently shown that maspin does not behave as a classical inhibitory serpin (PEMBERTON et al. 1995). Maspin is not an inhibitor of serine proteinases including trypsin, chymotrypsin, elastase, plasmin, thrombin, and the plasminogen activators, tissue type plasminogen activator (tPA) and uPA, but it does act as a substrate for several of them. Importantly, maspin does not undergo the

(S-R) transition predictive of inhibitory function. These findings support the view that the tumor suppressing activity of maspin is not based on a latent or intrinsic trypsin-like serine proteinase inhibitory activity. The question of its mode of action as a membrane-bound or cytoplasmic serpin remains to be determined.

In view of the close linkage between maspin, SCCA, and PAI-2 on chromosome 18q, as well as their sequence similarities, it is likely that these serpins evolved from a common precursor. The hinge region of the RSL represents one important site of evolutionary diversity, which has apparently led to functional diversity, since both SCCA and PAI-2 act as typical inhibitory serpins, whereas maspin apparently does not.

3 Production and Characterization of Recombinant Maspin Proteins Produced in Bacteria, Baculovirus, and Yeast Systems

The rGST-maspin protein was produced in *E. coli* using the pGEX-2T/Mas vector (Sheng et al. 1994). The fusion protein contains two polypeptides covalently linked by a thrombin cleavage site plus a short sequence resulting from *Bam*HI/*Bcl*I ligation. rGST-maspin accounted for 30%–40% of the total cellular protein. The recombinant protein was purified to near 100% homogeneity in a single step on a glutathione affinity column (for details see Sheng et al. 1994). The glutathione S-transferase polypeptide could not be removed by thrombin cleavage without degrading maspin, and therefore the fusion protein was used in further studies.

rMaspin(i) was produced in insect cells (*Spodoptera frugiperda*) infected with the recombinant viral vector pVL 1393/mas. rMaspin(i) was purified by a combination of anion exchange and heparin affinity chromatography, and accounted for 30%–40% of the total extractable protein. The identity of the two recombinant proteins (molecular 42 mass kDa) was confirmed by western blot analysis using the polyclonal antibody preparation AbS4A (Zou et al. 1994).

The identification of the probable reaction center was determined by cleavage of rMaspin(i) with limiting concentrations of trypsin (Sheng et al. 1994), leading to loss of the 42 kDa band and appearance of a 38 kDa band. Concomitant appearance of a 4227 Da fragment was detected by mass spectometry. The NH_2-terminal sequence of the 4227 Da fragment was ILQHKDELNAD, in agreement with the published cDNA sequence of amino acids located 3' to arginine (Sheng et al. 1994). These results show that the reactive center of the recombinant protein is readily accessible to trypsin cleavage, and suggest that maspin is an Arg-serpin.

rMaspin (y) was produced in yeast cells transformed with the expression vector pYMV4 (Pemberton et al. 1995). Recombinant maspin comprised at least 40% of the total soluble yeast protein. After purification by anion exchange chromatography on Q-sepharose, size fractionation S100HR columns, and affinity chromatography on heparin superflow, the average yield was 13.5 mg per gram wet weight of yeast (PEMBERTON et al. 1995). Yeast maspin is fully active biologically in the invasion assay (SAGER et al. 1994; SHENG et al. 1994).

4 Antibodies to Maspin

Polyclonal antibodies were produced in rabbits directed to three poorly conserved sequences, using conjugation of the corresponding synthetic oligopeptides to keyhole limpet hemocyanin. The selected sequences are underlined in Fig. 1 as S1A, S3A, and S4A. AbS4A recognizes the RSL encompassing the reactive site. All three antibody preparations react with a 42 kDa band present in proteins in extracts of normal cells, tumor transfectants that express maspin mRNA, and recombinant maspin protein on western blots. No 42 kDa protein was detected in extracts of mammary carcinoma cell lines that do not express maspin mRNA. All three antibody preparations are effective in protocols for immunostaining of formalin-fixed, paraffin-embedded tissue sections. Because of its high specificity, AbS4A is being used in retrospective and prospective studies of maspin expression in tumors for application to diagnostic and prognostic clinical evaluation.

5 Biological Properties of Maspin

5.1 Inhibition of Invasion in the Matrigel Assay

Classically, invasion is measured in cell culture by use of the Boyden chamber assay, in which cells are introduced into upper wells of the double-welled device and allowed to invade through a defined membrane into lower wells, where they are counted at suitable time points. In the maspin studies, MICS (membrane invasion culture system) manifolds comprising sets of 12 Boyden-like chambers were used. The perforated polycarbonate support membrane separating the upper and lower chambers was coated with growth factor reduced Matrigel as described (HENDRIX et al. 1987). Cells were removed from the bottom chambers at 24 or 48 h, fixed, stained, and counted microscopically.

The inhibitory effects of recombinant maspin added to tumor cells 1 h before transfer to MICS chambers are shown in the dose response graphs of Fig. 3. When the fusion protein rGST-maspin was used, a linear response curve was seen in the range from 0.04 to 0.77 μ*M* maspin. The control, recombinant GST alone, showed no effect. However, using the recombinant maspins from

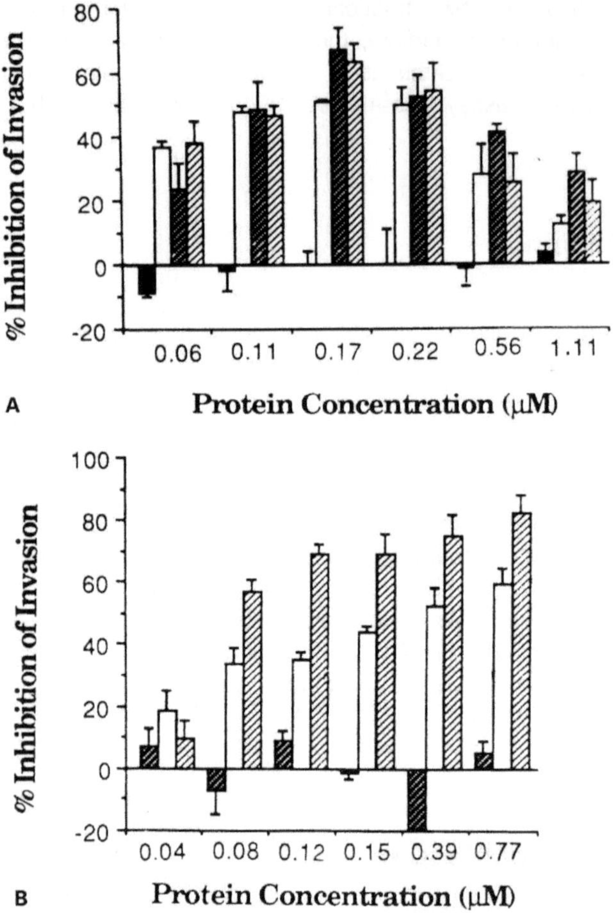

Fig. 3A,B. The effects of r-GST-maspin (**A**) and rMaspins (**B**) on invasion by two mammary carcinoma cell lines, MDA-435 and MDA-231. The invasion data from untreated tumor cells were normalized as 100% and invasion data from treated tumor cells are expressed as a percentage of this control. The values of "% inhibition of invasion" were obtained by substracting the invasion percentage of treated cells from 100%. Each value represents the average of the triplicate results and the error bars represents the S.E.M. **A** ■, invasion by MDA-435 in the presence of trypsin-cleaved rMaspin(y); ▨, invasion by MDA-435 in the presence of rMaspin(i); □, invasion by MDA-435 in the presence of rMaspin(y); and ▨, invasion by MDA-231 in the presence of rMaspin(i). **B** ■, invasion by MDA-435 in the presence of recombinant gluthathione S-transferase; □, invasion by MDA-435 in the presence of rGST-maspin; ▨, invasion by MDA-231 in the presence of rGST-maspin

insects (rMaspin(i)) or from yeast (rMaspin(y)), a bell-shaped curve was obtained with its peak at 0.17 μ*M*. Maspin cleaved within the RSL by trypsin was totally inactive. The inhibitory effects of maspin were fully reversible by antibody AbS4A (Sheng et al. 1994). The basis for the bell-shaped curve is not known, but it is suggested that interaction between maspin and its target on

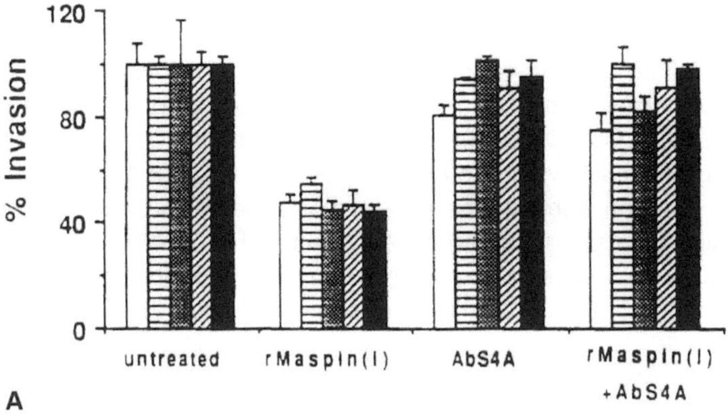

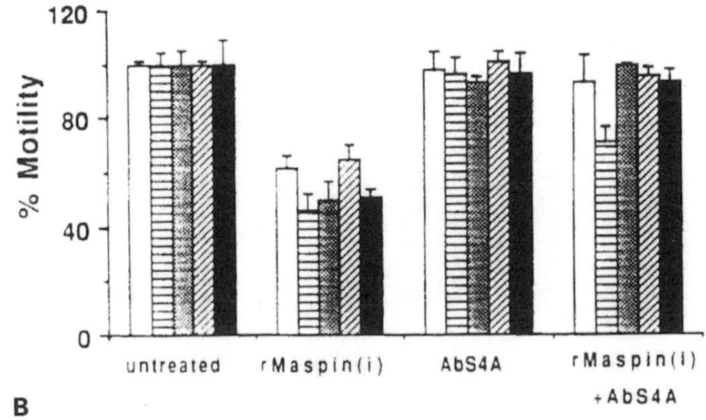

Fig. 4A,B. Effect of rMaspin(i) and AbS4A on invasion (**A**) and motility (**B**) of breast carcinoma cell lines 70N (□), ZR75-1 (⊟), 21NT (▩), MDA-MB-231 (▨) and MDA-MB-435 (■). Invasiveness and the motility rate of the untreated cells in each corresponding set of data were normalized as 100%, respectively, and the data from treated cells are expressed as a percentage of these controls. The data represent the average of three parallel experiments and the error bars represent the standard error

the membrane may induce a conformational change leading to polymerization and concomitant decreased activity. Polymerization is favored by a decreased pH, as occurs in cell culture. The GST-maspin fusion protein may by restricted in its ability to undergo this conformational change.

In further studies, the effect of rMaspin(i) on invasion was examined in a series of mammary carcinoma cell lines (ZR-75-1, 21NT, MDA-MB-231, and MDA-MB-435) and in normal cells (70N) (Sheng et al. 1995). Although the normal cells are not invasive in vivo, they show low level invasiveness in culture, which is inhibited by maspin. Fig. 4A shows inhibition of invasion by maspin and reversion by treatment with antibody.

5.2 Inhibition of Motility

Motility can be examined qualitatively by the Burk method in which a patch of cells in a confluent monolayer are removed, e.g., by scraping, and the movement of surrounding cells into the denuded area is observed. For quantitation, a modified Boyden assay or video time lapse recording can be used. All three methods have been used in maspin studies. The Burk assay readily revealed the inability of maspin treated cells to move. The modified Boyden assay gave quantitative results (Fig. 4B) showing the inhibition of motility measured as movement through a minimally coated Matrigel barrier in 6 h (SHENG et al. 1995).

For time-lapse video microscopy, cells were seeded onto coverslips coated with a combination of human laminin, collagen IV, and gelatin, and grown overnight, then treated with rMaspini(i) (or untreated) for 1 h prior to videotaping for 24 h. A Zeiss Axiovert 135 microscope equipped with a Focht environmental chamber and DIC optics was used for videotaping. Tapes were analyzed as described (SHENG et al. 1995).

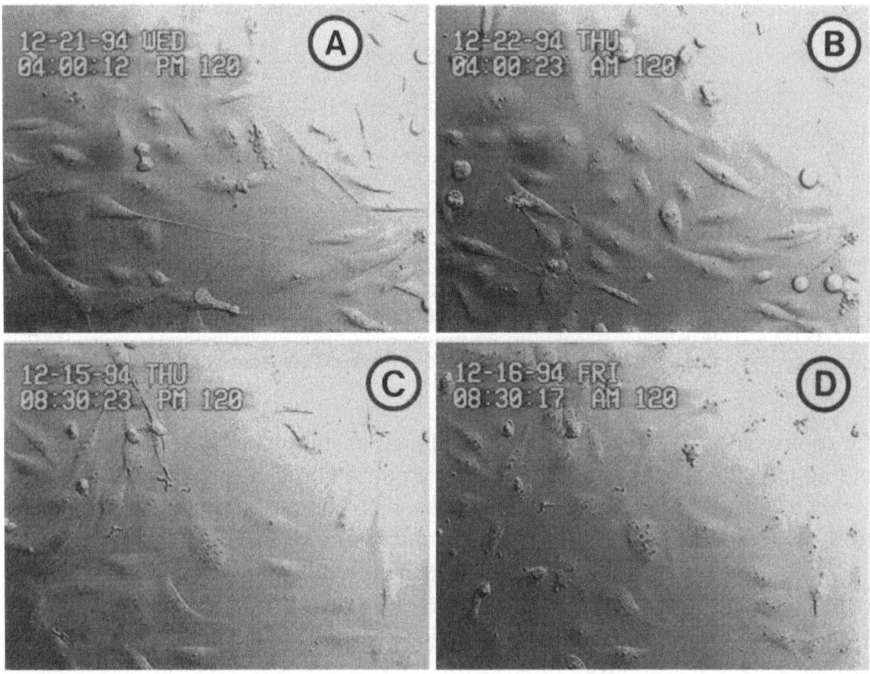

Fig. 5A–D. Photographs of video frames from time-lapse cinematography of MDA-MB-231 cells (motility data presented in Table 1) without rMaspin(i) (**A** and **B**) or with rMaspin(i) (**C** and **D**), recorded over a 12h interval in the Focht's environmental chamber. Note the relative inactivity of the cells treated with exogenous rMaspin(i) (0.47μ*M*). x 580

Table 1. Migration rate on extracellular matrix[a]

Cell line	–Maspin	+Maspin
MDA-MB-435	1.42±0.14 (*n*=6)	0.23±0.15 (*n*=9)
MDA-MB-231	4.48±0.38 (*n*=7)	0.19±0.12 (*n*=6)

[a]Data represent the average cell movement measured by video cinematography in μm/h with standard error(s)

As shown in Fig. 5 and summarized in Table 1, the migratory ability of tumor cell lines was reduced up to 75% by maspin compared with untreated controls over a 24 h period. The actual migration rate of cell line MDA-MB-231 was about three fold higher than that of MDA-MB-435, which is similar to their comparative metastatic potential in vivo (Thompson et al. 1992).

Two-dimensional migration across an extracellular matrix (ECM) involves attachment, detachment, and ultimately motility. Thus our data suggest that maspin, perhaps in combination with other proteins, inhibits this entire process.

5.3 Inhibition of Invasion and Motility in Prostate Cells

Because of the similarities of mammary and prostate tissues, both being secretory glands under steroid hormonal control, it seemed of interest to examine the effects of maspin on prostate cells in culture. Accordingly, invasion and motility assays were performed with three prostate cell lines: LNCaP, DU 145, and PC 3. The results (Fig. 6) show that invasion is strongly reduced by rMaspin(i) and that this inhibition is partially reversed by the AbS4A antibody. Motility is reduced about 40%–60%, similar to results with breast carcinoma cells, and reversible by antibody.

These results raise the novel possibility that maspin will have similar therapeutic potential in breast and prostate cancers. To follow up this lead, investigations have been initiated using recombinant maspin in animal testing. Preliminary studies in nude mice with breast tumor transfectants expressing maspin have been encouraging and will be presented below.

5.4 Inhibition of Tumor Growth and Metastasis by Maspin in Nude Mice

In initial studies (Zou et al. 1994) it was reported that tumor transfectants (MDA-MB-435 tumor cells expressing transfected maspin) were inhibited in tumor cell growth and metastasis compared with mock transfectants not

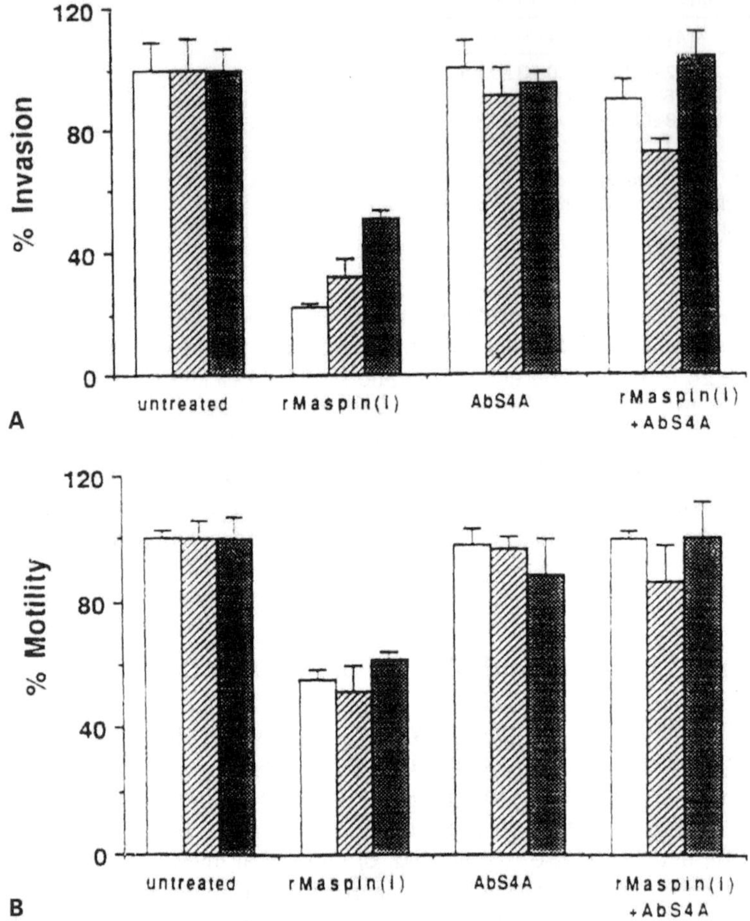

Fig. 6A,B. Effect of rMaspin(i) and AbS4A on the invasion (**A**) and motility (**B**) of prostate cancer cell lines LNCaP (□), DU145 (▨) and PC3 (■) Invasiveness and the motility rate of the untreated cells in each corresponding set of data were normalized as 100%, respectively, and the data from treated cells are expressed as a percentage of these controls. The data represent the average of three parallel experiments and the error bars represent the standard error

expressing maspin in the nude mouse assay. In the initial study, three out of four transfectants expressed this inhibition. The fourth cell line showed increased tumor growth in this study and elevated invasion in MICS chamber studies, suggesting an elevated invasive potential in this clone. These initial studies have now been repeated with a new set of five transfectants of similar origin, and four out of five of these cell lines have shown inhibition of tumor growth compared with controls (Table 2). Thus it seems evident that expression of maspin in tumor transfectants inhibits the growth rate of the primary tumors as well as their metastatic potential.

Table 2. Tumorigenicity of maspin-transfected MDA-MB-435 cells[a]

Cells	Median weight (g)[b]	*p* value[c]
Nn12	1.60	
Tn8	0.60	0.019
Tn13	0.45	0.024
Tn15	0.75	0.016
Tn16	0.36	0.0072
Tn45	0.60	0.190

[a]Transfected cell (5×10^6) were injected into the mammary fat pads of nude mice. Each mouse was injected at two sites and the tumor development was monitored weekly

[b]Tumor weights were measured 12 weeks postinjection

[c]*p*-values were obtained by Wilcoxon rank sum test using null hypothesis

5.5 Loss of Maspin Expression in Mammary Tumor Tissue

In initial studies of benign and malignant breast cancer tissues immunostained with maspin antibody, benign breast tissues (n=6) and benign epithelium adjacent to invasive carcinomas were maspin positive, with intense staining in myoepithelial cells and more heterogeneous staining in luminal cells. Inflammatory and stromal cells were negative. Twenty specimens of invasive primary carcinoma, lymph node and distant metastases were also evaluated. Most malignant cells in invasive carcinomas failed to express maspin but a minority of cells in well differentiated tumors expressed maspin focally. Maspin was undetectable or very weakly expressed in lymph node and distant metastases (Zou et al. 1994).

6 Summary

Maspin, a serpin found in mammary epithelial cells, has been shown to have tumor suppressor activity. The gene is expressed in normal human mammary epithelial cells but down-regulated in invasive breast carcinomas. Similar patterns of expression at the RNA and protein levels are seen by Northern analysis with cells grown in culture and by immunostaining of tissues. Biological assays of invasion by tumor cells through matrigel membranes and of motility have shown that recombinant maspin inhibits both processes, and that its inhibitory action is totally lost by a single cleavage at the reaction center. Tumor transfectants expressing maspin are inhibited in growth and invasion in nude mice. Maspin is located in the cell membrane and extracellular matrix, and does not behave as a classical inhibitory serpin against any known target protease. Its mode of action is presently unknown.

Acknowledgments. This work was supported by grant CA61253 to R.S. from the National Institutes of Health, and by LXR Biotechnology, Inc., Richmond, CA.

References

Carrell RW, Evans DLI (1992) Serpins: mobile conformations in a family of proteinase inhibitors. Cur Opin Cell Biol 50: 438–446

Coughlin P, Sun J, Cerruti L, Salem HH, Bird P (1993) Cloning and molecular characterization of a human intracellular serine protease inhibitor. Proc Natl Acad Sci USA 90: 9417–9421

Hendrix MJC, Seftor EA, Seftor REB, Fidler IJ (1987) A simple quantitative assay for studying the invasive potential of high low human metastatic variants. Cancer Lett 38: 137–147

Hopkins PCR, Whisstock J, Sager R (1994) Function of maspin. Science 265: 1893–1894

Huber R, Carrell RW (1989) Implications of the three-dimensional structure of al-antitrypsin for structure and function of serpins. Biochemistry 28: 8951–8966

Pemberton PA, Wong DT, Gibson HL, Kiefer MC, Fitzpatrick PA, Sager R, Barr PJ (1995) The tumor suppressor maspin does not undergo the S-R transition or inhibit trypsin-like serine proteases. J Biol Chem (in press)

Potempa J, Korzus E, Travis J (1994) The serpin superfamily of proteinase inhibitors: structure, function, and regulation. J Biol Chem 269: 15957–15960

Sager R, Sheng S, Anisowicz A, Sotiropoulou G, Zou Z, Stenman G, Swisshelm K, Chen Z, Hendrix MJC, Pemberton P, Rafidi K, Ryan K (1994) RNA genetics of breast cancer: maspin as paradigm. Cold Spring Harb Symp Quant Biol 59: 537–546

Schneider SS, Schick C, Fish KE, Miller E, Pena JC, Treter SD, Hui SM, Silverman GA (1995) A serine proteinase inhibitor locus at 18q21.3 contains a tandem duplication of the human squamous cell carcinoma antigen gene. Proc Natl Acad Sci USA 92: 3147–3151

Sheng S, Pemberton PA, Sager R (1994) Production purification and characterization of recombinant maspin proteins. J Biol Chem 269: 30988–30993

Sheng S, Carey J, Hendrix MJC, Seftor EA, Dias L, Sager R (1995) Maspin inhibits the invasion of mammary and prostatic tumor cell lines in culture by inhibiting cell motility (submitted)

Thompson EW, Paik S, Brunner N, Sommers CL, Zugmaier G, Clarke R, Shima TB, Torri J, Donahue S, Lippman ME, Martin GR, Dickson RB (1992) Association of increased basement membrane invasiveness with absence of estrogen receptor and expression of vimentin in human breast cancer cell lines. J Cell Physiol 150: 534–544

Wei A, Rubin H, Cooperman BS, Christianson DW (1994) Crystal structure of an uncleaved serpin reveals the conformation of an inhibitory reactive loop. Struct Biol 1: 251–258

Zou Z, Anisowic A, Hendrix MJC, Thor A, Neveu M, Sheng S, Rafidi K, Seftor E, Sager R (1994) Identification of a novel serpin with tumor suppressing activity in human mammary epithelial cells. Science 263: 526–529

Different Roles for Plasminogen Activators and Metalloproteinases in Melanoma Metastasis

B.M. Mueller

1 Introduction

The architecture of normal tissues is maintained by interactions between cells and the extracellular matrix (ECM). In this cooperation, cells produce and assemble the components of the ECM whereas the ECM stabilizes the physical structure of tissues and also regulates the behavior of cells (McDonald 1988, Adams and Watt 1993). Transformed cells can alter the composition and structural integrity of adjacent matrix promoting tumor progression. Enzymes produced by tumor cells can degrade different components of the ECM, such as collagens, glycoproteins and proteoglycans, and such matrix-degrading enzymes were shown to contribute to tumor invasion and metastasis in various experimental systems (for review of the extensive literature see Mareel et al. 1991; Mignatti and Rifkin 1993). The current view emerging from these studies is that the concerted action of several different enzymes, whose activities are modulated by complex control mechanisms, is required in malignancy. The functional role of the enzymatic degradation of ECM by tumor cells has been thought to be the dissolution of occluding matrix, thus enabling the cells to invade into the interstitial matrix of the mesenchyme at the site of a primary tumor, to cross the basement

---

The Scripps Research Institute, Department of Immunology, 10666 N. Torrey Pines Road, IMM13, La Jolla, CA 92037, USA

membrane of blood vessels in the process of intra- and extravasation and lastly to invade into the mesenchyme of a secondary site. More recently, it has been recognized that enzymatically altered matrix also provides a complex source of molecular information that can modulate tumor cell behavior including adhesion, survival, growth and migration. It appears now that spatially and temporally limited matrix degradation rather than continuous, widespread enzymatic activity is required for the successful dissemination of the tumor (LIOTTA et al. 1991; MAREEL et al. 1991; MIGNATTI and RIFKIN 1993; DECLERCK and IMREN 1994).

Cutaneous melanoma is an invasive and metastatic tumor that has received considerable interest. From an epidemiological and clinical point of view, malignant melanoma is important, because the rate of incidence is continuously rising and because metastatic melanoma is resistant to conventional radio- and chemotherapy. For the study of tumor biology, malignant melanoma is important, because tumor progression is well defined starting from benign melanocytic nevi, dysplastic nevi, radial growth phase melanoma, invasive vertical growth phase melanoma to melanoma metastasis (CLARK 1991). Also, a large number of human melanoma cell lines with varying invasive and metastatic potential are available for studies in vitro and in experimental animals. Current information on the molecular and cellular biology of human melanoma has been reviewed by HERLYN (1993). Here, we focus on matrix-degrading activities elaborated by human melonama cells and discuss our recent studies that address the synergism between different proteases in a human melanoma model.

2 Proteases Produced by Human Melanoma Cells

Human melanoma cells produce the same matrix-degrading enzymes that have been implicated in normal invasive processes, such as trophoblast invasion or leukocyte diapedesis and in the biology of other malignant tumors. These enzymes include proteolytic enzymes that are classified according to the nature of their active centres as serine proteases, metalloproteases, and cysteine proteases, as well as glycosidases that can degrade the carbohydrate moieties of proteoglycans.

2.1 Plasminogen Activators

The plasminogen activator (PA)/plasmin system has been biochemically well characterized and its role in tissue remodeling, particularly in malignancy, has been the subject of several reviews (DANØ et al. 1985; VASSALLI et al. 1991; KWAAN 1992). The two known mammalian PAs, tissue type PA (tPA) and urokinase type PA (uPA), are serine proteases with narrow specificity cleaving the ubiquitous zymogen plasminogen and thus generating plasmin. Plasmin is a broad acting serine protease that can degrade fibrin and the matrix glycoproteins fibronectin and laminin and is also able to contribute to matrix degradation by activating some latent metalloproteases. The PA/plasmin system is tightly controlled by specific inhibitors, including the PA inhibitors (PAI) and the plasmin inhibitor α2 antiplasmin. Pericellular plasminogen activation is enhanced on the cell surface and is regulated by cell surface receptors, as has been most clearly demonstrated for the uPA receptor (uPAR) (ELLIS et al. 1992; BLASI 1993).

Human melanoma cell lines have been shown to produce both tPA (RIJKEN and COLLEN 1981; HOAL-VAN HELDEN et al. 1986; VAHERI et al.1994) and uPA (QUAX et al. 1991; MEISSAUER et al. 1991; MONTGOMERY et al. 1993). It appears that the production of tPA is somewhat unique to human melanoma cells, whereas the production of uPA has been described for a large number of cell lines derived from malignant tumors of different histological origins.

Similar to other malignancies (TESTA and QUIGLEY 1990), the elaboration of uPA activity by human melanoma cell lines correlates with an highly invasive and metastatic phenotype. In a study of six melanoma cell lines by QUAXet al. (1991), all were found to express both tPA mRNA and protein, and grew to form subcutaneous tumors in immunodeficient mice. Only two cell lines, MV3 and BLM, which produced uPA, PAI-1 and PAI-2 and were able to degrade ECM in vitro in a uPA dependent fashion, grew in mice as invasive subcutaneous tumors that consistently developed spontaneous metastasis (QUAX et al. 1991). Similarly, two human melanoma cell lines, M24met and C8161, which are highly invasive in vitro and aggressively metastatic in mice, produce uPA (Table 1). In contrast, low net PA activity, which was mostly tPA activity, was detected for a cell line, WM35, established from a primary melanoma and another cell line, M21, originally derived from a melanoma metastasis, but poorly metastatic in mice. It is a noteworthy that C8161 cells secreted roughly 50 times more uPA activity into the conditioned media than M24met cells, while these two cell lines are quite similar in their metastatic behavior. The cell surface bound PA activity of both cell lines, detected in acidic eluates from the cell surface, was very similar (Table 1), suggesting that cell surface rather than total uPA activity determines the metastatic capability of these cells.

The importance of cell surface proteolytic activity for melanoma invasion has been emphasized by MEISSAUER et al. (1992). These authors show that MelJuso cells, which use cell surface bound uPA for plasminogen activation, are more invasive through ECM or layers of keratinocytes than MeWo cells, which activate plasminogen by secreted tPA much less efficiently. In vitro invasiveness of both cell lines is enhanced in the presence of cell-associated plasmin. In this regard, plasminogen (MEISSAUER et al. 1992) and tPA (BIZIK et al. 1993) can be eluted with lysine analogues from the surface of melanoma cells suggesting the presence of cell surface receptors. Candidates for such receptors are α-enolase (MILES et al. 1991) and annexin II (HAJJAR et al. 1994), which

Table 1. Activity of plasminogen activators in human melanoma cell lines

Origin	Cell line		PA activity	Inhibition (%)			
			(IU/10E6 cells)	PAI-2 (1 μM)	PAI-2 (100 nM)	anti-uPA (50 nM)	anti-tPA (50 nM)
Primary melanoma	WM35	Secreted	3.8	89.4	72.4	5.7	88.2
		cell surface	9.8	84.6	60.8	1.7	81.6
Melanoma metastasis	M21	Secreted	3.8	93.3	71.2	4.9	82.1
		cell surface	2.9	n.d.	n.d.	8.1	71.5
Melanoma metastasis	M24met	Secreted	42.8	88.0	73.3	85.2	13.4
		cell surface	37.2	91.9	83.9	80.7	0
Melanoma metastasis	C8161	Secreted	2828.4	89.7	70.6	87.9	0
		cell surface	31.5	n.d	n.d.	86.7	4.3

PA activity was measured in a solid phase fibrinolytic assay. Secreted activity was measured in 3 day cell conditioned media and cell surface activity was measured in pH3 eluates

n.d., not determined

have been demonstrated to function as cell surface binding proteins for plasminogen and tPA on monocytic and endothelial cells.

Metastatic human melanoma cell lines, including M24met cells, express high levels of a receptor for uPA that is biochemically and immunologically identical with uPAR described on monocytic cells and other tumor cells (STAHL and MUELLER 1994). M24met cells utilize uPA for matrix degradation and invasion efficiently only when it is bound to its receptor. uPA binding to uPAR also stimulates haptotactic migration of M24met cells independent of uPA's catalytic activity (STAHL and MUELLER 1994). Similarly, uPA can act as an autocrine growth factor for the human melanoma cell line GUBSB, which also expresses the high affinity uPAR (KIRCHHEIMER et al. 1989). Thus, uPAR may contribute to melanoma metastasis by enhancing pericellular proteolysis of occluding ECM and by acting as a signaling receptor to activate tumor growth and motility. We have recently demonstrated that uPAR on the surface of M24met melanoma cells is mainly localized in caveolae (STAHL and MUELLER 1995), flask-shape micro-invaginations of the plasma membrane that cluster cell surface receptors and signal transduction molecules in a characteristic lipid environment (ANDERSON 1993; LISANTI et al. 1994). The function of caveolae is currently not well understood, but the structural integrity of caveolae on melanoma cells appears to be required for efficient plasmin generation on the cell surface (STAHL and MUELLER 1995) possibily by virtue of clustering uPAR with other protease receptors. Furthermore, caveolae may also provide a specialized compartment that allows for efficient signal transduction through the uPAR.

Recently, two studies examined the expression of plasminogen activators in melanoma tumor progression. DE VRIES et al. (1994) examined benign and dysplastic nevi, early and advanced primary melanomas, and melanoma metastases by immunohistochemistry and RNA in situ hybridization. They found tPA in the endothelial cells of all lesions and in the tumor cells of a few metastatic tumors. In contrast, uPA, uPAR, PAI-1 and PAI-2 was not detected in benign and early stages, but appeared frequently in tumor cells of advanced primary melanoma and metastasis, indicating that plasminogen activation is a late event in melanoma tumor progression. In a similar study, DELBALDO et al. (1994) also detected uPA mRNA in some atypical nevi, but found uPA activity, as measured by histological zymography, exclusively in malignant melanomas where it was preferentially localized with tumor cells at borders of malignant nodules.

2.2 Matrix Metalloproteinases

The matrix metalloproteinases (MMPs) are a family of zinc-dependent matrix-degrading enzymes. The structure and function of MMP have been reviewed in detail by WOESSNER (1991) and MATRISIAN (1992). Based on substrate specificity, MMPs can be subdivided into three subclasses: collagenases, gelatinases and stromelysins. Interstitial collagenase and neutrophil collagenase are characterized by their ability to cleave fibrillar collagens, i.e., type I, II and III collagen. Both 72 kDa gelatinase A and 92 kDa gelatinase B have specificity for denatured collagens (gelatins) and for type IV basement membrane collagen. The stromelysin subclass consisting of stromelysin 1, stromelysin 2 and matrilysin has a broader specificity with major substrates including glycoproteins, proteoglycans and type IV collagen.

MMP synthesis and secretion is inducible by a variety of biologically active reagents, such as growth factors, oncogenes and tumor promoters (Frisch and Werb 1989). The activity of secreted MMPs is regulated by their requirement for proteolytic activation and by specific inhibitors (Murphy et al. 1994) Latent MMP can be activated by plasmin and other proteases. For example, a recently discovered cell surface MMP (Sato et al. 1994) specifically cleaves and activates gelatinase A. MMPs are inhibited by tissue inhibitors of metalloproteinase (TIMP), a family of glycoproteins with two well characterized members, namely TIMP-1 and TIMP-2 (Denhardt et al. 1993).

MMPs are secreted by normal cells, including leukocytes and fibroblasts, and are frequently overexpressed in malignant tumors. The experimental evidence implicating MMP activity in tumor invasion and metastasis has been reviewed by Stetler-Stevension et al. (1993). In this regard it has been demonstrated that: (a) MMP production is positively correlated with invasive and metastastic behavior of tumor cell lines; (b) growth factors and oncogenes upregulate the production of MMP; (c) cell-matrix interactions which promote cellular invasiveness also enhance MMP production; (d) MMP expression correlates with malignancy in various human tumors; and (e) TIMPs inhibit tumor invasion and experimental metastasis (Stetler-Stevenson et al. 1993).

Human melanoma cell lines were reported to produce gelatinase A, gelatinase B, interstitial collagenase and stromelysin. Expression of gelatinase A is found in virtually all cell lines studied, whereas expression of gelatinase B appears to be restricted to more invasive and metastatic cells lines (Hendrix et al. 1990; Welch et al. 1991; Montgomery et al. 1993; MacDougall et al. 1995). The metastatic potential of three human melanoma cell lines and their ability to invade through reconstituted basement membrane correlated positively with the expression of gelatinases and type IV collagenolytic activity (Welch et al. 1991). Information on the expression of MMPs in different stages of melanoma progression is currently not available.

Complex interactions between MMP expression and activity and cell adhesion, particularly mediated by integrins, have been described for human melanoma cells. Integrin signaling can regulate MMP expression. For example, binding of the vitronectin receptor $\alpha v\beta_3$ or the fibronectin receptor $\alpha_5\beta_1$ to their cognate ligands increases gelatinase A expression in A375M and C8161 human melanoma cells, respectively (Seftor et al. 1992; 1993). Alternatively, collagenolytic activity can expose or destroy adhesive substrates and thus modulate cell function. When A2058 human melanoma cells were transfected to overexpress TIMP-2 in order to lower their collagenolytic activity, the resulting transfectants showed increased cellular adhesion to matrix proteins and decreased chemotactic migration (RAY and STETLER-STEVENSON 1994). Controlled collagenolysis may expose cryptic Arg-Gly-Asp (RGD) sequences in dermal type I collagen and thereby promote RGD-dependent adhesion by integrin $\alpha_v\beta_3$ (MONTGOMERY et al. 1994b). Ligation of this integrin may in turn promote melanoma cell survival, growth and metastasis (MONTGOMERY et al. 1994b).

2.3 Other Matrix-Degrading Enzymes

Cathepsins B, L and H are lysosomal cysteine proteinases that are active at acidic pH. Tumor cells frequently release cathepsins and extracellular cathepsins can act as endo- or ectopeptidases on various proteins in the ECM. Cathepsin B activity has been positively correlated with the metastatic ability of tumor cell lines and with tumor progression (SLOANE et al. 1990). Extracellular or plasma membrane-associated cathepsins have been described in human melanoma cell lines including cathepsin B (TSUSHIMA et al. 1991a), cathepsin L (ROZHIN et al. 1989) and a cathepsin H-like protein (TSUSHIMA et al. 1991b). Cysteine proteinase inhibitors were described to inhibit melanoma cell motility (BOIKE et al. 1991) implying a role for cathepsins in cell migration.

A 170 kDa membrane protease complex with gelatinase activity, named seprase, has been described in invasive melanoma cell lines (AOYAMA and CHEN 1990; MONSKY et al. 1994). Consistent with a proposed role in melanoma invasion, seprase is localized in a specialized membrane region, invadopodia, at the invasive front of melanoma cells.

Aminopeptidase N was identified as another cell surface protease on human melanoma cells (MENRAD et al. 1993). Aminopeptidase N activity contributes to melanoma cell invasion through reconsistuted basement membrane and appears to specifically degrade nidogen/entactin.

A unique endo-β-D-glucuronidase, heparanase, with specificity for heparan sulfate proteoglycans is produced by many human melanoma cell lines. Heparanase activity has been implicated in melanoma metastasis in mice and elevated levels of heparanase have been detected in sera from malignant melanoma patients (for review see NAKAJIMA et al. 1988).

3 Melanoma-Mediated Dissolution of Extracellular Matrix In Vitro

Malignant tumor cells usually produce several matrix-degrading activities with different substrate specificities that are thought to cooperate in the degradation of complex matrices. For example, invasion of mouse B 16 melanoma cells through amniotic membrane can be inhibited with either inhibitors of uPA and plasmin or by collagenase inhibitors (Mignatti et al. 1986). Similarly, the ability of mouse melanoma cells and human fibrosarcoma or ovarian carcinoma cells to invade through reconstituted basement membrane can be inhibited by serine protease inhibitors, uPA inbibitors or type IV collagenase (gelatinase) inhibitors, but not by an elastase inhibitor (Reich et al. 1988). Human squamous carcinoma cell lines have to express uPA, the uPA receptor and interstitial collagenase to effectively invade through the chorioallantoic membrane of chick embryos (Ossowski 1992).

The human melanoma cell line M24met was demonstrated to produce uPA, and several MMPs, including gelatinase A, gelatinase B, interstitial collagenase and small amounts of stromelysin (Montgomery et al. 1993). While the cells also express protease inhibitors, including PAI-1, PAI-2, TIMP-1 and TIMP-2, the endogenously produced protease activities were found to exceed these of the inhibitors. Thus, uPA and MMP-dependent activities enable M24met cells to degrade various matrix proteins, including the major components of the interstitial stroma, fibronectin and type 1 collagen, as well as laminin and type IV collagen, major components of basement membranes.

Degradation of interstitial matrix by M24met cells was examined with ^{3}H-proline-labeled matrices produced by rat smooth muscle cells. In these matrices, approximately 49% of the labeled proline was incorporated into glycoprotein, 27% into elastin and 29% into collagen. The kinetics of interstitial matrix degradation by M24met cells in the presence or absence of the specific recombinant inhibitors PAI-2 and TIMP-2 is shown in Fig. 1. Untreated M24met cells caused significant accumulated degradation over the 14-day culture period. In the presence of PAI-2 or TIMP-2, the total accumulated degradation was inhibited by 50% and 30%, respectively. During the first 5 days of culture, there was a relatively rapid hydrolysis of the matrix by untreated M24met cells. During this early phase, less than 10% of the observed hydrolysis was inhibited by TIMP-2. In contrast, upto 50% of this early activity was inhibited by PAI-2. These findings suggest an early and rapid uPA-dependent hydrolysis, with a relatively minor contribution from a metalloproteinase-dependent pathway. After day 5, untreated M24met cells degraded the matrix at a slower, but constant, rate. During this late phase, TIMP-2 inhibition increased inhibiting approximately 80% of the degradation occurring between days 9 and 14. These results suggest the delayed activity of a metalloproteinase-dependent proteolytic pathway that predominates during the later stages of matrix hydrolysis. Analysis of the matrix composition at different time points confirmed that the majority of the matrix glycoprotein was digested in the first 7 days, while collagen degradation mainly took place between days 7 and 14. These data suggest that the interstitial glycoproteins function to protect the fibrillar collagen from proteolytic degradation.

The degradation of subendothetlial matrix by M24met cells demonstrated a similar temporal relationship between uPA and MMP activities. These results led to the conclusion that, in vitro, both uPA and MMP activities are

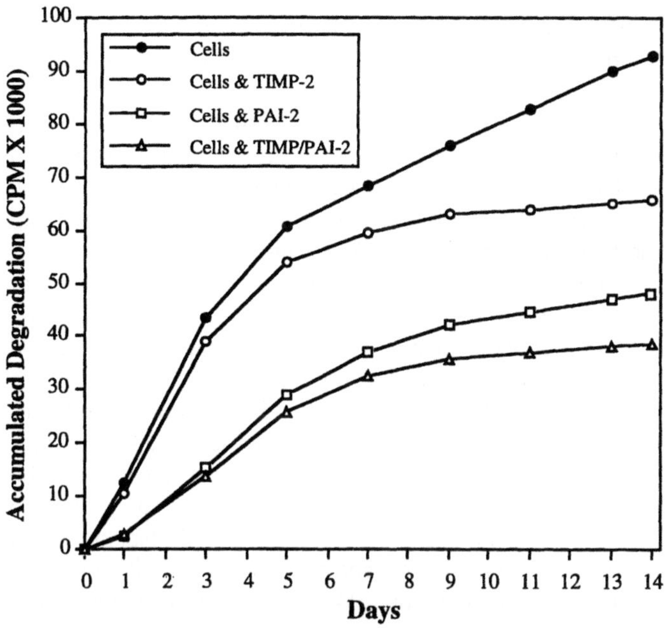

Fig. 1. Contribution of melanoma cell-associated urokinase type plasminogen activator (uPA)- and matrix metalloproteinase (MMP)-dependent proteolysis to the degradation of interstitial matrix M24met cells in RPMI and 10% acid-treated fetal bovine serum were plated on smooth muscle cell matrices labeled with [^{3}H] proline. Tissue inhibitor of metalloproteinase (TIMP)-2 and plasminogen activator inhibitor (PAI)-2 were added to this media at 20 or 45 µg/ml, respectively. Aliquots of 200 µl supernatant were harvested for counting every 2 days and the culture media were replaced with fresh media and inhibitors. Results were expressed as accumulated degradation and represent the mean release of triplicate samples over background. (From MONTGOMERY et al. 1993)

involved in the degradation of occluding extracellular matrix by malignant human melanoma cells and that uPA-dependent removal of matrix glycoprotein must precede MMP-dependent collagenolysis as a prerequisite rate-limiting step (MONTGOMERY et al. 1993). Thus, cooperation between different proteolytic activities may be imposed by the structural architecture of the ECM in which interstitial glycoproteins can function to protect the fibrilar collagen from proteolytic degradation.

4 In Vivo Studies Implicate Different Roles for Proteolytic Activities

Human M24met melanoma cells metastasize very efficiently from subcutaneous tumors to lymph nodes and lungs of SCID mice (MULLER et al. 1991). To study the contribution of proteolytic activities to melanoma invasion and metastasis in vivo. M24met cells were transfected to overexpress protease inhibitors. Specifically, PAI-2 was overexpressed to address the role of the uPA/plasmin system and TIMP-2 was overexpressed to address the role of MMP in melanoma metastasis.

Clonal M24met cell lines stably transfected with human PAI-2 cDNA produced two to nine times more PAI-2 protein in cell lysates and cell-conditioned media than either parental M24met cells or mock transfectants (Mueller et al. 1995). The clone with the highest PAI-2 expression, PAI-2-4, exhibited complete inhibition of soluble and cell-surface bound PA activities. The level of PAI-2 overexpression for the clonal cell lines correlated positively with the inhibition of their ability to degrade complex matrix in vitro. Parental, mock-transfected and PAI-2 transfected M24met cells produced rapidly growing tumors when injected into the flank of SCID mice. The tumors producing the highest levels of PAI-2 were surrounded by a dense tumor capsule. Both parental and mock-transfected cells invariably metastasized from the subcutaneous tumors to lymph nodes and lungs of the mice. PAI-2-transfected cell lines produced little or no macroscopic foci with the degree of the inhibition of spontaneous metastasis correlating well with the overexpression of PAI-2. These data indicate that PAI-2 mediated inhibition of uPA activity suppresses human melanoma invasion and metastasis in SCID mice and provide direct evidence for a requirement of uPA activity in this animal model (Mueller et al. 1995).

Degradation of occulding ECM as a major function of tumor associated PA activity has been previously implicated in different experimental systems in vivo. Anticatalytic anti-uPA antibodies inhibit invasion and metastasis of a human squamous carcinoma cell line in chick embryos (Ossowski and Reich 1983); local invasion of human tumors grown in mice (Ossowski et al. 1991), and experimental metastasis of murine melanoma cells in mice (Hearing et al. 1988). Overexpression of uPA in poorly metastatic murine melanoma cells (Yu and Schultz 1990) and in H-*ras*-transformed NIH 3T3 fibroblasts (Axelrod et al. 1989) enable these cells to form experimental metastases in mice. Human fibrosarcoma cells that overexpress PAI-2 are less invasive in vivo and form encapsulated tumors (Laug et al. 1993). Overexpression of enzymatically inactive uPA reduced spontaneous metastasis of prostate carcinoma cells thus indicating a requirement for cell surface bound uPA activity in this system (Crowley et al. 1993).

Stable transfection of M24met with human TIMP-2 cDNA resulted in clonal cell lines producing up to eight times more TIMP-2 than the parental cell line (Montgomery et al. 1994a). Transfectants overexpressing TIMP-2 demonstrated an effective inhibition of collagenolytic activity. Overexpression of TIMP-2 markedly reduced melanoma growth in the skin of SCID mice, but did not prevent these cells from spontaneously metastasizing to the lungs and lymph nodes of inoculated mice. No difference in growth rates of parental and TIMP-2 overproducing M24met cells was observed when the cells were maintained on untreated or gelatin coated plastic. In contrast, TIMP-2 overexpressing melanoma cells assumed a differentiated morphology and were significantly growth-inhibited compared to the parental cells when plated in three-dimensional interstitial collagen gels. Similarly, the growth of parental M24met cells in three-dimensional collagen gels was inhibited in the presence of recombinant TIMP-2. These data suggest that in a three-dimensional collagen matrix and in the dermal collagen of the mouse skin TIMP-2 has a growth inhibitory effect on human melanoma cells indicating a requirement for MMP activities and collagenolysis to promote melanoma growth (MONTGOMERY et al. 1994a).

MMP activities have been linked to metastatic spread. For example, mouse B16-F10 melanoma cells overexpressing TIMP-1 are less invasive in vitro (KHOKHA et al. 1992a) and have a lower metastatic potential in chick embryos (KHOKHA et al. 1992b) and in mice (KHOKHA 1994). Transfection of TIMP-2 into a transformed rate embryo cell line partially inhibited experimental metastasis (DECLERCK et al. 1992). Overexpression of TIMP in these model system; however, also resulted in a profound suppression of tumor growth (DECLERCK et al. 1992; KHOKHA 1994). Interestingly, TIMP-1 overexpressing B16-F10 cells were shown by intravital vidoemicroscopy of chick embryo choriallantoic membranes to be not affected in their ability to extravasate, but rather in tumor growth post extravasation (KOOP et al. 1994). A role for MMP activity in promoting tumor growth is also supported by the in vivo effects of the synthetic MMP inhibitor batimastant. In the B16BL5 mouse melanoma model, batimastast inhibits the growth of subcutaneous tumors and spontaneous metastases, but not the number of pulmonary metastases in the spontaneous metastasis assay (CHIRIVI et al. 1994). Similarly, the growth of orthotopic human colon tumors in nude mice was inhibited by batimastat (WANG et al. 1994). Furthermore, expression of matrilysin in human colon carcinoma cell lines was not sufficient for tumor invasion and metastasis, but contributed to tumorgenicity and progression of colorectal tumors (WITTY et al. 1994). Thus, different in vivo models support the conclusion that the promotion of tumor growth is clearly an important function of MMP activities.

It is currently not clear by what mechanism MMPs promote tumor growth. Blocking of endothelial MMP by MMP inhibitors might result in suppression of tumor neovascularization and, indirectly, inhibition of tumor growth. In this regard, metalloproteinase inhibitors have been shown to inhibit angiogenesis (MOSES and LANGER 1991; JOHNSON et al. 1994). Alternatively, proteolytic alteration of occluding ECM might be a requirement for tumor cell growth, a mechanism that is supported by the finding that a similar degree of growth inhibition was observed for TIMP-2 overexpressing M24met cells in the subcutaneous environment and in three-dimensional collagen gels in vitro (MONTGOMERY et al. 1994a). MMP activity may release matrix-bound growth factor or fragments of ECM proteins with growth factor activity. Limited proteolysis of matrix proteins may also expose new adhesive ligands for cellular adhesion molecules and, in turn, promote proliferation and tumorigenesis.

The effect of PAI-2 and TIMP-2 overexpression on the survival of M24met metastases bearing SCID mice was compared directly. Compared to mice inoculated with M24met tumors, PAI-2 overexpression resulted in markedly

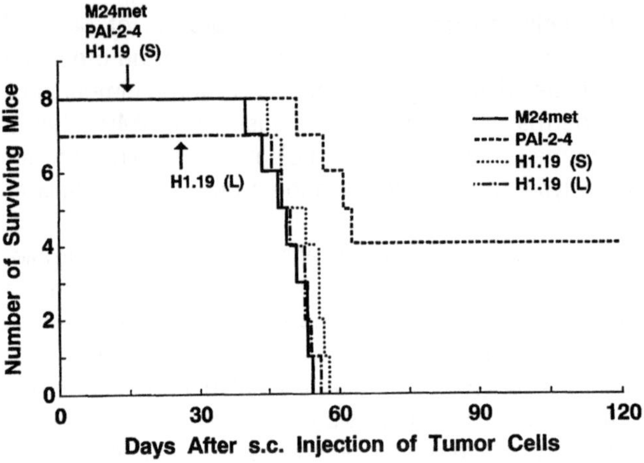

Fig. 2. Survival of SCID mice bearing M24met or transfectants in the spontaneous metastasis assay. Groups of 6 week old, female SCID mice were injected subcutaneously into the flank with 2 × 10 cells of either M24met human melanoma cells, plasminogen activator inhibitor (PAI)-2 overexpressing M24met (PAI-2-4) or tissue inhibitor of metalloproteinase (TIMP)-2 overexpressing M24met (H1.19). On day 15, subcutaneous tumors were removed surgically with mean weights for M24met tumors being 0.80 + 0.19 g, for PAI-2-4 tumors being 0.56 + 0.14 g and for H1.19(S) being 0.12 + 0.03 g. TIMP-2-overexpressing tumors were excised from a second group, H1.19(L), on day 26 with a mean tumor weight of 0.78 + 0.22 g

prolonged survival for all mice and in 50% long-term survivors (Fig. 2). To adjust for the slower growth rate of subcutaneous TIMP-2 overexpressing M24met tumors, two experimental groups were tested. In one group, small subcutaneous tumors were excised at the same time as parental and PAI-2 expressing tumors. In a second group, TIMP-2 overexpressing tumors were allowed to grow to the same volume as the control groups. Interestingly, in either situation, the survival of mice with TIMP-2 overexpressing tumors was not prolonged compared to mice bearing M24met tumors (Fig. 2) and all animals had massive pulmonary metastasis at the time of their death. These data confirm the earlier observation that TIMP-2 overexpression does not interfere with spontaneous metastasis in the M24met model (Montgomery et al. 1994a).

From our results in the spontaneous melanoma metastasis model in SCID mice, we conclude that there are different roles for proteolytic activities produced by melanoma cells. MMP activities are responsible for the restructuring of the local environment, thus allowing tumor growth. PA/plasmin activities are involved in the generation of distant metastasis, presumably by facilitating tumor invasion. While MMP activities can assist in the degradation of occluding ECM, their contribution to tumor invasion may not be required and can be overcome by other collagenolytic proteases (Pauli et al. 1986; Mackay et al. 1990).

5 Summary

Malignant human melanoma cells produce many matrix-degrading enzymes, including plasminogen activators and matrix metalloproteinases. These enzymes have substrate specificity for different components of ECM and most of them have been demonstrated to contribute to melanoma cell-mediated dissolution of matrices and to melanoma cell invasion. The degradation of complex matrices in vitro requires the cooperation of proteases with specificity for glycoproteins and collagens.

The contribution of proteases to spontaneous melanoma metastasis was studied by overexpressing specific protease inhibitors in human melanoma cells. Overexpression of PAI-2 inhibited the spread of distant metastasis indicating a role for uPA/plasmin in melanoma invasion. Overexpression of TIMP-2, in contrast, reduced the growth rate of subcutaneous tumors, but did not inhibit metastasis, indicating that MMP activities promote melanoma growth in the skin and may not be required for metastatic dissemination. Thus, uPA and MMP activities are involved in different processes, but they both contribute to melanoma malignancy.

Acknowledgments. The author would like to acknowledge the contributions of Ralph Reisfeld, Anthony Montgomery, Andreas Stahl, Walter Laug and Yves DeClerck to the work presented here. The author is supported by a Junior Faculty Research Award from the American Cancer Society JFRA-550 and by NIH grant CA 59692. This communication is The Scripps Research Institute's Manuscript Number 9309-IMM.

References

Adams JC, Watt FM (1993) Regulation of development and differentiation by the extracellular matrix. Development 117: 1183–1198

Anderson RGW (1993) Caveolae: where incoming and outgoing messengers meet. Proc Natl Acad Sci USA 90: 10909–10913

Aoyama A, Chen WT (1990) A 170-kDa membrane-bound protease is associated with the expression of invasiveness by human malignant melanoma cells. Proc Natl Acad Sci USA 87: 8296–8300

Axelrod JH, Reich R, Miskin R (1989) Expression of human recombinant plasminogen activators enhances invasion and experimental metastasis of H-ras-transformed NIH 3T3 cells. Mol Cell Biol 9: 2133–2141

Bizik J, Stephens RW, Grofova M, Vaheri A (1993) Binding of tissue-type plasminogen activator to human melanoma cells. J Cell Biochem 51: 326–335

Blasi F (1993) Urokinase and urokinase receptor: a paracrine/autocrine system regulating cell migration and invasiveness. Bioessays 15: 105–111

Boike G, Lah T, Sloane BF, Rozhin J, Honn K, Guirguis R, Stracke ML, Liotta LA, Schiffman E (1991) A possible role for cysteine proteinase and its inhibitors in motility of malignant melanoma and other tumor cells. Melanoma Res 1: 333–340

Chirivi RGS, Garofalo A, Crimmin MJ, Bawden LJ, Stoppacciaro A, Brown PD, Giavazzi R (1994) Inhibition of the metastatic spread and growth of B16–BL6 murine melanoma by a synthetic matrix metalloproteinase inhibitor. Int J Cancer 58: 460–464

Clark WH (1991) Tumour progression and the nature of cancer. Br J Cancer 64: 631–644
Crowley CW, Cohen RL, Lucas BK, Liu G, Shuman MA, Levinson AD (1993) Prevention of metastasis by inhibition of the urokinase receptor. Proc Natl Acad Sci USA 90: 5021–5025
Danø K, Andreasen PA, Grondahl-Hansen J, Kristensen P, Nielsen LS, Skriver L (1985) Plasminogen activation, tissue degradation and cancer. Adv Cancer Res 44: 139–266
DeClerck YA, Imren S (1994) Protease inhibitors: role and potential therapeutic use in human cancer. Eur J Cancer 30A: 2170–2180
DeClerck YA, Perez N, Shimada H, Boone TC, Langley KE, Taylor SM (1992) Inhibition of invasion and metastasis in cells transfected with an inhibitor of metalloproteinases. Cancer Res 52: 701–708
Delbaldo C, Massouye I, Saurat J-H, Vassalli J-D, Sappino A-P (1994) Plasminogen activation in melanocytic neoplasia. Cancer Res 54: 4547–4552
Denhardt DT, Feng B, Edwards DR, Cocuzzi ET, Malyankar UM (1993) Tissue inhibitor of metalloproteinases (TIMP, aka EPA): structure, control of expression and biological functions. Pharmacol Ther 59: 329–341
De Vries TJ, Quax PHA, Denijn M, Verrijp KN, Verheijen JH, Verspaget HW, Weidle UH, Ruiter DJ, Van Muijen GNP (1994) Plasminogen activators, their inhibitors, and urokinase receptor emerge in late stages of melanocytic tumor progression. Am J Pathol 144: 70–81
Ellis V, Pyke C, Eriksen J, Solberg H, Danø K (1992) The urokinase receptor: Involvement in cell surface proteolysis and cancer invasion. Ann NY Acad Sci 667: 13–31
Frisch SM, Werb Z (1989) Molecular biology of collagen degradation. In: Olsen BR, Nimni ME (eds) Collagen: molecular biology, vol IV. CRC Press, Boca Raton, Florida, pp 85–105
Hajjar KA, Jacovina AT, Chacko J (1994) An endothelial cell receptor for plasminogen/tissue plasminogen activator. I. Identity with annexin II. J Biol Chem 269: 21191–21197
Hearing VJ, Law JW, Corti A, Appella E, Blasi F (1988) Modulation of metastastic potential by cell surface urokinase of murine melanoma cells. Cancer Res 48: 1270–1278
Hendrix MJC, Wood WR, Seftor EA, Lotan D, Nakajima D, Misiorowski RL, Seftor REB, Stetler-Stevenson WG, Bevacqua SJ, Liotta LA, Sobel ME, Raz A, Lotan A (1990) Retinoic acid inhibition of human melanoma cell invasion through a reconstituted basement membrane and its relation to decreases in the expression of proteolytic enzymes and motility factors. Cancer Res 50: 4121–4130
Herlyn M (1993) Molecular and cellular biology of melanoma. Landes. Austin, Texas
Hoal-Van Helden EG, Wilson EL, Dowdle EB (1986) Characterization of seven human melanoma cell lines: melanogenesis and secretion of plasminogen activators. Br J Cancer 54: 287–295
Johnson MD, Kim HR, Chesler L, Tsao WG, Bouck N, Polverini PJ (1994) Inhibition of angiogenesis by tissue inhibitor of metalloproteinase. J Cell Physiol 160: 194–202
Khokha R (1994) Suppression of the tumorigenic and metastatic abilities of murine B16-F10 melanoma cells in vivo by the overexpression of the tissue inhibitor of the metalloproteinases-1. J Natl Cancer Inst 86: 299–304
Khokha R, Zimmer MJ, Graham CH, Lala PK, Waterhouse P (1992a) Suppression of invasion by inducible expression of tissue inhibitor of metalloproteinases-1 (TIMP-1) in B16-F10 melanoma cells. J Natl Cancer Inst 84: 1017–1022
Khokha R, Zimmer MJ, Wilson SM, Chambers A (1992b) Up-regulation of TIMP-1 expression in B16-F10 melanoma cells supresses their metastatic ability in chick embryo. Clin Exp Metastasis 10: 365–304
Kirchheimer JC, Wojta J, Christ G, Binder BR (1989) Functional inhibition of endogenously produced urokinase decreases cell proliferation in a human melanoma cell line. Proc Natl Acad Sci USA 86: 5424–5428
Koop S, Khokha R, Schmidt EE, MacDonald IC, Morris VL, Chambers AF, Groom AC (1994) Overexpression of metalloproteinase inhibitor in B16-F10 cells does not affect extravasation but reduces tumor growth. Cancer Res 54: 4791–4797
Kwaan HC (1992) The plasminogen-plasmin system in malignancy. Cancer Metastasis Rev 11: 291–311
Laug WE, Cao XR, Yu YB, Shimada H, Kruithof EKO (1993) Inhibition of invasion of HT1080 sarçoma cells expressing recombinant plasminogen activator inhibitor 2. Cancer Res 53: 6051–6057
Liotta LA, Steeg PS, Stetler-Stevenson WG (1991) Cancer metastasis and angiogenesis: An imbalance of positive and negative regulation. Cell 64: 327–336
Lisanti MP, Scherer PE, Tang Z, Sargiacomo M (1994) Caveolae, caveolin and caveolain-rich membrane domains: a signalling hypothesis. Trends Cell Biol 4: 231–235

MacDougall JR, Bani MR, Lin Y, Rak J, Kerbel RS (1995) The 92-kDa gelatinase B is expressed by advanced stage melanoma cells: suppression by somatic cell hybridization with early stage melanoma cells. Cancer Res 55: 4174–4181

Mackay AR, Corbitt RH, Hartzler JL, Thorgeirsson UP (1990) Basement membrane type IV collagen degradation: evidence for the involvement of a proteolytic cascade independent of metalloproteinases. Cancer Res 50: 5997–6001

Matrisian LM (1992) The matrix-degrading metalloproteinasess. Bioessays 14: 455–463

Mareel MM, De Baetselier P, Van Roy FM (1991) Cellular activities implicated in invasion and metastasis. In: Mareel MM, De Baetselier P, Van Roy FM (eds) Mechanisms of invasion and metastasis. CRC Press, Boca Raton, Florida, pp 73–219

McDonald J (1988) Extracellular matrix assembly. Annu Rev Cell Biol 4: 183–207

Meissauer A, Kramer MD, Hofmann M, Erkell LJ, Jacob E, Schirrmacher V, Brunner G (1991) Urokinase-type and tissue type plasminogen activators are essential for in vitro invasion of human melanoma cells. Exp Cell Res 192: 453–459

Meissauer A, Kramer MD, Schirrmacher V, Brunner G (1992) Generation of cell surface-bound plasmin by cell associated urokinase-type or secreted tissue-type plasminogen activator: A key event in melanoma invasiveness in vitro. Exp Cell Res 199: 179–190

Menrad A, Speicher D, Wacker J, Herlyn M (1993) Biochemical and functional characterization of aminopeptidase N expressed by human melanoma cells. Cancer Res 53: 1450–1455

Mignatti P, Rifkin DB (1993) Biology and biochemistry of proteinases in tumor invasion. Physiol Rev 73: 161–195

Mignatti P, Robbins E, Rifkin DB (1986) Tumor invasion through the human amniotic membrane: requirement for a proteinase cascade. Cell 47: 487–498

Miles LA, Dahlberg CM, Plescia J, Felez, Kato K, Plow EF (1991) Role of cell surface lysines in plasminogen binding to cells: identification of α-enolase as a candidate plasminogen receptor. Biochemistry 30: 1682–1686

Monsky WL, Lin CY, Aoyama A, Kelly T, Akiyama SK, Mueller SC, Chen WT (1994) A potential marker of invasiveness, seprase, is localized on invadopodia of human malignant melanoma cells. Cancer Res 54: 5702–5710

Montgomery AMP, De Clerck YA, Langley KE, Reisfeld RA, Mueller BM (1993) Melanoma-mediated dissolution of extracellular matrix. Contribution of urokinase-dependent and metalloproteinase dependent proteolytic pathways. Cancer Res 53: 693–700

Montgomery AMP, Mueller BM, Reisfeld RA, Taylor SM, De Clerck YA (1994a) Effect of tissue inhibitor of the matrix metalloproteinases-2 expression on the growth and spontaneous metastasis of a human melanoma cell line. Cancer Res 54: 5467–5473

Montgomery AMP, Reisfeld RA, Cheresh DA (1994b) Integrin $\alpha_v\beta_3$ rescues melanoma cells from apoptosis in three-dimensional dermal collagen. Proc Natl Acad Sci USA 91: 8856–8860

Moses MA, Langer R (1991) A metalloproteinase inhibitor as an inhibitor of neovascularization. J Cell Biochem 47: 230–235

Mueller BM, Romerdahl CA, Trent JM, Reisfeld RA (1991) Suppression of spontaneous melanoma metastasis in SCID mice with an antibody to the epidermal growth factor receptor. Cancer Res 51: 2193–2198

Mueller BM, Yu YB, Laug WE (1995) Overexpression of plasminogen activator inhibitor 2 in human melanoma cells inhibits spontaneous metastasis in scid/scid mice. Proc Natl Acad Sci USA 92: 205–209

Murphy G, Willenbrock F, Crabbe T, O'Shea M, Ward R, Atkinson S, O'Connell J, Reynolds JJ (1994) Regulation of matrix metalloproteinase activity. Ann NY Acad Sci 732: 31–41

Nakajima M, Irimura T, Nicolson GL (1988) Heparanases and tumor metastasis. J Cell Biochem 36: 157–167

Ossowski L (1992) Invasion of connective tissue by human carcinoma cell lines: requirement for urokinase, urokinase receptor, and interstitial collagenase. Cancer Res 52: 6754–6760

Ossowki L, Reich E (1983) Antibodies to plasminogen activator inhibit human tumor metastasis. Cell 35: 611–619

Ossowski L Russo-Payne H, Wilson EL (1991) Inhibition of urokinase-type plasminogen activator by antibodies: the effect on dissemination of a human tumor in the nude mouse. Cancer Res 51: 274–281

Pauli BU, Arsenis C, Hohberger LH, Schwart DE (1986) Connective tissue degradation by invasive rat bladder carcinomas: action of nonspecific proteinases on collagenous matrices. Cancer Res 46: 2005–2012

Quax PHA, Van Muijen GNP, Weening-Verhoeff EJD, Lund LR, Danø K, Ruiter DJ, Verheijen JH (1991) Metastatic behaviour of human melanoma cell lines in nude mice correlates with urokinase-type plasminogen activator, its type-1 inhibitor, and urokinase-mediated matrix degradation. J Cell Biol 115: 191–199

Ray JA, Stetler-Stevenson WG (1994) TIMP-2 expression modulates human melanoma cell adhesion and motility. Ann NY Acad Sci 732: 233–247

Reich R, Thompson EW, Iwamoto Y, Martin GR, Deason JR, Fuller GC, Miskin R (1988) Effects of inhibitors of plasminogen activator, serine proteinases, and collagenase IV on the invasion of basement membranes by metastastic cells. Cancer Res 48: 3307–3312

Rijken DC, Collen D (1981) Purification and characterization of the plasminogen activator secreted by human melanoma cells in culture. J Biol Chem 256 : 7035–7041

Rozhin J, Wade R, Honn KV, Sloane BF (1989) Membrane associated cathepsin L: a role in metastasis of melanomas. Biochem Biophys Res Commun 164: 556–561

Sato H, Takino T, Okada Y, Cao J, Shinagawa A, Yamamoto E, Seiki M (1994) A matrix metalloproteinase expressed on the surface of invasive tumor cells. Nature 370: 61–65

Seftor REB, Seftor EA, Gehlsen KR, Stetler-Stevenson WG, Brown PD, Ruoslathi E, Hendrix MJC (1992) Role of the $\alpha_v\beta_3$ integrin in human melanoma invasion. Proc Natl Acad Sci USA 89: 1557–1561

Seftor REB, Seftor EA, Stetler-Stevenson WG, Hendrix MJC (1993) The 72 kDa type IV collagenase is modulated via differential expression of $\alpha_v\beta_3$ and $\alpha_5\beta_1$ integrins during human melanoma cell invasion. Cancer Res 53: 3411–3415

Sloane BF, Moin K, Krepela E, Rozhin J (1990) Cathepsin B and its endogenous inhibitors: The role in tumor malignancy. Cancer Metastasis Rev 9: 333–352

Stahl A, Mueller BM (1994) Binding of urokinase to its receptor promotes migration and invasion of human melanoma cells in vitro. Cancer Res 54: 3066–3071

Stahl A, Mueller BM (1995) The urokinase-type plasminogen activator receptor, a GPI-linked protein, is localized in caveolae. J Cell Biol 129: 335–344

Stetler-Stevenson WG, Aznavoorian S, Liotta LA (1993) Tumor cell interactions with the extracellular matrix during invasion and metastasis. Annu Rev Cell Biol 9: 541–573

Testa JE, Quigley JP (1990) The role of urokinase-type plasminogen activator in aggressive tumor cell behavior. Cancer Metastais Rev 9: 35–367

Tsushima H, Hyodoh F, Yoshida E, Ueki A, Hopsu-Havu VK (1991a) Inacative cathepsin B-like enzyme in human melanoma culture medium. Melanoma Res 1: 341–347

Tsushima H, Ueki A, Matsuoka Y, Mihara H, Hopsu-Havu VK (1991b) Characterization of a cathepsin-H-like enzyme from a human melanoma cell line. Int J Cancer 48: 726–732

Vaheri A, Tapiovaara H, Myohanen H, Bizik J (1994) α_2-Macroglobulin in the regulation of pericellular plasminogen activation of human tumor cells. Ann NY Acad Sci 737: 409–418

Vassalli J-D, Sappino A-P, Belin D (1991) The plasminogen activator/plasmin system. J Clin Invest 88: 1067–1072

Wang X, Fu X, Brown PD, Crimmin MJ, Hoffman RM (1994) Matrix metalloproteinase inhibitor BB-94 (batimastat) inhibits human colon tumor growth and spread in a patient-like orthotopic model in nude mice. Cancer Res 54: 4726–4728

Welch DR, Bisi JE, Miller BE, Conaway D, Seftor EA, Yohem KH, Gilmore LB, Seftor REB,Nakajima M, Hendrix MJC (1991) Characterization of a highly invasive and spontaneously metastastic human malignant melanoma cell line. Int J Cancer 47: 227–237

Witty JP, McDonnell S, Newell KJ, Cannon P, Navre M, Tressler RJ, Matrisian LM (1994) Modulation of matrilysin levels in colon carcinoma cell affects tumorigenicity in vivo. Cancer Res 54: 4805–4812

Woessner JF Jr (1991) Matrix metalloproteinases and their inhibitors in connective tissue remodeling. FASEB J 5: 2145–2154

Yu, H, Schultz RM (1990) Relationship between secreted urokinase plasminogen activator activity and metastatic potential in murine B16 cells transfected with human urokinase sense and antisense genes. Cancer Res 50: 7623–7633

SPARC and the Extracellular Matrix: Implications for Cancer and Wound Repair

M.J. Reed[1] and E.H. Sage[2]

1 Introduction

Complex interactions between cells and the extracellular matrix are requisite during development, wound repair, and neoplasia. Proteins that regulate cell-cell and cell-matrix contacts are likely to influence both normal responses (e.g. wound healing) and abnormal processes (e.g. malignant transformation). The glycoprotein SPARC (secreted protein, acidic and rich in cysteine), also termed osteonectin, BM-40, and 43 K protein, has been classified, along with tenascin and thrombospondin, as an "anti-adhesin" (for review see Sage and Bornstein 1991). For many cell types, this function promotes cell rounding, a decrease in focal adhesions, and diminished cell-cell and cell-surface contacts. The attendant consequences for tissue remodeling and cellular migration implicate

[1]Department of Medicine, Box 359755, University of Washington, Seattle, WA 98195, USA
[2]Department of Biological Structure, Box 357420, University of Washington, Seattle, WA 98195, USA

SPARC as a regulatory protein in both wound healing and cancer. The purpose of this chapter is to review the properties of this protein and the functional implications for SPARC in neoplasia and tissue repair.

2 Expression of SPARC

SPARC is a 43 kDa glycoprotein first identified as a noncollagenous component of bone by TERMINE et al. (1981). It was later isolated as a protein secreted by endothelial cells in vitro (SAGE et al. 1984), a product of fibroblasts in culture (OTSUKA et al. 1984), and a major secreted protein in cells derived from mouse embryonic endoderm (MASON et al. 1986b). SPARC is produced by a diverse array of other cell types that include platelets, smooth muscle cells, and many tumor cell lines (STENNER et al. 1986; MANN et al. 1987; VILLARREAL et al. 1991). The variety of cell types that express this protein, coupled with a myriad of conditions under which they secrete SPARC, has led to a lack of consensus regarding its biological activity. Determination of functions for SPARC has been assisted by enhanced expression of this protein under certain experimental conditions. For example, expression of SPARC is increased when cells, including fibroblasts, smooth muscle cells, and endothelial cells, are placed in culture at sparse plating density or when cells are exposed to endotoxin (SAGE et al. 1986). Moreover, SPARC is secreted by chondrocytes in response to "heat shock" (NERI et al. 1992). Although SPARC is not a structural component of the extracellular matrix (MASON et al. 1986b), this secretion during conditions of "culture shock" is analogous to the enhanced production of proteins when tissues undergo remodeling. Not all cell types are consistent in the expression of SPARC; macrophages and several tumor cell lines do not produce SPARC in culture, whereas these cell types do express SPARC in injured and neoplastic tissue (RAINES et al. 1992; REED et al. 1993; PORTER et al. 1995).

Studies of the expression of a protein during development can elucidate potential functions of the protein in growth and remodeling. SPARC is regulated in a temporal and spatial manner during murine development (HOLLAND et al. 1987; SAGE et al. 1989a). Localization by in situ hybridization and immunocytochemistry has demonstrated that SPARC is expressed early in embryogenesis in Reichert's membrane, trophoblastic giant cells, and in areas of chondrogeneis, osteogenesis, and somitogenesis. Later in development, expression was noted in the thymus, heart, kidney, lung, gut, testis, and all mineralizing tissues (HOLLAND et al. 1987; SAGE et al. 1989a,b). In adult mice, SPARC has demonstrated an association with epithelium exhibiting high rates of turnover, such as glandular tissues, gut, and skin (SAGE et al. 1989a,b). Disruption of the expression of SPARC in *Xenopus* embryos via microinjection with affinity-purified antibodies directed against this protein resulted in severe external developmental anomalies by the early tadpole stage. These perturbations included a lack of visible eye pigment and deformities in eye development as well as disorganized myotome patterns and loss of segmental boundaries (PURCELL et al. 1993). Overexpression of SPARC in *C. elegans* was associated with an uncoordinated (*Unc*) phenotype in the F1 generation. Transcripts for SPARC were restricted to muscle cells of the body wall and vulva. Adults demonstrated abnormal movements or paralysis and were unable to produce viable offspring (SCHWARZBAUER and SPENCER 1993). These data, obtained from several different species, implicate SPARC in the regulation of synthesis and turnover of the extracellular matrix during embryonic development.

3 Functional Domains

The SPARC gene is highly conserved among bovine, mouse, and human species, a feature consistent with positive selection for the preservation of the structure of this protein (MCVEY et al. 1988; VILLARREAL et al. 1989). SPARC cDNA consists of a 17 amino acid signal sequence followed by a secreted protein of 283–287 amino acids (ENGEL et al. 1987). Peptide sequences derived from SPARC have permitted the study of functional consequences of specific domains that are not accessible in the native protein (LANE and SAGE 1990). These studies have utilized synthetic peptides as well as specific antibodies directed at portions of these domains. SPARC has at least four distinct domains based on the predicted secondary structure (Fig. 1).

Domain I (amino acids 3–51) is a highly acidic region with clusters of glutamic acid residues. It binds calcium with low affinity (ENGEL et al. 1987) and contains a sequence, designated 1.1 (amino acids 5–23), that is an highly efficient inhibitor of endothelial cell spreading (LANE and SAGE 1990).

Domain II (amino acids 52–132) is a cysteine-rich region that is homologous to a repeated domain in follistatin and agrin. The NH_2-terminal portion, termed 2.1 (amino acids 54–73), exhibits regulatory effects on the cell cycle (FUNK and SAGE 1991, 1993). The COOH-terminal region, termed 2.3 (amino acids 113–130), contains two cationic binding sites and the sequence (K)GHK, and has been implicated in the regulation of cell proliferation and angiogenesis (LANE et al. 1994).

Domain III (amino acids 133–227), an extended series of α-helical segments, contains an endogenous protease-sensitive site between amino acids 197 and 198. Cysteine 137, at the NH_2-terminal of this region, is disulfide-bonded to Cys-247 in domain IV. This linkage is one of only two disulfide bonds that have been definitively identified among the 14 cysteine residues of SPARC.

Domain IV (amino acids 228–285) contains an EF hand-like loop that has a high affinity binding site for Ca^{2+} (ENGEL et al. 1987). Moreover, a synthetic peptide, designated 4.2, (amino acids 254–273), derived from this region mediates the interaction between SPARC and collagens (LANE and SAGE 1990) as well as endothelial cells (YOST and SAGE 1993). A disulfide bond between Cys-225 and Cys-271 stabilizes the conformation of this domain (MANN et al. 1987).

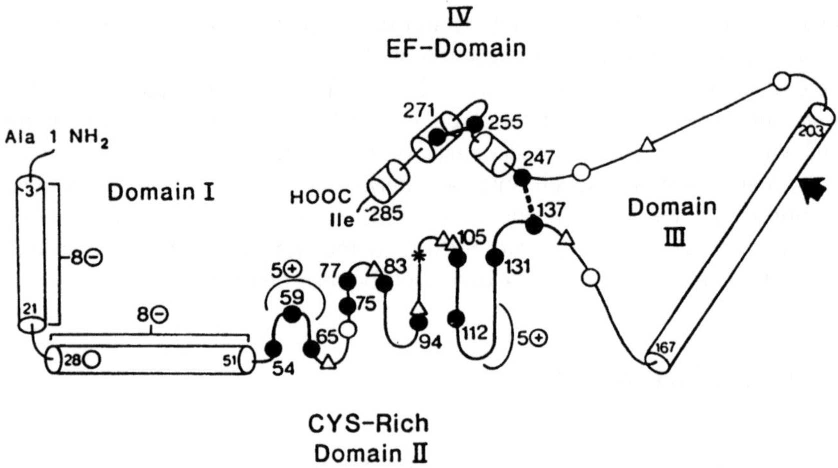

Fig. 1. The SPARC protein. *Numbers* are designated according to MASON et al. (1986), without the signal peptide. Structural domains of the protein are designated I–IV. In domain I, the two regions with eight negatively charged residues are noted. Cysteine-rich domain II has 11 cysteine residues. One of these (denoted by Cys-137) is thought to be disulfide bonded with Cys-247; all of the other Cys residues are presumed to be disulfide bonded within domain II. Domain III is predicted to have an α-helical structure as depicted by the cylinders. Three-dimensional modeling of the EF-hand domain IV indicates that Cys-255 and Cys-271 form another disulfide bond. (From ENGEL et al. 1987)

Peptide sequences from these domains and specific antibodies against these regions have permitted the elucidation of structure-function relationships. These studies have demonstrated that specific sequences of SPARC modulate interactions between cells and the extracellular matrix in a differential manner (for review see LANE and SAGE 1994). Disparate roles for SPARC and peptides derived from proteolysis of this protein are further supported by the demonstration that SPARC is only present intracellularly (SAGE et al. 1989a,b; REED et al. 1993; PORTER et al. 1995). This limited expression is attributed to rapid degradation of the protein after secretion, a process expected to preclude detection of the native protein in the extracellular milieu. Sequences with specific functions could then modulate cell behavior during wound healing and neoplasia, integral aspects of which include adhesion, proliferation, interaction with the extracellular matrix, degradation by proteases, and angiogenesis. The effect of SPARC on these processes will be examined in further detail.

4 Cell Shape and Adhesion

SPARC was first described as a secreted protein that was produced by endothelial cells in response to conditions of "culture shock" (SAGE et al. 1984). However, the introduction of exogenous SPARC inhibits cell spreading and induces cellular rounding, effects which do not support cell attachment in vitro (SAGE et al. 1989c). Moreover, stably transfected F9 embryonal carcinoma cell lines that overexpressed SPARC were aggregated and rounded to a greater extent than control cells transformed with vector alone (EVERITT and SAGE 1992). This function as an anti-adhesin has implicated SPARC in the enhancement of detachment and subsequent migration of cells in vivo. Functional mapping studies have localized this property to peptide sequences in domains I and IV (LANE and SAGE 1990). Further studies have determined that peptides 2.1 and 4.2 mediate the effect of SPARC on the disassembly of focal adhesions (MURPHY-ULLRICH et al. 1995). In cultured pulmonary vascular endothelial cells, exogenous SPARC induces changes in cell shape and increases in endothelial cell permeability that result in barrier dysfunction (GOLDBLUM et al. 1994). The ability of SPARC to diminish cell-cell and cell-matrix contacts concomitant with permeabilization of cell junctions is consistent with a role for this protein in facilitating cellular migration, a process necessary for the progression of tumors and the development of metastases.

5 Cellular Proliferation

The effect of SPARC on cellular proliferation is distinct from that of peptide sequences derived from the native molecule. SPARC protein that is secreted by endothelioma cells transformed by the polyoma middle T-oncogene inhibits the growth of normal bovine aortic endothelial cells, but not of established murine capillary endothelial cells (SAGE 1991). The addition of exogenous SPARC inhibits the entry of bovine aortic endothelial cells into S-phase in a concentration-dependent manner. Peptide 2.1 produced a similar, but less marked, reduction of cell entry into S-phase. In contrast, peptide 1.1, a sequence that inhibits cell spreading, had no activity in the modulation of cell cycle progression (FUNK and SAGE 1991). Human umbilical vein endothelial cells, fetal bovine endothelial cells, and bovine capillary endothelial cells exhibited a response to SPARC and peptide 2.1 similar to that of bovine aortic endothelial cells. Human foreskin fibroblasts and fetal bovine ligament fibroblasts exhibited an increase in thymidine incorporation in the presence of SPARC and peptide 2.1 at lower concentrations, but were inhibited at higher concentrations (FUNK and SAGE 1993). Recent studies have determined that peptide 4.2 is a more potent inhibitor of DNA synthesis that acts synergistically with peptide 2.1 to diminish the incorporation of tritiated thymidine (SAGE et al. 1995). In contrast, peptide 2.3, a sequence implicated in angiogenesis, increased thymidine incorporation in both bovine aortic endothelial cells and fibroblasts (FUNK and SAGE 1993). The differential effect of SPARC and peptides derived from SPARC on cell proliferation implicates this protein in the regulation of cell growth during wound healing and neoplasia.

6 Interaction with the Extracellular Matrix

SPARC binds to divalent cations, several types of collagens, albumin, thrombospondin, and to the cell membranes of platelets and endothelial cells (SAGE et al. 1989c; CLEZARDIN et al. 1988; YOST and SAGE 1993). SPARC and type I collagen are coexpressed and coordinately regulated in cultured fibroblasts derived from human gingiva and skin (VUORIO et al. 1991; WRANA et al. 1991; REED et al. 1994), and in tissues during remodeling and wound repair (SALONEN et al. 1990; REED et al. 1993). SPARC also binds to type IV collagen and is expressed in specialized basement membranes during development and neoplastic transformation (MASON et al. 1986b; WEWER et al. 1988). Tissue-specific alterations in the structure of SPARC can alter the affinity of the protein for the extracellular matrix. For example, partial degradation of bone SPARC (osteonectin) by a protease enhances the binding of this protein to type I collagen (TYREE 1989).

The addition of exogenous SPARC to endothelial cells modulates the synthesis of a variety of proteins associated with the extracellular matrix. Examples include the enhanced synthesis of plasminogen activator inhibitor-1 and metalloproteases, and the inhibition of synthesis of fibronectin and thrombospondin-1 (LANE et al. 1992; TREMBLE et al. 1993). These data support the hypothesis that the transient expression of SPARC early in wound repair and at sites of tissue remodeling might regulate fibrinolysis and thrombosis as well as the deposition or assembly of other matrix proteins (RAINES et al. 1992; REED et al. 1993).

7 Interaction with Growth Factors

Transcription of SPARC is increased by retinoic acid, progesterone, and dexamethasone (MASON et al. 1986a; NOMURA et al. 1989; KASUGAI et al. 1991). Transforming growth factor beta-1 (TGFβ-1), a cytokine with regulatory activity in wound repair, stimulates the production of several matrix proteins such as type I collagen (MUSTOE et al. 1987; PENTTINEN et al. 1988). TGF-β1 also increases the expression of SPARC mRNA and protein in osteoblasts and fibroblasts in vitro as well as during wound repair (MASAKI and RODAN 1989; WRANA et al. 1991; REED et al. 1994; PUOLAKKAINEN et al. 1995).

SPARC might also directly or indirectly modulate the interaction of cells with a growth factor. SPARC binds to platelet-derived growth factor (PDGF)-AB and PDGF-BB, but not to PDGF-AA. Moreover, SPARC inhibited the binding of PDGF to its cognate receptor(s) on fibroblasts (RAINES et al. 1992). In the absence of serum, exogenous SPARC antagonizes the migration of bovine aortic endothelial cells to the chemokine basic fibroblast growth factor (bFGF).

In contrast to its affinity for PDGF, SPARC did not bind bFGF or block the binding of bFGF to its receptors. The antagonistic effect was mimicked by peptides 1.1 and 4.2, but not by peptides 2.1 and 3.4 (HASSELAAR and SAGE 1992), data consistent with specificity of peptides from the calcium-binding regions with respect to the interaction of SPARC with bFGF.

8 Interaction with Proteases

Proteases and protease inhibitors act in concert to facilitate cell turnover and remodeling in tissues (DUFFY 1992). Although a secreted product in culture media, SPARC protein has only been noted intracellularly during development, wound repair, and neoplasia (REED et al. 1993; PORTER et al. 1995). Moreover, SPARC is expressed and degraded in a temporal and spatial manner that is coincident with angiogenesis in the chicken chorioallantoic membrane (IRUELA-ARISPE et al. 1995). These data support the contention that interactions with proteases, metalloproteases, and serine proteases result in rapid proteolytic degradation of SPARC in the extracellular space in vivo. This possibility is supported by studies in vitro. SPARC induces the expression of type I plasminogen activator inhibitor, a regulator of serine proteases, in cultured bovine aortic endothelial cells (HASSELAAR et al. 1991). Moreover, in fibroblasts, SPARC has been demonstrated to induce the expression of three different metalloproteases involved in morphogenesis and tissue remodeling: collagenase, stromelysin, and 92 kDa gelatinase (TREMBLE et al. 1993), which are, in turn, activated by serine proteases. Therefore, SPARC is a substrate and a modulator of proteolytic events in remodeling tissues.

9 Angiogenesis

The expression of SPARC is induced in endothelial cells derived from the bovine aorta and rat cerebral microvessels during cord formation, a model of angiogenesis in vitro whereby endothelial cells spontaneously form tube-like structures with patent lumens (IRUELA-ARISPE et al. 1991a). During development of the chicken chorioallantoic membrane and during wound repair, SPARC mRNA and protein are expressed by endothelial cells in newly formed blood vessels (REED et al. 1993; IRUELA-ARISPE et al. 1995) (Fig. 2A).

In wound healing, expression of SPARC in fibroblasts was shown to be maximal during the phase in which there is active angiogenesis in the wound bed (Fig. 2B). This result contrasted with that obtained for thrombospondin-1, an inhibitor of angiogenesis in vitro and in vivo (BAGANVADOSS and WILKS 1990;

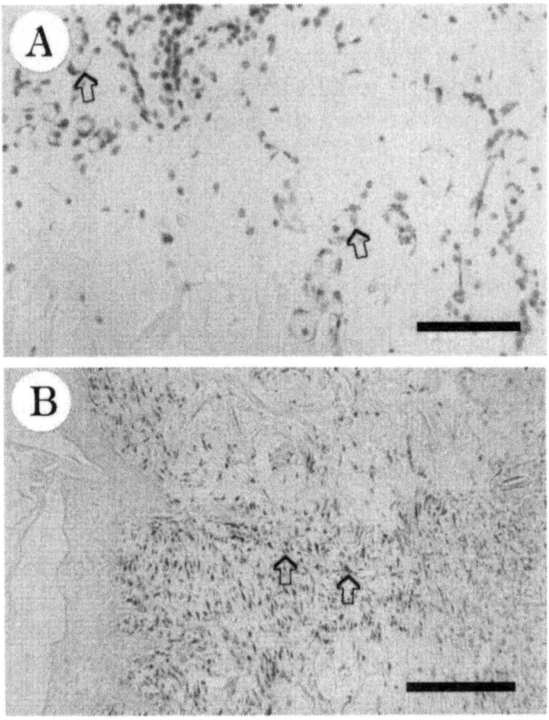

Fig. 2A,B. Expression of SPARC in endothelial cells and fibroblasts during wound repair. Shown are representative photographs of tissue at 7 days after wounding. Immuncytochemistry was performed with an affinity-purified antibody directed against SPARC. The dark stain represents SPARC protein (*arrows*) in endothelial cells (**A**) and in fibroblasts (**B**). *Bar* = 50μm (**A**), 200μm (**B**)

GOOD et al. 1990; IRUELA-ARISPE et al. 1991b), which was expressed only in the early phase of wound repair (REED et al. 1993). Moreover, endothelial cells treated with exogenous SPARC exhibited a fourfold decrease in thrombospondin production (LANE et al. 1992).

Synthetic peptides of SPARC have domains that regulate endothelial cell proliferation and angiogenesis. Peptide 2.3, in contrast to peptide 2.1, increased the proliferation of endothelial cells and angiogenesis in vitro and in vivo (LANE et al. 1994). However, peptide 2.3 was not as potent a stimulator of angiogenesis as smaller cleavage products derived from the proteolysis of either peptide 2.3 or intact SPARC. For example, digestion of SPARC with trypsin and peptide 2.3 with plasmin resulted in the release of the sequence GHK (LANE et al. 1994). This tripeptide is a growth factor for differentiated cells and is chemotactic for several cell types (PICKART 1981; POOLE and ZETTER 1983; ZETTER et al. 1985). Moreover, GHK stimulated angiogenesis in vitro and in vivo (RAJU et al. 1982; LANE et al. 1994). Recently, GHK has been shown to stimulate the accumulation of connective tissue and enhance wound repair in animal models (MAQUART et al. 1993). In studies of the proteolytic degradation of SPARC in the chicken chorioallantoic membrane assay, extracellular proteolysis of SPARC occurred during the middle stages (days 9–15) of chorioallantoic membrane development, but not at later stages (days 17–21) (IRUELA-ARISPE et al. 1995).

This cleavage might be mediated, in part, by plasmin. Moreover, although intact SPARC was inactive, synthetic peptides ([K]GHK) derived from peptide 2.3, a plasmin-sensitive region of SPARC, stimulated angiogenesis in a concentration-dependent manner (LANE et al. 1994; IRUELA-ARISPE et al. 1995). These studies support the proposal that SPARC, and peptides derived from proteolytic degradation of SPARC, regulate angiogenesis in vitro and in vivo.

10 Wound Healing

The early stages of wound healing are characterized by coagulation and inflammation. Later stages of tissue repair consist of cell migration, proliferation, angiogenesis, synthesis of extracellular matrix, and, finally, tissue remodeling (FALANGA et al. 1988). In studies of wound repair in animal models, it has been demonstrated that maximal expression of SPARC mRNA and protein were noted in the wound bed from the middle through late stages of repair (REED et al. 1993). This expression was coincident with the phase of granulation tissue formation and was consistent with a role for this protein in remodeling and the facilitation of angiogenesis. Since SPARC was not detected in the extracellular space, the degradation of this protein by proteases expressed during wound healing might serve as a mechanism whereby fragments of SPARC are released into the extracellular space. To test this hypothesis, we have studied the effect of matrix metalloproteases from the three classes of this extended family (collagenases, gelatinases, and stromelysins) on the proteolytic degradation of SPARC. Whereas matrix metalloprotease (MMP)-1 (interstitial collagenase), MMP-2 (72 kDa gelatinase), and MMP-9 (92 kDa gelatinase) did not degrade SPARC, MMP-3 (stromelysin) cleaved SPARC into discrete fragments. Analysis of the cleavage products has shown a protected region that encompasses domain II (REED et al., in preparation). Wound fluid also contains an activity (or activities) that cleaves SPARC in a concentration-dependent manner. Of interest, cleavage is enhanced in fluids that are obtained from the later stages of wound healing. This result could be due to facilitated proteolysis during the stages of repair that require the formation of granulation tissue or the absence of inhibitors during the later stages of wound healing. The effect of wound fluid on the degradation of SPARC is not obviated by tissue inhibitor of matrix metalloprotease-1 (TIMP-1) alone, but required the addition of TIMP, leupeptin, and PMSF. It is not surprising that degradation of SPARC is mediated by several classes of proteases. Since proteases are temporally and differentially expressed during wound repair, there are multiple points of regulation in the cleavage of SPARC. Future experiments will investigate whether cleavage of SPARC results in the formation of angiogenic peptides in vivo and will characterize the enzymes that are responsible for the proteolytic degradation of this protein during wound repair.

11 Cancer

Much of the evidence implicating SPARC in cancer has been indirect and arises from the presence of this protein in numerous neoplastic and malignant tissues. BM-40 (basement membrane-40) was isolated from the basement membrane of a murine tumor and was subsequently found to be identical to SPARC/osteonectin (MANN et al. 1987). WEWER et al. (1988) found SPARC mRNA and protein in human decidua and several types of carcinomas. In poorly differentiated carcinoma, SPARC was present in the pericellular stroma of cells, whereas in decidua and well-differentiated tumors SPARC was restricted to newly deposited basement membrane and the cytoplasm of some cells. In human decidua, the presence of SPARC in a distribution similar to laminin was consistent with an association of this protein with active formation of the basement membrane.

BELLAHCÉNE and CASTRONOVO (1995) have shown increased expression of SPARC in malignant, compared to benign, lesions of the breast. The localization of SPARC with microcalcifications underscores the affinity of this protein for hydroxyapatite and the binding of Ca^{2+} to SPARC in the formation of ectopic microcalcifications associated with malignancies of the breast. Moreover, the high expression of both SPARC and osteopontin, two bone matrix proteins, in breast carcinoma may underlie the predilection of this form of cancer for metastasis to bone (BELLAHCÉNE and CASTRONOVO 1995). SPARC protein has also been noted in endothelial cells and fibrocytes in invasive malignant tumors of the gastrointestinal tract, breast, lung, kidney, adrenal cortex, ovary, and brain (Fig. 3) (PORTER et al. 1995). This immunoreactivity was greatest in malignancies of mesenchymal origin. The majority of epithelial tumors showed high levels of SPARC, although some specimens were completely negative. A high tumor grade did not appear to be consistent with high levels of SPARC. There were low levels of expression in normal human tissues in a pattern that was similar to that observed in studies of adult murine tissues. Further studies of SPARC in neoplasia will focus on apparent correlations among the expression of SPARC, the extent of angiogenesis, and the clinical outcome of breast cancers and other malignancies.

12 Conclusions

Studies in vitro and in vivo support a regulatory role for SPARC in cellular processes required for both wound healing and neoplasia. In fibroblasts and endothelial cells, SPARC diminishes focal contacts and prevents cell spreading. SPARC inhibits proliferation of endothelial cells, but peptide sequences

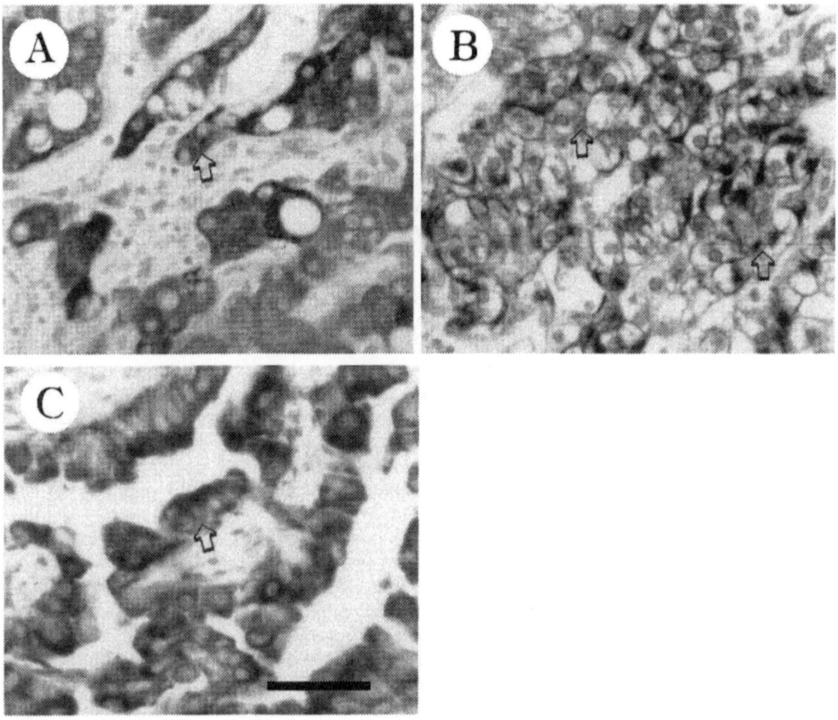

Fig. 3A–C. Expression of SPARC in human carcinomas. Shown are sections of tumors exposed to a monoclonal antibody directed against domain I of SPARC. Immunostain (*arrows*) for SPARC is noted in malignant cells of adenocarcinoma of the breast (**A**), renal cell carcinoma (**B**), and adenocarcinoma of the lung (**C**). *Bar* = 50μm. (Photographs provided by Dr. P. PORTER).

derived from this protein enhance endothelial cell proliferation and angiogenesis. SPARC has an affinity for Ca^{2+} and Cu^{2+} and binds to multiple ECM components including collagens, thrombospondin, and the basement membrane. SPARC stimulates the production of plasminogen activator inhibitor-1 as well as matrix metalloproteases and is also a substrate for certain proteases. Expression of SPARC is induced by TGF-β1, but SPARC abrogates the cellular effect of other growth factors such as PDGF and bFGF. SPARC is expressed at high levels in tissues that undergo morphogenesis, remodeling, and repair. Although many characteristics of SPARC are known, further studies are necessary to elucidate precise functions. Future goals include: (1) the characterization of cell-surface receptors for SPARC, (2) determining the interactions of SPARC with specific cytokines, morphogens, and inflammatory mediators, and (3) understanding the mechanisms of signal transduction that are mediated by SPARC. These areas of study provide the basis for further investigation of the role of SPARC in the processes of wound repair and malignant transformation.

References

Bagavandoss P, Wilks JW (1990) Specific inhibition of endothelial cell proliferation by thrombospondin. Biochem Biophys Res Commun 170: 867–872

Bellahcéne A, Castronovo V (1995) Increased expression of osteonectin and osteopontin, two bone matrix proteins, in human breast cancer. Am J Pathol 146: 95–100

Clezardin P, Malaval L, Ehrensperger A, Delmas PD, Dechavanne M, McGregor JL (1988) Complex formation of human thrombospondin with osteonectin. Eur J Biochem 175: 275–284

Duffy MJ (1992) The role of proteolytic enzymes in cancer invasion and metastasis. Clin Exp Metastasis 10: 145–155

Engel J, Taylor W, Paulsson M, Sage EH, Hogan B (1987) Calcium binding domains and calcium-induced conformational transition of SPARC/BM-40/osteonectin, an extracellular glycoprotein expressed in mineralized and nonmineralized tissues. Biochemistry 26: 6958–6965

Everitt E, Sage EH (1992) Expression of SPARC is correlated with altered morphologies in transfected F9 embryonal carcinoma cells. Exp Cell Res 199: 134–146

Falanga V, Zitelli JA, Eaglstein WH (1988) Wound healing. J Am Acad Dermatol 19: 559–563

Funk SE, Sage EH (1991) The Ca^{2+}-binding glycoprotein SPARC modulates cell cycle progression in bovine aortic endothelial cells. Proc Natl Acad Sci USA 88: 2648–2652

Funk SE, Sage EH (1993) Differential effects of SPARC and cationic SPARC peptides on DNA synthesis by endothelial cells and fibroblasts. J Cell Physiol 154: 53–63

Goldblum SE, Ding X, Funk SE, Sage EH (1994) SPARC (secreted protein acidic and rich in cysteine) regulates endothelial cell shape and barrier function. Proc Natl Acad Sci USA 91: 3448–3452

Good DJ, Polverini PJ, Rastinejad F, LeBeau MM, Lemons RS, Frazier WA, Bouck NP (1990) A tumor suppressor-dependent inhibitor of angiogenesis is immunologically and functionally indistinguishable from a fragment of thrombospondin. Proc Natl Acad Sci USA 87: 6624–6628

Hasselaar P, Sage EH (1992) SPARC antagonizes the effect of basic fibroblast growth factor on the migration of bovine aortic endothelial cells. J Cell Biochem 49: 272–283

Hasselaar P, Loskutoff DJ, Sawdey M, Sage EH (1991) SPARC induces the expression of type I plasminogen activator inhibitor in cultured bovine aortic endothelial cells. J Biol Chem 266: 13178–13184

Holland PWH, Harper SJ, McVey JH, Hogan BLM (1987) In vivo expression of mRNA for the Ca^{++}-binding protein SPARC (osteonectin) revealed by in situ hybridization. J Cell Biol 105: 473–482

Iruela-Arispe ML, Hasselaar P, Sage EH (1991a) Differential expression of extracellular proteins is correlated with angiogenesis in vitro. Lab Invest 64: 174–186

Iruela-Arispe ML, Bornstein P, Sage EH (1991b) Thrombospondin exerts an antiangiogenic effect on cord formation by endothelial cells in vitro. Proc Natl Acad Sci USA 88: 5026–5030

Iruela-Arispe ML, Lane TF, Redmond D, Reilly M, Bolender RP, Kavanagh TJ, Sage EH (1995) Expression of SPARC during development of the chicken chorioallantoic membrane: evidence for regulated proteolysis in vivo. Mol Biol Cell 6: 327–343

Kasugai S, Todescan R, Nagata T, Yao K-L, Butler WT, Sodek J (1991) Expression of bone matrix proteins associated with mineralized tissue formation by adult rat bone marrow cells in vitro: inductive effects of dexamethasone on the osteoblastic phenotype. J Cell Physiol 147: 111–120

Lane TF, Sage EH (1990) Functional mapping of SPARC: peptides from two distinct Ca^{++}-binding sites modulate cell shape. J Cell Biol 111: 3065–3076

Lane TF, Sage EH (1994) The biology of SPARC, a protein that modulates cell-matrix interactions. FASEB J 8: 163–173

Lane TF, Iruela-Arispe ML, Sage EH (1992) Regulation of gene expression by SPARC during angiogenesis in vitro: changes in fibronectin, thrombospondin-1, and plasminogen activator inhibitor-1. J Biol Chem 267: 16736–16745

Lane TF, Iruela-Arispe ML, Johnson RS, Sage EH (1994) SPARC is a source of copper-binding peptides that stimulate angiogenesis. J Cell Biol 4: 929–943

Mann K, Deutzmann R, Paulsson M, Timpl R (1987) Solubilization of protein BM-40 from a basement membrane tumor with chelating agents and evidence for its identity with osteonectin and SPARC. FASEB J 218: 167–172

Maquart F-X, Bellon G, Chaqour B, Wegrowski J, Patt LM, Trachy RE, Monboisse JC, Chastang F, Birembaut P, Gillery P, Borel JP (1993) In vivo stimulation of connective tissue accumulation by the tripeptide-copper complex glycyl-L-histidyl-L-lysine-Cu^{2+} in rat experimental wounds. J Clin Invest 92: 2368–2376

Masaki N, Rodan GA (1989) Type β transforming growth factor regulates expression of genes encoding bone matrix proteins. Connect Tissue Res 21: 71–75
Mason IJ, Murphy D, Munke M, Francke U, Elliott RW, Hogan BLM (1986a) Developmental and transformation-sensitive expression of the SPARC gene on mouse chromosome 11. EMBO J 5: 1831–1837
Mason IJ, Taylor A, William JG, Sage EH, Hogan BLM (1986b) Evidence from molecular cloning that SPARC, a major product of mouse embryo parietal endoderm, is related to an endothelial cell "culture shock" glycoprotein. EMBO J 5: 1465–1472
McVey JH, Nomura S, Kelly P, Mason IJ, Hogan BLM (1988) Characterization of the mouse SPARC/osteonectin gene: intron/exon organization and an unusual promoter region. J Biol Chem 263: 11111–11116
Murphy-Ullrich JE, Lane TF, Pallero MA, Sage EH (1995) SPARC mediates focal adhesion disassembly in endothelial cells through a follistatin-like region and the Ca^{2+}-binding EF-hand. J Cell Biochem 57: 341–350
Mustoe TA, Pierce GF, Thomason A, Gramates P, Sporn MB, Deuel TF (1987) Accelerated healing of incisional wounds in rats induced by transforming growth factor β. Science 237: 1333–1335
Neri M, Descalzi-Cancedda F, Cancedda R (1992) Heat-shock response in cultured chick embryo chondrocytes: osteonectin is a secreted heat-shock protein. Eur J Biochem 205: 569–574
Nomura S, Hashmi S, McVey JH, Ham J, Parker M, Hogan BLM (1989) Evidence for positive and negative regulatory elements in the 5'-flanking sequence of the mouse SPARC (osteonectin) gene. J Biol Chem 264: 12201–12207
Otsuka K, Yao KL, Wasi S, Tung PS, Aubin JE, Sodek J, Termine JD (1984) Biosynthesis of osteonectin by fetal porcine calvarial cells in vitro. J Biol Chem 259: 9805–9812
Penttinen RP, Kobayashi S, Bornstein P (1988) Transforming growth factor β increases mRNA for matrix proteins both in the presence and in the absence of changes in mRNA stability. Proc Natl Acad Sci USA 85: 1105–1108
Pickart L (1981) The use of glycyl-histidyl-lysine in culture systems. In Vitro Cell Dev Biol 17: 459–466
Poole TJ, Zetter BR (1983) Stimulation of rat peritoneal mast cell migration by tumor-derived peptides. Cancer Res 43: 5857–5861
Porter PL, Sage EH, Lane TF, Funk SE, Gown AM (1995) Distribution of SPARC in normal and neoplastic human tissue. J Histochem Cytochem 43: 791–800
Purcell L, Gruia-Gray J, Scanga S, Ringuette M (1993) Developmental anomalies of Xenopus embryos following microinjection of SPARC antibodies. J Exp Zool 265: 153–164
Puolakkainen P, Reed MJ, Gombotz WR, Twardzik DR, Abrass IB, Sage EH (1995) Acceleration of wound healing in aged rats by topical application of transforming growth factor beta 1. Wound Repair Regen 3: 330–339
Raines EW, Lane TF, Iruela-Arispe ML, Ross R, Sage EH (1992) The extracellular glycoprotein SPARC interacts with platelet-derived growth factor (PDGF)-AB and -BB and inhibits the binding of PDGF to its receptors. Proc Natl Acad Sci USA 89: 1281–1285
Raju KS, Alessandri J, Ziche M, Gullino PM (1982) Ceruloplasmin, copper ions and angiogenesis. J Natl Cancer Inst 69: 1183–1188
Rastinejad R, Polverini PJ, Bouck NP (1989) Regulation of the activity of a new inhibitor of angiogenesis by a cancer suppressor gene. Cell 56: 345–355
Reed MJ, Puolakkainen P, Lane TF, Dickerson D, Bornstein P, Sage EH (1993) Differential expression of SPARC and thrombospondin 1 in wound repair: immunolocalization and in situ hybridization. J Histochem Cytochem 41: 1467–1477
Reed MJ, Vernon RB, Abrass IB, Sage EH (1994) TGF-β1 induces the expression of type I collagen and SPARC, and enhances contraction of collagen gels, by fibroblasts from young and aged donors. J Cell Physiol 158: 169–179
Reed MJ, Johnson RS, Sage EH. Degradation of SPARC by stromelysin and wound fluid: Evidence for regulation of angiogenesis in vivo (manuscript in preparation)
Sage EH (1991) Secretion of SPARC by endothelial cells transformed by polyoma middle T oncogene inhibits the growth of normal endothelial cells in vitro. Biochem Cell Biol 70: 579–592
Sage EH, Bornstein P (1991) Extracellular proteins that modulate cell-matrix interactions. J Biol Chem 266: 14831–14834
Sage EH, Johnson C, Bornstein P (1984) Characterization of a novel serum albumin-binding glycoprotein secreted by endothelial cells in culture. J Biol Chem 259: 3993–4007
Sage EH, Tupper J, Bramson R (1986) Endothelial cell injury in vitro is associated with increased secretion of an Mr 43,000 glycoprotein ligand. J Cell Physiol 127: 373–387

Sage EH, Decker J, Funk SE, Chow M (1989a) SPARC, a Ca^{2+}-binding extracellular protein associated with endothelial cell injury and proliferation. J Mol Cell Cardiol 21: 13–22
Sage EH, Vernon RB, Decker J, Funk S, Iruela-Arispe ML (1989b) Distribution of the calcium-binding protein SPARC in tissues of embryonic and adult mice. J Histochem Cytochem 37: 819–829
Sage EH, Vernon RB, Funk SE, Everitt EA, Angello J (1989c) SPARC, a secreted protein associated with cellular proliferation, inhibits cell spreading in vitro and exhibits Ca^{+2}-dependent binding to the extracellular matrix. J Cell Biol 109: 341–356
Sage EH, Bassuk JA, Yost JC, Folkman JM, Lane TF (1995) Inhibition of endothelial cell proliferation by SPARC is mediated through a Ca^{2+}-binding EF-hand sequence. J Cell Biochem 57: 127–140
Salonen J, Domenicucci C, Goldberg HA, Sodek J (1990) Immunohistochemical localization of SPARC (osteonectin) and denatured collagen and their relationship to remodelling in rat dental tissues. Arch Oral Biol 35: 337–346
Schwarzbauer J, Spencer CS (1993) The Caenorhabditis elegans homolog of the extracellular calcium binding protein SPARC/osteonectin affects nematode body morphology and mobility. Mol Biol Cell 4: 941–952
Stenner DD, Tracy RP, Riggs BL, Mann KG (1986) Human platelets contain and secrete osteonectin, a major protein of mineralized bone. Proc Natl Acad Sci USA 83: 6892–6896
Termine JD, Kleinman HK, Whitson SW, Conn KM, McGarvery ML, Martin GR (1981) Osteonectin, a bone-specific protein linking mineral to collagen. Cell 26: 99–105
Tremble PM, Lane TF, Sage EH, Werb Z (1993) SPARC, a secreted protein associated with morphogenesis and tissue remodeling, induces expression of metalloproteinases in fibroblasts through a novel extracellular matrix-dependent pathway. J Cell Biol 121: 1433–1444
Tyree B (1989) The partial degradation of osteonectin by a bone-derived metalloprotease enhances binding to type I collagen. J Bone Miner Res 4: 877–883
Villarreal XC, Mann KG, Long GL (1989) Structure of human osteonectin based upon analysis of cDNA and genomic sequences. Biochemistry 28: 6483–6491
Villarreal XC, Grant BW, Long GL (1991) Demonstration of osteonectin mRNA in megakaryocytes: the use of the polymerase chain reaction. Blood 78: 1216–1222
Vuorio T, Kähäri V-M, Black C, Vuorio E (1991) Expression of osteonectin, decorin, and transforming growth factor-β1 genes in fibroblast cultures from patients with systemic sclerosis and morphea. J Rheumatol 18: 247–251
Wewer U, Albrechtsen R, Fisher L, Young M, Termine J (1988) Osteonectin/SPARC/BM-40 in human decidua and carcinoma, tissues characterized by de novo formation of basement membrane. Am J Pathol 132: 345–355
Wrana JL, Overall CM, Sodek J (1991) Regulation of the expression of a secreted acidic protein rich in cysteine (SPARC) in human fibroblasts by transforming growth factor β. Eur J Biochem 197: 519–528
Yost JC, Sage EH (1993) Specific interaction of SPARC with endothelial cells is mediated through a carboxyl-terminal sequence containing a calcium-binding EF hand. J Biol Chem 268: 25790–25796
Zetter BR, Rasmussen N, Brown L (1985) Methods in laboratory investigation. An in vivo assay for chemoattractant activity. Lab Invest 53: 362–368

MUC18: A Cell Adhesion Molecule with a Potential Role in Tumor Growth and Tumor Cell Dissemination

J.P. Johnson, M.M. Rummel, U. Rothbächer[1], and C. Sers[2]

1 Introduction

The development of metastatic capacity in a tumor is a multistep process characterized both by the loss of normal behavior and by the acquisition of new properties (Fidler and Hart 1982; Hart and Saini 1992). The identification of molecules which are newly expressed, altered or lost during this process can lead to an understanding of the molecular basis of tumor progression and metastasis formation and can also lead to the development of diagnostic and therapeutic tools. We have used monoclonal antibodies to define, in human melanoma, molecules that change their expression patterns in concert with the appearance of parameters known to be predictive for the occurrence of metastatic disease (Clark et al. 1984; Herlyn et al. 1987; Johnson 1995). Using this approach the 113kDa cell surface glycoprotein MUC18 was identified as a human melanoma associated antigen that increases in expression as tumors increase in vertical thickness and begin to acquire metastatic potential (Lehmann et al. 1987).

Institute for Immunology, Goethestrasse 31, 80336 Munich, Germany
[1]Present Address: Developmental Biology Center, Department of Developmental and Cell Biology, University of California, Irvine, CA 92717, USA
[2]Present Address: Unispital Zürich, Departement Pathologie, Schmelzbergstrasse 24, 8091 Zürich, Switzerland

2 MUC18 Belongs to a Subgroup of the Immunoglobulin Superfamily

Analysis of the MUC18 cDNA isolated from a melanoma expression library (LEHMANN et al. 1989) indicated that the molecule has all of the structural characteristics of a type I cell surface glycoprotein and that it belongs to the immunoglobulin superfamily (WILLIAMS and BARCLAY 1988). The predicted MUC18 protein consists of five immunoglobulin-like domains, a transmembrane region and a cytoplasmic tail of 63 amino acids. Eight potential N-glycosylation sites have been identified, consistent with biochemical evidence that MUC18 is a heavily glycosylated protein (LEHMANN et al. 1987). To date only a single mRNA of approximately 3.3 kb has been identified.

Isolation of the human MUC18 gene revealed that it consists of 16 exons extending over approximately 12 kb (SERS et al. 1993). The promoter region lacks classical CAAT and TATA boxes but is GC rich and shows consensus to transcriptional initiator sequences which have been identified in several other genes (ROY et al. 1991). Primer extension analysis revealed a single transcriptional start site localized within the putative initiator sequence (CTCACTTG). The intron-exon organization of the MUC18 gene demonstrates yet another variation within the growing immunoglobulin superfamily. Among members of this family, single immunoglobulin-like domains are most commonly encoded as single exons (WILLIAMS and BARCLAY 1988). However in the MUC18 gene, each of the five immunoglobulin-like domains is encoded by multiple exons. With the exception of the second domain, MUC18 resembles the neural cell adhesion molecules Po and NCAM in that the immunoglobulin-like domains are encoded by two exons with the intron splitting the domains between β strands, so that each exon encodes a half-domain (LEMKE et al. 1988). The penultimate NH_2-terminal immunoglobulin-like domain of MUC18 is encoded by three exons and the first 2 exons are separated by a large intron (> 2 kb).

A search of the Genbank and EMBL sequence databases led to the identification of several molecules with significant homology to MUC18 (Table 1). These include gicerin (TAIRA et al. 1994), SC1 (also known as DM-GRASP or BEN; BURNS et al. 1991; POUQUIE et al. 1992; TANAKA et al. 1991) and neurolin (LAESSING et al.

Table 1. Molecules with sequence similarity to MUC18

Molecule	Species	Identity %	Overlapping amino acids (*n*)	Score[a]
muMUC18	Mouse	75.4	625	2433
Gicerin	Chicken	33.3	519	809
Bcam	Human	29.4	521	586
SC1	Chicken	25.9	495	526
Neurolin	Goldfish	25.3	383	333

Evaluated with the FastA program;
[a]Optimized score

1994), molecules primarily expressed in the nervous system, and B-CAM, an epithelial cell surface antigen showing increased expression in carcinomas (CAMPBELL et al. 1994). The most closely related is gicerin isolated from the chicken. However since gicerin and human MUC18 show an identity of only 33%, it does not appear to be the MUC18 homologue. It is interesting to note that these proteins not only show the highest sequence homology to MUC18 using both the BLAST and FastA algorithms but that they also share overall structural similarities (CAMPBELL et al. 1994; LAESSING et al. 1994; TAIRA et al. 1994; TANAKA et al. 1991). All encode cell surface glycoproteins with five predicted extracellular immunoglobulin-like domains. In each case, the two most NH_2-terminal domains are larger and appear to belong to the V set of immunoglobulin-like sequences while the other three have been assigned to the C2 set of immunoglobulin-like sequences (WILLIAMS and BARCLAY 1988).

The murine homologue of the MUC18 gene has recently been isolated (ROTHBÄCHER et al. 1995). Comparison of the human and murine MUC18 predicted protein sequence reveals a conservation in both predicted protein structure and intron-exon organization. The predicted proteins have an overall identity of 75% on the amino acid level, a moderate level of conservation when compared to other immunoglobulin family members. Examination of the individual domains indicates that the most poorly conserved region (63% identity) is the NH_2-terminal domain, a region likely to be involved in ligand binding. In contrast the highest conservation (92%) is found in the cytoplasmic domain, suggesting that interactions with other molecules in the cytosol may be critical to MUC18 function. Interestingly this conserved region contains four potential sites for phosphorylation by serine/threonine kinases: three for protein kinase C (SXK) and one potential casein kinase II motif (SSGD).

3 MUC18 Expression by Human Tumors

MUC18 was first identified as a melanoma associated antigen and further studies have confirmed that its expression is characteristic of cutaneous melanoma tumors and cell lines. Northern analyses as well as antibody binding studies of cell lines derived from a variety of tumors indicate further that the expression of MUC18 is not limited to melanomas but also extends to other tumors of neuroectodermal origin. The majority of melanoma, glioma and neuroblastoma cell lines are MUC18 positive (Fig. 1). In contrast, it can neither be detected on hematopoietic cell lines nor on the vast majority of carcinoma cell lines. However, MUC18 expression (protein and mRNA) was observed in three of 24 tested carcinoma cell lines: one cervix carcinoma, one breast carcinoma and one colon carcinoma cell line.

The apparent tissue restricted expression observed on tumor cell lines is also seen in situ. The majority of primary cutaneous melanomas express MUC18 and a minority of breast carcinomas (4/42), and gastric carcinomas (4/54) were also

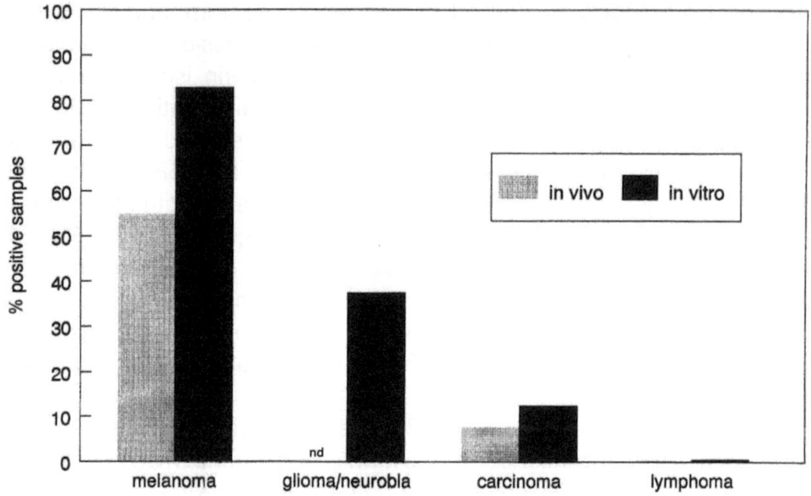

Fig. 1. Expression of MUC18 on human tumors in vivo and in vitro. The percent of MUC18 positive tumor samples or cell lines as determined by antibody binding. Melanomas in vivo, *n*=110; cell lines, *n*=12; gliomas/neuroblastoma cell lines, *n*=8; carcinomas in vivo, *n*=99, cell lines, *n*=24; lymphomas in vivo, *n*=10, cell lines *n*=5. *nd*, not done, *neurobla*, neuroblastoma

found to be MUC18 positive, a reactivity which was confirmed with antibodies directed against two independent epitopes. Although both the breast and stomach patient collectives were extensively characterized for histological and clinical parameters, no associations were evident between MUC18 expression and tumor characteristics or clinical course. In contrast, immunohistochemical analyses of cutaneous melanoma using monoclonal antibodies directed against different epitopes indicate that the expression of MUC18 increases with increasing tumor progression (Holzmann et al. 1987; Shih et al. 1994b). Epidermal melanocytes in situ do not express MUC18 and benign melanocytic tumors (i.e. melanocytic nevi) express it only weakly. Among primary tumors MUC18 expression increases with increasing vertical thickness, the main parameter of tumor progression in melanoma (Breslow 1970; Clark et al. 1984), and MUC18 expression is strongest and most frequent on metastatic lesions.

4 Regulation of MUC18 Expression: Evidence for a Loss of Sensitivity to Negative Regulation in Melanomas

There is no evidence that the expression of MUC18 by malignant tumors is associated either with chromosomal translocation, mutation or gene amplification. The observations that MUC18 expression is essentially restricted to tumors of the neuroectodermal lineage and that in cutaneous melanomas it appears to increase with tumor thickness and tumor progression must therefore reflect changes in the regulation of MUC18 expression.

As primary tumors increase in size and grow into the dermis, the melanoma cells become increasingly distant from their normal epidermal environment. Recent studies indicate that keratinocyte-melanocyte interactions are important in the regulation of MUC18 expression in benign melanocytes and perhaps in early primary tumors (SHIH et al. 1994a). In contrast to the observations in situ, nevus cells and epidermal melanocytes grown in culture express high levels of MUC18. Coculture with keratinocytes for 5 days led to an 80%–90% reduction in MUC18 expression on the melanocytic cells, an effect which was not observed when cocultured with fibroblasts or carcinoma cells. However, MUC18 expression by melanoma cells was not affected by coculture with keratinocytes, suggesting that melanoma cells are no longer sensitive to the keratinocyte-mediated effects. This keratinocyte effect and the resistance of advanced melanomas may partially explain the in vivo observations that MUC18 expression increases with increasing tumor progression.

The mechanisms by which keratinocytes are able to down-regulate MUC18 expression remain unknown although direct cellular contact appears to be required. This phenomenon may be related to the observation that the presence of phorbol ester also leads to MUC18 down-regulation in certain cells (RUMMEL et al., manuscript in preparation). Exposure of the glioma cell line LN215 to phorbol ester led to the loss of cell surface MUC18 within 24 h (Fig. 2). Immunoreactive MUC18 could not be detected in the supernatants of the treated cells, suggesting that proteolytic cleavage of the molecule from the membrane is probably not responsible for this down-regulation. Furthermore, a decrease in MUC18 mRNA was apparent at 4 h and was maximum at 18 h, indicating that down-regulation is occurring at the level of transcription. Interestingly, phorbol ester did not induce a similar down-regulation of MUC18 expression in all cells. MUC18 expression was able to be down-regulated in all glioma and MUC 18 positive carcinoma cell lines examined. In contrast, similar to the kertainocyte-mediated down-regulation, melanoma cell lines, regardless of the levels of MUC18 expression, were insensitive to the effects of phorbol ester. These results may point to a protein kinase C activated pathway that can repress MUC18 expression, a pathway that is no longer functional in malignant melanoma cells.

The elements responsible for constitutive MUC18 expression in melanoma cells have not yet been identified. Analyses of reporter gene constructs indicate that the preferential expression of MUC18 by neuroectodermal tumors is not determined by sequence elements close to the promoter. Reporter gene constructs containing 900 bp of human MUC18 5' region were tested for activity using transient transfection assays. This region does confer promoter activity and leads to levels of luciferase which are seven- to175-fold above those obtained in the absence of a promoter (Fig. 3). However, high levels of luciferase activity were observed both in the carcinoma cell lines which do not express MUC18 (LoVo, Colo320) and in the neuroectodermally derived cells (melanomas Mel JuSo,

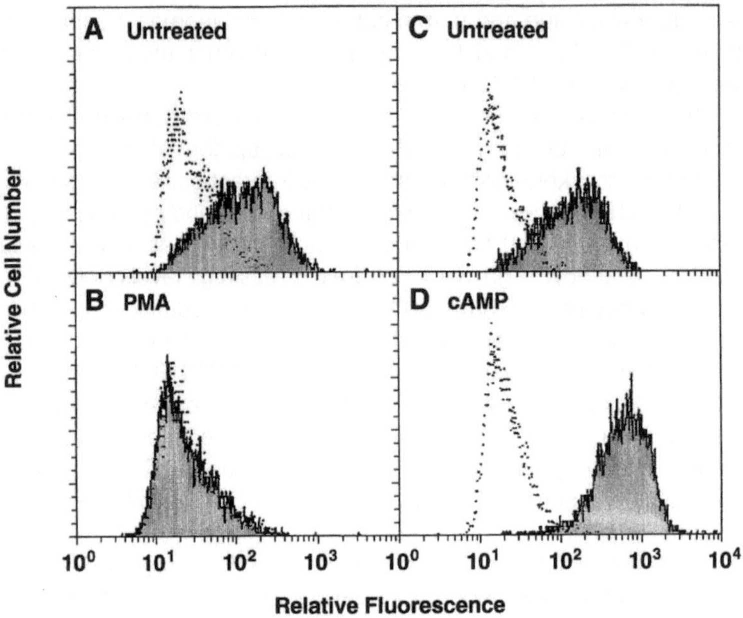

Fig. 2A–D. Modulation of MUC18 expression. Glioma cell line LN215 was exposed to phorbol ester (phorbol 12-myristate 13-acetate, 10 ng/ml, 24 h; (**A,B**) or to CT-cAMP (8-(4-chlorophenylthio)-adenosine 3':5'-cyclic monophosphate, 200uM, 48 h (**C,D**). FACS profiles showing isotype control binding (myeloma Upc10, IgG2a;*unfilled* curves and anti-MUC18 binding (monoclonal antibody MUC18, IgG2a; *filled curves*) of treated and untreated control cells

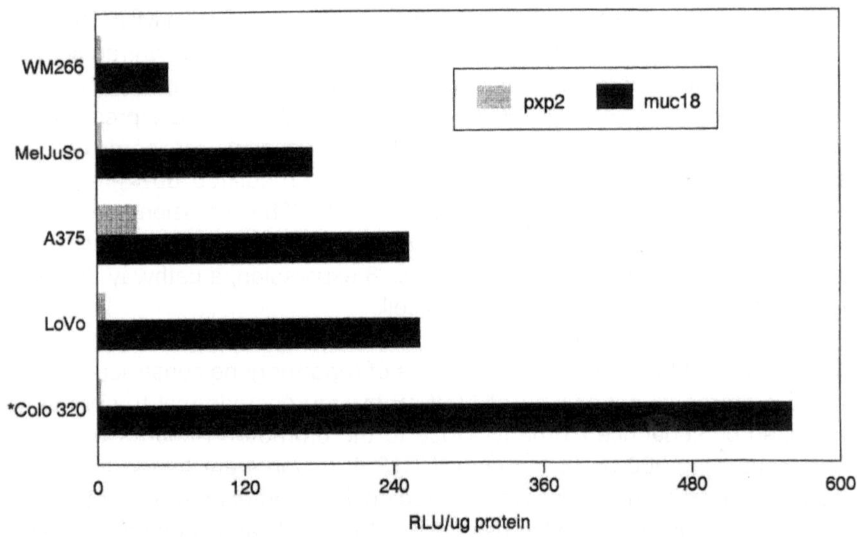

Fig. 3. Activity of MUC18 promotor in various cell lines. Luciferase reporter gene activity is shown in relative light units (RLU) per microgram of protein except for Colo 320 (star) where RLU per 0.1 ug protein are shown. *Black bars:* luciferase activity under the control of the MUC18 gene promoter region; *gray bars*: activity driven by the vector, pxp2 alone. WM266, Mel JuSo, A375, melanoma cell lines; LoVo, Colo320, colon carcinoma cell lines

WM-226; glioma LN215) which do express this molecule. While tissue restricted expression of genes is frequently controlled by elements 5' to the gene, elements 3' as well as within the gene may be responsible. In the case of MUC18, such elements may reside in the large introns.

5 Constitutive and Induced Expression of MUC18 in the Vascular System

In contrast to the restricted expression observed in cell lines, northern analysis of various normal human and murine tissues indicate that MUC18 expression in vivo is ubiquitous (ROTHBÄCHER et al. 1995; SERS et al. 1994). In human tissues, the highest levels relative to mRNA encoding the enzyme GAPDH were observed in placenta and lung, while low levels were observed in kidney and skeletal muscle. Immunohistochemical analysis of many of these tissues with monoclonal antibodies directed against three different epitopes of the MUC18 glycoprotein revealed that the primary site of expression of this molecule in normal tissue is smooth muscle (LEHMANN et al. 1987; SERS et al. 1994). Smooth muscle in the lung, skin, and gastrointestinal tract are MUC18 positive as is the smooth muscle of small and large blood vessels. Expression of MUC18 in the vessels is not limited to smooth muscle but can occasionally be seen on endothelial cells (SHIH et al. 1994b; SERS et al. 1994). MUC18 reactive endothelia are generally seen on only a fraction of the vessels present in a section as estimated by CD31 (a pan-endothelial marker) expression. In contrast, many of the vessels observable in tumors are MUC18 reactive. These observations suggest that MUC18 expression may be associated with activation of the endothelial cells.

In order to test whether MUC18 expression can be induced in vitro, MUC18 positive and MUC18 negative cell lines were examined for their response to a variety of cytokines including interleukin-1, interleukin-6, interferon-γ, and tumor necrosis factor (TNF)-α, as well as to modulators of cAMP (CPT-cAMP, forskolin). Although none of the cytokines influenced MUC18 expression, increasing cAMP levels led to increased surface expression of MUC18 in several MUC18 positive cell lines (Fig. 2; RUMMEL et al., manuscript in preparation). Northern analyses indicates that the increase in surface MUC18 expression is due to increased mRNA, which could be observed as early as 4 h following treatment. An examination of the sequence of the human MUC18 promoter region revealed the presence of a potential cAMP responsive element, further supporting a direct effect on MUC18 expression (SERS et al. 1993).

6 MUC 18: A Cell Adhesion Molecule with a Potential Role in Tumor Growth and Tumor Cell Dissemination

Solely on the basis of structural characteristics and homology with other molecules, MUC18 was originally postulated to function as a cell adhesion molecule (Lehmann et al. 1989). The molecules most closely related to MUC18 include gicerin, neurolin, and SC1, all of which have been shown to be cell adhesion molecules (Taira et al. 1994; Laessing et al. 1994; Tanaka et al. 1991). Furthermore, MUC18 can carry a posttranslational carbohydrate modification (a sulfated glucuronic acid which forms an epitope denoted CD57 or HNK-1) that has been implicated in the adhesive interactions of several cell adhesion molecules including neural cell adhesion molecule (NCAM) and myelin associated glycoprotein (MAG) (Riopelle et al. 1986; Shih et al. 1994b). The first evidence that MUC18 may actually mediate intercellular adhesion was obtained from studies using antigen purified from melanoma cells (Shih et al. 1994b). Plates coated with MUC18 antigen were shown to bind to melanoma cell lines, a binding that was inhibited by soluble MUC18 protein and by an antibody to CD 57. Studies using both mouse L cells transfected with MUC18 cDNA and soluble MUC18 Ig constructs (unpublished observations) suggest that MUC18 does not function as a homophilic cell adhesion molecule and implicate a heterophilic ligand on the melanoma cells.

In order to evaluate the role of MUC18 in intercellular adhesion between melanoma cells, stable MUC18 expressing transfectants were produced using MUC18 negative melanoma cells. Transfection of the cell line SBCL2 with MUC18 cDNA in the eukaryotic expression vector pcDNA1 led to stable expression of MUC18 by most of the cells (Fig. 4A). Intercellular adhesion was assessed using aggregation rate assays (Tanaka et al. 1991; Taira et al. 1994) in which single cell suspensions are prepared and the cells are incubated at 37°C for 1 h. At various timepoints the number of single cells and the number of cells in clusters (>2 cells) are determined. No differences were observed between the MUC18 positive and MUC18 negative cells under standard assay conditions (Fig. 4B, solid lines). Both cell lines formed large clusters (> 10 cells) at a similar rate. When the assays were carried out in the presence of 3 m*M* EDTA, much of the non-MUC18 mediated aggregation was inhibited (SBCL2-E, Fig. 4B). Under these conditions, significant aggregation of the MUC18 expressing cell was still observed (SBCL2-M18-E) indicating that MUC18 can mediate intercellular adhesion and that this is independent of divalent cations. Mixing experiments indicated that the aggregates contained both SBCL2 and SBCL2-MUC18 cells, confirming that MUC18 is interacting with a heterophilic ligand on the melanoma cells.

These studies confirm that MUC18 can function as an intercellular adhesion molecule and that it binds to a ligand which can itself be expressed on melanoma cells. As MUC18 is expressed on the endothelia of certain blood vessels, in particular those found in tumors, interaction via the MUC18 ligand on melanoma cells may play a role in the ability of the tumor cells to enter (and exit) the vascular

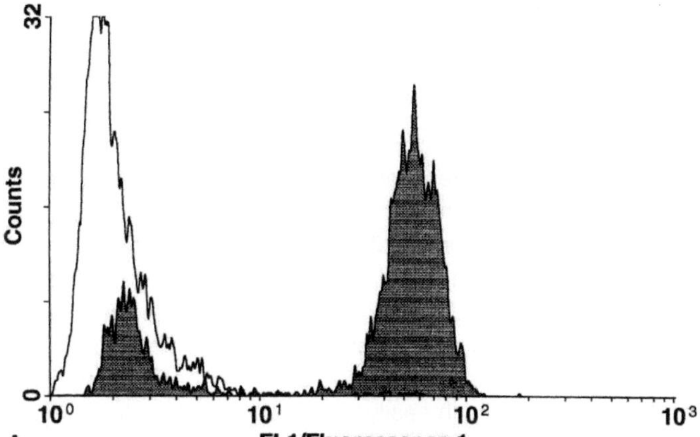

A

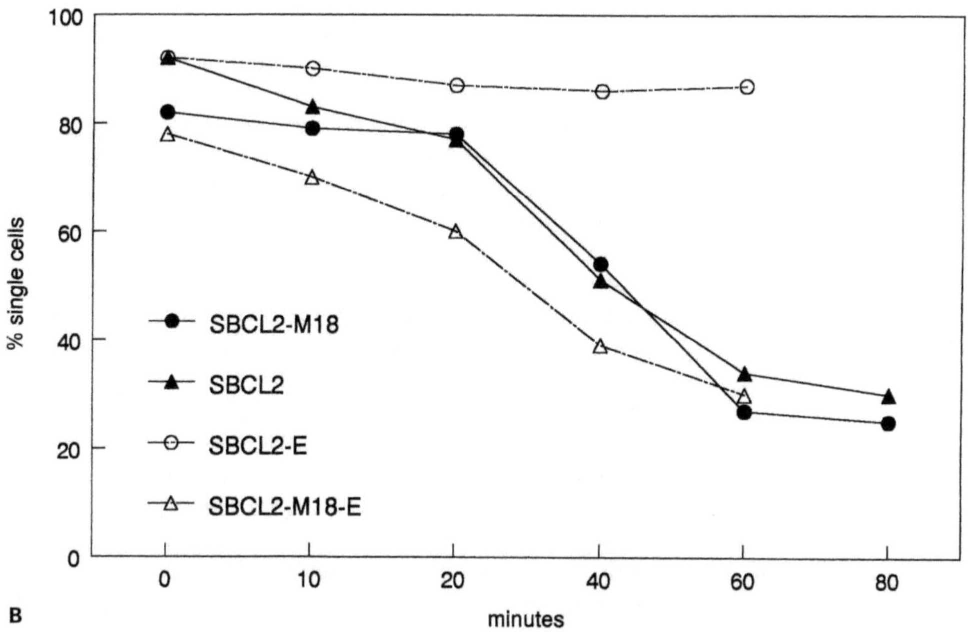

B

Fig. 4A,B. A Expression of MUC18 by transfected melanoma cells. Binding of monoclonal antibody MUC18 to SBCL2 melanoma cells (*unfilled curve*) and to SBCL2 cells transfected with MUC18 cDNA (SBCL2-M18; *filled curve*). **B** Contribution of MUC18 expression to melanoma cell aggregation. Kinetics of aggregation (TANAKA et al. 1991) of SBCL2 and SBCL2-M18 at 37°C in normal culture medium (*solid lines, filled markers*) and in the presence of 3 m*M* EDTA (*dotted lines, unfilled markers*). Aggregation was allowed to occur without rotation and at 10^5 cells/ml

system. However, it is the expression of MUC18 by the melanoma cells themselves which has been associated with tumor progression. In an attempt to determine whether MUC18 expression by melanoma cells is associated with metastasis development in vivo, nine human melanoma cell lines previously characterized for their ability to form lung colonies following intravenous injection into nude mice were evaluated for MUC18 expression (Luca et al. 1993). All of the cells which were able to metastasize in this system showed high surface expression of MUC18. Of the five cell lines incapable of forming experimental metastases in nude mice, three had no surface MUC18 expression while two expressed high levels of this molecule. These observations suggest that the expression of MUC18 by melanoma cells may be necessary for lung colony formation although it is not sufficient. Experimental metastasis assays measure the ability of cells to leave the vascular system and to establish foci of growth in the parenchyma of the lung. Thus the role of MUC18 in this process could involve adhesion, migration through the endothelium, or establishment of tumor growth. A comparsion of two sublines of a melanoma which differ primarily in their tumorigenicity showed that the subline which easily produced local tumors after subcutaneous inoculation expressed MUC18 while the subline which was not able to reliably give rise to tumors in this assay was negative for both MUC18 protein and mRNA. These observations suggest that MUC18 expression may also contribute to tumor growth. Determination of whether MUC18 actually plays a role the formation of tumors by human melanoma cells in nude mice will require inactivating the gene in MUC18 expressing tumor cells as well as transfecting it into the nonexpressing tumor cells. Nevertheless, these studies provide the first suggestion that MUC18 expression by the tumor cells themselves may be important in tumor growth and metastasis formation.

Metastasis formation requires the cells to enter the vascular system. The presence of MUC18 on blood vessel endothelia suggests a possible pathway for melanoma cells expressing the MUC18 ligand to enter vessels. MUC18-MUC18 ligand interactions among the melanoma cells may lead to the formation of tumor cell clusters. These clusters can result in vessel blockage and subsequent anchoring to the endothelium. Invasion into the surrounding tissue could again be mediated by endothelial MUC18 expression. In addition, MUC18 mediated tumor cluster formation may support growth promoting interactions between the tumor cells. Aspects of particular future interest include identifying the heterophilic ligand responsible for MUC18 mediated adhesion as well as the factors regulating endothelial MUC18 expression. Overall, the recent findings regarding MUC18 open numerous possibilities for further study of its role in tumor growth and metastasis.

Acknowledgments. This work was supported by the Deutsche Krebshilfe, Mildred Scheel Stiftung, Bonn, and by the Deutsche Forschungsgemeinschaft, SFB 217 (A3).

References

Breslow A (1970) Thickness, cross-sectional areas and depth of invasion in the prognosis of cutaneous melanoma. Ann Surg 172: 902–908

Burns FR, von Kannen S, Guy L, Raper JA, Kamholz J, Chang S (1991) DM-GRASP, a novel immunoglobulin superfamily axonal surface protein that supports neurite extension. Neuron 7: 209–220

Campbell IG, Foulkes WD, Senger G, Trowsdale J, Garin-Chesa P, Rettig WJ (1994) Molecular cloning of the B-CAM cell surface glycoprotein of epithelial cancers: a novel member of the immunoglobulin family. Cancer Res 54: 5761–5765

Clark WH, Elder DE, Guerry D, Epstein MN, Greene MH, van Horn M (1984) A study of tumor progression. The precursor lesions of superficial spreading and nodular melanoma. Hum Pathol 15: 1147–1165

Fidler IJ Hart IR (1982) Biological diversity in metastatic neoplasms: origins and implications. Science 217: 998–1003

Hart IR, Saini A (1992) Biology of tumour metastasis. Lancet 339: 1453–1457

Herlyn M, Clark WH, Rodeck U, Mancianti ML, Jambrosic J, Koprowski H (1987) Biology of tumor progression in human melanocytes. Lab Invest 56: 461–474

Holzmann B, Bröcker EB, Lehmann JM, Ruiter DJ, Sorg C, Riethmüller G, Johnson JP (1987) Tumor progression in human malignant melanoma: five stages defined by their antigenic phenotypes. Int J Cancer 39: 466–471

Johnson JP (1995) The identification of molecules associated with metastasis formation in human melanoma. Invasion Metastasis (in press)

Laessing U, Giordano S, Stecher B, Lottspeich F, Stuermer CAO (1994) Molecular characterization of fish neurolin: a growth-associated cell surface protein and member of the immunoglobulin superfamily in the fish retinotectal system with similarities to chick protein DM-GRASP/SC-1/BEN. Differentiation 56: 21–29

Lehmann JM, Holzmann B, Breitbart EW, Schmiegelow P, Riethmüller G, Johnson JP (1987) Discrimination between benign and malignant cells of melanocytic lineage by two novel antigens, a glycoprotein with a molecular weight of 113,000 and a protein with a molecular weight of 76,000. Cancer Res 47: 841–845

Lehmann JM, Riethmüller G, Johnson JP (1989) MUC18, a marker of tumor progression in human melanoma shows sequence similarity to the neural cell adhesion molecules of the immunoglobulin superfamily. Proc Natl Acad Sci USA 86: 9891–9895

Lemke G, Lamar E, Patterson J (1988) Isolation and analysis of the gene encoding peripheral myelin protein zero. Neuron 1: 73–83

Luca M, Hunt B, Bucana CD, Johnson JP, Fidler IJ, Bar Eli M (1993) Direct correlation between MUC 18 expression and metastatic potential of human melanoma cells. Melan Res 3: 35–41

Pourquie O, Corbel C, La Caer JP, Rossier J, LeDouarin NM (1992) BEN, a surface molecule of the immunoglobulin superfamily expressed in a variety of developing systems. Proc Natl Acad Sci USA 89: 5261–5265

Riopelle RJ, McGarry RC, Roder JC (1986) Adhesion properties of a neuronal epitope recognized by the monoclonal antibody HNK-1. Brain Res 367: 20–25

Rothbächer U, Sers C, Riethmüller G, Johnson JP (1995) Cloning and characterization of the murine homolog of the human melanoma associated molecule MUC18 (submitted)

Roy AL, Meisterernst M, Pognonec P, Roeder RG (1991) Cooperative interaction of an initiator-binding transcription initiator factor and the helix-lool-helix activator USF. Nature 354: 245–248

Sers C, Kirsch K, Rothbacher U, Riethmuller G, Johnson JP (1993) Genomic organization of the melanoma-associated glycoprotein MUC18: implications for the evolution of the immunoglobulin domains. Proc Natl Acad Sci USA 90: 8514–8518

Sers C, Riethmuller G, Johnson JP (1994) MUC18, a melanoma-progression associated molecule, and its potential role in tumor vascularization and hematogenous spread. Cancer Res 54: 5689–5694

Shih IM, Elder DE, Hsu MY, Herlyn M (1994a) Regulation of Mel-CAM/MUC18 expression on melanocytes of different stages of tumor progression by normal keratinocytes. Am J Pathol 145: 837–845

Shih IM, Elder DE, Speicher D, Johnson JP, Herlyn M (1994b) Isolation and functional characterization of the A32 melanoma-associated antigen. Cancer Res 54: 2514–2520

Taria E, Takaha N, Taniura H, Kim CH, Miki N (1994) Molecular cloning and functional expression of gicerin, a novel cell adhesion molecule that binds to neurite outgrowth factor. Neuron 12: 861–872

Tanaka H, Matsui T, Agata A, Tomura M, Kubota I, McFarland KC, Kohr B, Lee A, Philips HS, Shelton DL (1991) Molecular cloning and expression of a novel adhesion molecule. SC1. Neuron 7: 535–545

Williams AF, Barclay AN (1988) The immunoglobulin superfamily-domains for cell surface recognition. Annu Rev Immunol 6: 381–405

Lu-ECAM-1 and DPP IV in Lung Metastasis

R.C. Elble and B.U. Pauli

1 Introduction

The development of the metastatic cell is a Darwinian process involving random genetic changes and selection over many generations of tumor growth. Early events include a loosening of cell-cycle control and disabling of checkpoints and failsafe mechanisms such as cell-cell constraints on proliferation and motility (Rustgi et al. 1992; Fearson and Vogelstein 1990). Successfully metastatic cells arising from this population display an altered repertoire of cell adhesion molecules, allowing escape from the primary tumor, adhesion and penetration of the extracellular matrix (ECM) and entry into the microvasculature (Nicholson 1988; Fidler 1990; Hart and Saini 1992; Lafrenie et al. 1993). Most such cells are destroyed by geometric and hemodynamic forces in their first encounter with the narrow bore capillary net, usually in the lungs (Weiss et al. 1988). The few survivors may give rise to secondary colonies. The frequency of metastasis to the lungs has often been attributed solely to mechanical entrapment of tumor cell emboli. However, this theory does not explain the early observation that certain

Cancer Biology Laboratories, Department of Pathology, Cornell University College of Veterinary Medicine, Ithaca, NY 14853, USA

tumor cell types prefer to metastasize to specific target organs. This behaviour was first noticed by Stephen PAGET in 1889. He observed that breast carcinomas had very specific organ preferences for metastasis, and those preferences were not obviously related to circulatory patterns. About 70% of the patients had metastases to the liver, 20% to lungs, and only 10% to a few other organs. These observations inspired the "seed and soil" hypothesis which postulates that metastatic tumor cells are potentially unlimited in distribution but can only find purchase and prosper in a congenial soil. Experimental support for this hypothesis was obtained by KINSEY in 1960. He showed that an ectopic lung, implanted beneath the dermis of a mouse thigh, is metastasized as frequently as the normal lung following injection of lung-metastatic cancer cells, regardless of the site of injection (KINSEY 1960). No metastases were observed in other ectopically placed organs. The molecular basis of this result in undoubtedly complex but could be mediated by tumor cell/host cell interactions that promote implantation, invasion, and growth at the new site (ROSSI and ZETTER 1992; ZETTER 1990; CHACKAL-ROY et al. 1989; SARGENT et al. 1988; NAITO et al. 1987; FIDLER 1990).

There is increasing evidence that the selection of a target organ for metastasis is mediated by specific interactions between blood-borne cancer cells and the endothelium of that target organ (reviewed by ALBELDA 1993; DAMJANOVICH et al. 1992; PAULI et al. 1992; ALBELDA 1991; MCCARTHY et al. 1991; PAULI et al. 1990, 1991; BELLONI and TRESSLER 1990). Two groups of endothelial cell adhesion molecules have been implicated in this process. The first group consists of cytokine-inducible adhesion molecules belonging to the selectin and immunoglobulin (Ig) superfamilies (BEVILACQUA 1993; BEVILACQUA and NELSON 1993; BUTCHER 1991). Cytokine induction involves mobilization from intracellular stores, posttranslational modifications (e.g., phosphorylation, glycosylation), and de novo synthesis (MCEVER et al. 1989). The primary role of these molecules is to promote vascular arrest and facilitate extravasation of leukocytes during episodes of acute inflammation and immunologic disorders. The notion that inducible endothelial cell adhesion molecules also mediate metastasis has some appealing elements. Potential ligands for endothelial E- and P-selectins, sialyl Le^x and Le^a, are common on tumor cells, especially colon carcinoma (TAKADA et al. 1993; DEJANA et al. 1992; MATSUSHITA et al. 1990; WALZ et al. 1990). Moreover, ligands for Ig superfamily members VCAM-1 and ICAM-1 can be found constitutively expressed on certain tumor cell types (QIAN et al. 1994; OKAHARA et al. 1994; LAURI et al. 1991 a,b; WETERMAN et al. 1994; ALBELDA 1993). However, a major obstacle arises in explaining how these molecules come to be induced and how that induction could be organ-specific. A further problem is that high levels of inducible cell adhesion molecules are found in serum of cancer patients (BANKS et al. 1993; HARNING et al. 1991). Soluble selectins, and ICAM-1 and VCAM-1 would be expected to act as competitive inhibitors of tumor cell adhesion to endothelium (HARNING et al. 1991). A more plausible role for inducible cell adhesion molecules might be in immunosurveillance or in neovascularization. Cytokine production by tumor cells or tumor-infiltrating leukocytes at the primary tumor site should be adequate to activate endothelium. Indeed, the vasculatures of several late stage tumor types show expression of selectins and inducible Ig superfamily members (ALBELDA 1993; COULDWELL et al. 1992). In this context, inducible cell adhesion molecules might serve to recruit tumor-attacking immune and inflammatory cells. Conversely, activated angiogenic endothelial cells might find purchase on tumor cells expressing the appropriate ligand, facilitating endothelial cell migration and penetration of the tumor. Such interactions could also enable tumor cells to passively enter the bloodstream as they become engulfed into the lumen of invading vascular sprouts.

The second group of endothelial cell adhesion molecules is comprised of constitutively expressed membrane glycoproteins that have restricted distributions in the vasculature of one (or more) organ (ALBELDA 1991, 1993; DAMJANOVICH et al. 1992; PAULI et al. 1990, 1992; PILEWSKI and ALBELDA 1993). The likely role for such organ-specific endothelial cell adhesion molecules in metastasis was foreshadowed by the work of Auerbach and coworkers in the mid-1980s (AUERBACH et al. 1987; ALBY and AUERBACH 1984). They showed that tumor cells mimicked lymphocytes during the "homing" to particular lymphoid tissues (LASKY et al. 1992; SPRINGER 1990; BUTCHER et al. 1979; STAMPER and WOODRUFF 1976) adhering preferentially to endothelial cells derived from the metastasized organ. For example, ovary-metastatic teratoma cells bound preferentially to ovary-derived endothelial cells, and glioma cells adhered preferentially to brain endothelium. However, primary cultures of organ-specific microvascular endothelial cells did not prove useful for molecular analysis of tumor cell/endothelial cell interaction, chiefly because the cells tend to lose their organ-specific phenotype during culture in vitro (BELLONI and NICOLSON 1988; DEBONO and GREEN 1984; BORSUM et al. 1982). A second problem is that cells from postcapillary venules, a frequent site of adhesion and extravasation of cancer cells, are underrepresented in endothelial cell isolates (ZHU and PAULI 1991, 1993). To circumvent these difficulties, we have devised two in vitro systems to facilitate recovery of the predicted adhesion molecules. One involves the modulation of readily isolated and cultured aortic endothelial cells by organ-specific matrices in vitro. Modulation presumably causes the cells to express organ-specific traits including constitutive cell adhesion molecules. This system was used to isolate and characterize a novel endothelial cell surface molecule called lung-derived endothelial cell adhesion molecule-1 (Lu-ECAM-1) that mediates arrest and lung metastasis of melanoma cells (ZHU et al. 1991). The second system we have used relies not on recovery of whole cells from the vasculature but only membrane samples in the form of microvesicles. The microvesicles are harvested by in situ perfusion of the lungs with paraformaldehyde, yielding a highly antigenic representative of the endothelial cell surface for adhesion assays and for the production of monoclonal antibodies. The method was used in the identification of dipeptidyl peptidase IV (DPP IV) as a cell adhesion molecule that mediates the metastasis of breast carcinoma cells to the lung. Here we shall focus on our attention to work on Lu-ECAM-1 and, to a lesser extent, DPP IV.

2 The Lung-Derived Endothelial Cell Adhesion Molecule-1

2.1 Expression and Adhesion

Lung matrix-modulated bovine aortic endothelial cells (BAECs) were the first in vitro system successfully used in the isolation, purification, and partial characterization of a constitutive endothelial cell adhesion molecule involved in organ-specific metastasis. This molecule is the lung-derived, melanoma-cell binding adhesion receptor Lu- ECAM-1. It was first identified by a monoclonal antibody (mAb; 6D3; IgG2a) that had been generated against outside-out, luminal membrane vesicles from lung matrix-modulated BAECs and selected for inhibition of adhesion of lung-metastatic B16-F10 melanoma cells to these endothelial cells (ZHU and PAULI 1991; ZHU et al. 1991). Anti-Lu-ECAM-1 mAb 6D3 was used to immunoprecipitate the 90 kDa Lu-ECAM-1 from detergent extracts of lung matrix-modulated BAECs (ZHU et al. 1991). Lu-ECAM-1 can also be precipitated from these extracts with immobilized wheat germ agglutinin and *Ricinus communis* agglutinin-I, but it binds less efficiently to concanavalin A and not at all to *Ulex europaeus* agglutinin-I. Lu-ECAM-1 is not up-regulated when BAECs grown on plastic are incubated with tumor necrosis factor (TNF)-α (50 ng/ml), recombinant interleukin-(rIL)-1α(50 ng/ml), or lipopolysaccharide (LPS 10 µg/ml) for 4, 8, 12, or 16 h (ZHU et al. 1991). Tissue distribution of Lu-ECAM-1 is restricted to the endothelial lumenal membrane of postcapillary venules and small and medium size venules in bovine lungs (ZHU et al. 1991). The pattern and level of expression are unaltered in lung sections from bovines with acute bronchopneumonia (ZHU and PAULI 1991).

Immunoaffinity-purified bovine Lu-ECAM-1 binds high lung metastatic B16-F10 melanoma cells in significantly larger numbers than their intermediate and low lung metastatic counterparts B16-L8-F10 and B16-F0, respectively. Maximal binding is observed at a density of approximately 2.4×10^2 Lu-ECAM-1 molecules per μm^2 of Lu-ECAM-1 (300 ng/ml)-coated plastic dishes (ZHU et al. 1992). Binding of B16 melanoma cells is dependent upon the presence of Ca^{2+} in HBSS adhesion medium (1 mM), but is unaffected by Mg^{2+} (ZHU et al. 1992). Bovine Lu-ECAM-1 also binds lung metastatic human melanoma cells (cell line 1205) but fails to bind other lung metastatic tumor cells (e.g., KLN205 squamous carcinoma) or lymphocytes and neutrophils isolated from bovine peripheral blood.Thus, Lu-ECAM-1 appears to be melanoma-specific.

The preferential binding of lung-metastatic B16 melanoma cells to Lu-ECAM-1 is significantly inhibited by anti-Lu-ECAM-1 mAb 6D3 (ZHU et al. 1991, 1992). At each coating-density of Lu-ECAM-1, anti-Lu-ECAM-1 mAb 6D3 (10 µg/ml) reduces binding of high lung metastatic B16-F10 cells to slightly lower levels than the low lung metastatic B16-F0 is able to bind to Lu-ECAM-1. Control mAbs 3C6 and SE12 (both directed against endothelial cell surface epitopes other than Lu-ECAM-1) as well mAbs directed against murine intercellular adhesion molecule-1 (ICAM-1) (HORLEY et al. 1989) and keratinocyte surface antigen (SUTER et al.

1990) have no effect on the binding of B16 melanoma cells to Lu-ECAM-1. Binding of lung-metastatic B16 melanoma cells to Lu-ECAM-1 coated dishes is also competitively inhibited in a dose-dependent manner by soluble Lu-ECAM-1 supplemented in the assay medium (ZHU et al. 1992). Maximal inhibition is achieved at the same Lu-ECAM-1 concentration that has been used for the coating of the plastic dishes, namely 100 ng/ml. Higher Lu-ECAM-1 concentrations have no further inhibitory effect. The control protein glycophorin A, tested at the same detergent concentration used to solubilize Lu-ECAM-1, had no binding inhibitory effect.

2.2 The Role of Lu-ECAM-1 in Melanoma Lung Metastasis

The well-known formation of pleural and subpleural B16-F10 melanoma colonies is correlated quantitatively with prominent histochemical staining of endothelia from mouse pleural and subpleural venules with anti-Lu-ECAM-1 mAb 6D3 (80% of the Lu-ECAM-1-positive blood vessels and 57% of the B16-F10 experimental metastases are localized in the pleura) (ZHU and PAULI 1993). The less frequent endothelial staining of perivenous and peribronchial venules is associated with correspondingly fewer B16-F10 colonies in these locations, and the occasional segmental staining of pulmonary veins coincides with rare tumor nodules which usually expand in an asymmetric fashion around these veins (ZHU et al. 1991; ZHU and PAULI 1993). Lu-ECAM-1 is also expressed on endothelia of some tumor vessels indicating that these vessels are recruited from the same host blood vessels that originally caused the arrest of blood-borne B16-F10 melanoma cells.

The close association between the lung distribution patterns of Lu-ECAM-1-positive blood vessels and experimental melanoma metastases prompted us to evaluate the effect of Lu-ECAM-1-blocking on the formation of B16-F10 lung colonies, using anti-adhesive, anti-Lu-ECAM-1 mAb 6D3 in a standard lung colony assay. Syngeneic mice passively immunized with anti-Lu-ECAM-1 mAb 6D3 1 h prior to an intravenous challenge with B16-F10 cells exhibit a more than 90% reduction in the number of lung colonies when compared to mice that have received control mAbs of the same immunoglobulin isotype (6D8) or against endothelial cell surface determinants other than Lu-ECAM-1 (3C6 and 5E12) (ZHU et al. 1991, 1992). The anti-metastatic effect of mAb 6D3 is still noticeable, when B16-F10 are injected 6 or 24 h after treatment of mice with mAb 6D3, although inhibition of metastasis is now only 41% and 33%, respectively (ZHU et al. 1992). As expected, anti-Lu-ECAM-1 mAb 6D3 also efficiently inhibits the colonization of the lungs by B16-L8-F10 (an intermediate lung metastatic and moderate liver metastatic B16 melanoma variant), but has no inhibitory effect on the propensity of this cell line to colonize the liver (ZHU et al. 1992). Anti-Lu-ECAM-1 antibody mAb 6D3 has no effect on the metastatic performance of other lung metastatic tumor cell types (e.g., KLN205 squamous carcinoma cells). These findings imply that different tumor cell types recognize distinct Lu-ECAMs and that inhibition of lung colonization by B16-F10 is not mediated by nonspecific, steric blocking of adhesion, as indicated by the inability of control mAbs 3C6 and 5E12, both detecting endothelial cell surface determinants other than Lu-ECAM-1, to inhibit lung colonization of B16 melanoma cells.

To further substantiate that the interruption of the metastatic cascade occurs at the site of tumor cell adhesion to lung endothelium, we have compared lung clearance rates of tail vein injected B16-F10 melanoma cells in untreated C57B1/6 mice and in mice treated with anti-Lu-ECAM-1 mAb 6D3. Lung clearance is rapid in both animal groups and is accomplished primarily within the first 12 h after i.v. injection of B16-F10 (Zhu et al. 1992). At the end of this period only 2.11% of the initial tumor cell inoculum remains in the lungs of untreated mice compared to 1.32% in mice treated with anti-Lu-ECAM-1 mAb 6D3. The more rapid tumor cell clearance from the lungs of anti-Lu-ECAM-1 mAb 6D3-pretreated mice is paralleled by an expected increase in the number of tumor cells in liver, spleen, and kidney relative to untreated controls. By 3–5 days after i.v. inoculation of tumor cells, the number of lung-bound tumor cells has further decreased to correspond approximately with the number of lung colonies counted 3 weeks later in each of the animal groups (Zhu et al. 1992). The remaining organs are all cleared of unspecifically arrested tumor cells within 3– 5 days after tail vein injection of B16-F10 melanoma cells, although a slightly slower pace of clearance is observed in the treated animal group as illustrated by the clearance rates of B16-F10 from livers and kidneys.

Active immunization of mice against bovine Lu-ECAM-1 purified by immunoaffinity chromatography from detergent extracts of superconfluent lung matrix-modulated BAECs also prevents metastatic colonization of the lungs by B16-F10 melanomas (Zhu et al. 1992). The efficiency of inhibition of experimental metastases is dependent upon the anti-Lu-ECAM-1 serum titer of the immunized mice. At a titer of 1:1000 an average of 21±6 B16-F10 tumor colonies are counted, but only 9±3 colonies are observed in mice with serum titers of 1:2000 ($p < 0.05$) (Zhu et al. 1992). In contrast, mice immunized with saline alone and having no detectable anti-Lu-ECAM-1 serum titers exhibit over 200 B16-F10 melanoma lung colonies. Similar to passive immunization, active immunization against Lu-ECAM-1 has no effect on the metastatic performance of KLN205.

2.3 Cloning and Sequencing of Lu-ECAM-1

Lu-ECAM-1 cDNAs have been cloned from a library using a probe based on Lu-ECAM-1 internal amino acid sequences. The library was constructed using RNA from matrix-modulated BAECs that expressed Lu-ECAM-1 at high levels. The amino acid sequences of two tryptic peptides from Lu-ECAM-1 were used to plan degenerate oligonucleotide primers for PCR. A 540 base-pair product was obtained and used to probe the library. A complete 2.8 kilobase-pair cDNA clone was isolated and its sequenced determined. The open reading frame predicts an 820 amino acid type 1 transmembrane protein with a cleavable signal sequence (Elble et al. 1993). Comparison with existing signal peptides and NH_2-terminal peptide sequence data suggest that cleavage occurs between the two serine residues, resulting in a protein of 799 amino acids with a predicted size of 87 kDa (VAN HEIJNE 1986). Its putative extracellular domain contains five potential, N-linked glycosylation sites that are more than sufficient to account for the 3kDa loss in molecular weight observed after N-glycanase treatment of Lu-ECAM-1 (PAULI et al. 1992; PAULI 1995).

The Lu-ECAM-1 cDNA sequence does not reveal homology with any known polypeptide. Although it has a cysteine-rich region in the extracellular domain and displays lectin-like binding characteristics (see below), there is no recognizable homology with the group of C-type mammalian lectins (DRICKAMER 1987, 1988) and selectin-type endothelial cell adhesion molecules (BEVILACQUA et al. 1989; JOHNSTON et al. 1989; LASKY 1991; LARSEN et al. 1992). Lu-ECAM-1 also has no recognizable Ca^{2+}-binding region, albeit binding to B16-F10 is Ca^{2+} dependent. Conservation of the Lu-ECAM-1 gene in mammalian species is shown by Southern blot analysis, using a multispecies genomic DNA blot. *Eco*RI-digested DNAs from human, monkey, rat, mouse, dog, bovine, rabbit, chicken, and yeast were hybridized under reduced stringency conditions to DNA fragments containing all or part of the Lu-ECAM-1 cDNA. The data show that the entire open reading frame, not just a portion, is conserved in mammals but not chicken or yeast. The results also demonstrate that Lu-ECAM-1 is a single-copy gene and not part of a closely related gene family. Conservation of the Lu-ECAM-1 gene in mammalian species argues for an involvement of Lu-ECAM-1 in processes other than melanoma cell adhesion (ZHU et al. 1991, 1992).

2.4 Lu-ECAM-1 Sugar Chain Structure and Binding Specificities

The Lu-ECAM-1 molecule has five putative, N-linked glycosylation sites (see above). These sugar chains donate approximately 3% of the molecular weight of Lu-ECAM-1. Preliminary information of their structure has been obtained by lectin blotting of immunoaffinioty purified Lu-ECAM-1. These studies indicate that the major chain structure is a biantennary complex type. All of the sugar chains appear to be acidic, with sialic acids linked exclusively by α2–6 linkages. *N*-acetyllactosamine repeating structures were found to most likely compose the terminal chain part. There is some indication of an α-linked fucosylation of the most terminal lactosamine unit. The sugar chain core is Manα1–6(Manα1–3)Manβ1–4GlcNAcβ1–4 (Fucα1–6)GlcNAc.

The contribution of various carbohydrate moieties to B16-F10 adhesion and metastasis was analyzed in competitive adhesion and metastasis inhibition experiments. Using mono- and disaccharides covalently linked to BSA, B16-F10 adherence to Lu-ECAM-1 and colonization of the lungs were significantly impeded by *N*-acetylglucosamine and *N*-acetyllactosamine, but not by glucose, galactose, and *N*-acetylgalactosamine (all used at 1 mg/ml BSA carbohydrate containing approximately 0.5 m*M* carbohydrate). Preliminary characterization of the carbohydrate structure that most effectively blocked adhesion of B16-F10 to Lu-ECAM-1 identified lacto-*N*-fucopentose I (LNF I) [65% inhibition at 1 m*M*

LNF I] (PAULI et al. 1992; PAULI 1995). Minor inhibitory activities were also observed with lacto-*N*-tetrose (LNT), lacto-*N*-fucopentose II (Lewis a), lacto-*N*-fucopentose III (LNF III), and sialyl lacto-*N*-tetrose a.

2.5 Melanoma Adhesion to Lu-ECAM-1 Under Conditions of Flow

Monolayers of lung matrix-modulated BAECs (high Lu-ECAM-1 expressors) and unmodulated BAECs (moderate Lu-ECAM-1 expressors) have been used as substrates for studying B16-F10 adhesion under hydrodynamic conditions in a radial flow chamber (GOETZ and HAMMER 1994). At physiological shear stresses between 0.32 and 1.0 dynes/cm^2, B16-F10 cells adhere in significantly higher number to lung matrix-modulated BAECs than they do to unmodulated BAEC monolayers grown on type I collagen ($p < 0.05$) (GOETZ et al. 1995). Both endothelial cell types strongly support adhesion of B16-F10 cells at low shear stresses (< 0.52 dynes/cm^2). However, at higher shear stresses (> correct 0.52 dynes/cm^2) only lung matrix-modulated BACEs express a sufficient Lu-ECAM-1 density to further support adhesion of B16-F10 melanoma cells. Specific adhesion of B16-F10 to lung matrix-modulated BAECs is blocked when tumor cells are incubated with soluble Lu-ECAM-1 prior to conducting a flow chamber adhesion assay. No inhibitory activity is detected when B16-F10 cells are pre-incubated with the control membrane protein glycophorin, A solubilized in the same detergent concentration as Lu-ECAM-1 (GOETZ et al. 1995).

B16-F10 interactions with the endothelium can be dissected into two distinct phases. The first phase is characterized by tumor cells moving along the endothelial cell monolayer at a velocity comparable to the theoretical velocity of an unencumbered, 15 µm diameter particle moving at a distance of 50 nm from the endothelial surface, e.g., 300 µm/s at 0.7 dynes/cm^2. The second phase entails the actual arrest of the tumor cell to the endothelium. Arrest occurs within a fraction of a second and is permanent (GOETZ et al. 1995). None of the analyzed trajectories provide any evidence that tumor cells roll in the manner of neutrophils on stimulated endothelium (LAWRENCE and SPRINGER 1991). This pattern of B16-F10 arrest remains unchanged when lung matrix-modulated BAECs are substituted with LPS-stimulated endothelium. However, LPS-stimulated, lung matrix-modulated BAECs do support rolling of peripheral bovine neutrophils as expected.

2.6 Induction of Gap Junctional Communication and Extravasation by Lu-ECAM-1

Adhesion of blood-borne tumor cells to vascular endothelial cells is a prerequisite for tumor cell extravasation and the formation of metastatic colonies. Although the mechanisms by which tumor cells penetrate the endothelial cell lining have been associated with exposure of subendothelial matrix, tumor cell adhesion and destruction of basement membrane, and tumor cell migration (PAULI et al. 1990; NICHOLSON 1988), the initial steps of extravasation that follow tumor cell adhesion to the endothelium lining remain largely obscure. Morphological studies of tumor cell/endothelial cell interactions indicate that adhesion preferentially occurs at the apposition zone between neighboring endothelial cells and is followed by the retraction of endothelial cells and the recanalization of the blood vessel around the arrested tumor cell (PAULI and LEE 1988). Using digital video imaging microscopy and fluorescence activated cell sorting techniques to further study interactions between B16-F10 melanoma cells and endothelium, we have recently shown that an adhesion-dependent establishment of gap junctional communication between metastatic tumor cells and endothelium might be critically involved in extravasation. Highly lung metastatic mouse B16-F10 melanoma cells previously labeled with the membrane-impermeable, fluorescent dye BCECF begin to transfer dye to endothelial cell monolayers shortly after adhesion via the melanoma cell-binding endothelial cell adhesion molecule Lu-ECAM-1 is established (EL-SABBAN and PAULI 1991). The extent of BCECF transfer to endothelial monolayers is dependent upon the number of BCECF labeled tumor cells seeded onto the endothelial cell monolayers and the time of coculture of the two cell types, as visualized by an increase in the number of BCECF-positive cells among cells stained with an endothelial cell-specific monoclonal antibody. The transfer of dye is bidirectional and sensitive to inhibition by 1-heptanol (EL-SABBAN and PAULI 1991). In contrast, poorly metastatic B16-F10 melanoma cells do not efficiently couple with vascular endothelium. It is inferred from these experiments and from the amounts of connexin43 (cx43) mRNA expressed by tumor cells that tumor cell/endothelial cell communication is mediated by gap junctional channels. Similar observations have been made with high lung metastatic R3230AC-MET rat mammary carcinoma cells and the nonmetastatic, lectin-resistant counterpart R3230AC-LR (EL-SABBAN and PAULI 1991).

Gap junction assembly and communication between B16 melanoma cells and endothelium appear to depend upon the expression levels of cx43 protein and Lu-ECAM-1 (EL-SABBAN 1993). In accordance, high levels of cx43 message are observed only in B16 melanoma cell variants (e.g., B16-F10) that efficiently communicate with endothelial cells. However, a high cx43 message alone is not sufficient to initiate efficient gap junction communication between B16 melanoma cells and endothelium. The establishment of functional gap junctions appears to depend critically upon the coexpression of tumor cell/endothelial cell adhesion receptor/ligand. This notion is supported by the inhibition of gap junctional assembly and communication between B16-F10 melanoma cells and lung matrix-modulated endothelial cells with anti-Lu-ECAM-1 mAb 6D3. The importance of adhesion in gap junctional assembly and communication is experimentally verified by transfection of poorly coupled S180 cells with cDNA encoding the liver cell adhesion molecule L-CAM. L-CAM transfection greatly increases cell-to-cell coupling between S180 cells without markedly affecting production of cx43 (MEGE et al. 1988; MUSLI et al. 1990). In the processs of establishing communication, cx43 becomes hyperphosphorylated and concentrated in regions of cell contact in L-CAM transfected cells, implicating an adhesion-mediated signaling event in cell-to-cell communication.

To further investigate the role of tumor cell/endothelial cell gap junctional communication in metastasis, low lung metastatic B16-F0 melanoma cells were transfected with the plasmid DOLG2A containing the entire coding region of rat heart cx43 (clone G2) and a selectable marker for neomycin resistance, using the lipofectamine method (El-Sabban 1993). A stably transfected cell line termed B16-F0H3 and a mock-transfected (DOL vector alone) cell line B16-F0D have been established. Northern blot analyses of these cells reveal that the transfected cell line B16-F0H3 expresses five times more cx43 message than B16-F0 and two times more than B16-F10 (El-Sabban 1993). The mock-transfected B16-F0D cells have message levels comparable to those of B16-F0. The same results are obtained when these cell lines are analyzed for cx43 protein expression using FACS analysis. When these transfectants are tested for their ability to produce lung colonies after tail vein inoculation, we find that cx43 transfection of low lung metastatic B16-F0 melanoma cells significantly increases the metastatic performance. However B16-F0H3 transfectants do not produce the same high number of lung colonies that are generated by the high lung metastatic B16-F10 melanoma cell line, although their cx43 protein expression was approximately two fold higher than that of B16-F10. This finding is explained by a much lower ability of cx43 transfected B16-F0H3 to adhere to endothelium than B16-F10. Therefore, the metastatic colony efficiency appears to be the result of an optimal expression of adhesion ligand and cx43 protein. Metastatic performance is also significantly affected by antisera directed against the extracellular domains of cx43 (EL-46 and EL-186). Our preliminary findings are supportive of the hypothesis that gap junctional communication is an essential, postadhesion event that facilitates extravasation of tumor cells and establishment of tumor colonies at secondary organ sites (El-Sabban 1993).

Functionally, cell-to-cell communication between blood-borne cancer cells (e.g., lung metastatic mouse B16-F10 melanoma cells and rat R3230AC-MET breast carcinoma cells) and endothelium might prepare the endothelium for tumor cell exit from the vasculature. This may be accomplished by the direct transfer of tumor metabolites to the endothelium, e.g. by inducing local retraction of endothelial cells and exposure of subendothelial matrix to adherent tumor cells. A candidate for such transfer is the arachidonic acid metabolite 12-(S)-HETE, recently proposed by Honn et al. (1993). This molecule has been shown to cause endothelial cell retraction in vitro and is produced in large amounts in the cytoplasm of malignant tumor cells from where it might be efficiently transferred to endothelium in vivo through gap junctional channels. A further possibility regarding the function of tumor cell/endothelial cell gap junctional communication currently pursued in our laboratory is the induction of proteolytic degradation of subendothelial basement membrane by tumor-stimulated endothelial proteases and glycosidases, which may aid tumor cells in their quest to gain access to the subendothelial extracellular matrix (Pauli et al. 1992; El-Sabban 1993).

3 Dipeptidyl Peptidase IV: An Endothelial Cell Adhesion Molecule for Lung Metastatic Rat Breast Carcinoma Cells

Dipeptidyl peptidase IV (DPP IV; CD24) is an integral membrane sialoglycoprotein of 110 kDa (HARTEL et al. 1988; PIAZZA et al. 1989). Its functions are those of an exopeptidase, cleaving x-pro dipeptides from the NH_2-terminus of polypeptides (HARTEL et al. 1988), and a fibronectin binding protein (PIAZZA et al. 1989). In addition to its expression in bile cannalicular membranes, brush borders of proximal kidney tubular epithelium, and intestinal villus epithelium, DPP IV is expressed on endothelia of distinct vascular branches including the vasa recta of the renal medulla, splenic red pulp venules, and lung capillaries (JOHNSON et al. 1991). DPP IV has a high binding affinity for lung metastatic rat breast carcinoma cells in vitro, but not for their low or nonmetastatic counterparts (JOHNSON et al. 1991, 1993). The DPP IV binding affinity parallels the propensity with which these tumor cells metastasize to the lungs. Tumor cell binding to immobilized DPP IV is inhibited by anti-DPP IV mAb 6A3 and by soluble DPP IV, both in a dose-dependent manner. No effect on tumor cell binding to DPP IV is observed with peptide substrates to the enzymatic portion of this exopeptidase, the fibronectin-derived peptide Gly-Arg-Gly-Asp-Ser, or with control mAbs of the same Ig class as anti-DPP IV mAb 6A3 or mAbs that are directed against lung endothelial epitopes other than DPP IV (JOHNSON et al. 1993). To probe whether tumor cell binding to DPP IV is mediated by fibronectin, lung metastatic breast carcinoma cells were analyzed by flow cytometry for cell surface-associated fibronectin. Our data show a high correlation between cell surface expression of fibronectin, adhesion to DPP IV, and lung metastatic propensity. Fibronectin appears to be bound to tumor cell surfaces by integrin receptors, as increased expression of $\alpha5$, $\beta1$ and $\beta3$ integrins indicate (JOHNSON et al. 1993). Thus, fibronectin may serve as a binding intermediary between tumor cell integrins and endothelial cell DPP IV to promote arrest and metastasis of rat breast carcinoma cells in the lungs. This conclusion is consistent with recent reports that associate increased expression of tumor cell surface fibronectin and/or integrins with higher binding affinity for endothelium and higher propensity for metastasis (AUGUSTIN-VOSS et al. 1991; JOHNSON et al. 1991, 1993; NICOLSON et al. 1989). Moreover, it sheds new light on the mechanisms by which RGD peptides inhibit tumor metastasis. Since RGD peptides cause rapid clearance of tumor cells from the lungs after i.v. inoculation and have an extremely short half-life in circulation (8 min) (HUMPHRIES et al. 1988), it is unlikely that these peptides exert their inhibitory role on metastasis, as originally proposed, by blocking tumor cell adhesion to subendothelial matrix, an event which starts approximately 4 h after i.v. inoculation of tumor cells (JOHNSON et al. 1993). Instead, RGD peptides may compete with the binding of fibronectin to tumor cell integrin receptors, thereby preventing surface assembly of an insoluble fibronectin matrix and tumor cell binding to endothelial DPP IV.

4 Conclusions

The recognition of endothelial cell adhesion molecules by tumor cell ligands is a crucial step in the metastatic cascade and one that is vulnerable to therapeutic attack. A considerable number of these molecules have been identified, mostly having high level expression and very broad distribution, thus accounting for their early and facile isolation. These molecules are important in invasion, migration and angiogenesis and may play a supporting role in vascular arrest, but their distribution precludes a primary role in organ-specific metastasis. More difficult to study have been those adhesion molecules whose expression is restricted by organ and vascular compartment and which are far more likely to account for site-specific metastasis. Lu-ECAM-1 expression, for example, is found in postcapillary and small to medium-sized venules, the vascular segments favored by melanoma cells for extravasation and metastasis. Doubtless more such molecules will be isolated as methodologies for isolation and culture of distinct endothelial subtypes continue to improve. Isolating the tumor cell ligands for such molecules may depend upon closely mimicking the conditions experienced by the intravascular tumor cell. As the study of integrins and their ligands has illustrated, conformational shifts in cell surface molecules may dramatically affect the ability of those molecules to bind others (Hynes 1992). These shifts are caused by inside-out signalling which, in the case of the migrating tumor cell, might occur in response to changes in matrix environment or the transition from a spread to a rounded state and back again (Roskelley et al. 1994). Similarly, the arcane conformational shifts undergone by fibronectin in response to inter- and intramolecular interactions (Aguirre et al. 1994; Huhtala et al. 1995) have been of paramount importance in understanding DPP IV interaction with carcinoma cells. Clearly the potential for such cryptic transitions in receptor and ligand activity will be a challenge in determining which interactions are most relevant to the attachment and arrest phase of metastasis.

Another largely unexplored territory is in the events downstream from receptor-ligand binding that lead to the formation of secondary colonies. These events include the secretion of active proteinases and subsequent scission of endothelial bonds to extracellular matrix and to other endothelial cells thus allowing retraction and exposure of basement membrane (Nicholson 1988; Lafrenie et al. 1993). The unexpected existence of heterotypic gap junctions between tumor cell and endothelial cell provides an additional avenue for transmission of signalling molecules such as 12-[S]-HETE that may expedite this process. Further understanding of this complex interplay of events should provide new and more specific targets for therapeutic intervention against these steps in the metastatic cascade.

Acknowledgments. The authors gratefully acknowledge the many contributions from the laboratory staff (Dr. Mossaad Abdel-Ghany, Ms. Joanne Widom, Ms. Renée Eff, Dr. Andrew Goodwin, Ms. Haigun Lin, and Mr. Hung-Chi Cheng) as well as from collaborators at Cornell University School of Chemical Engineering (Dr. Daniel A. Hammer and Dr. Douglas Goetz). This work was supported in part by Public Health Service Grant CA47668 from the National Cancer Institute, a grant from the Council for Tobacco Research, Inc., USA, and a grant from the Cornell Biotechnology Program.

References

Aguirre KM, McCormick RJ, Schwarzbauer JE (1994) Fibronectin self-association is mediated by complementary sites within the amino-terminal one-third of the molecules. J Biol Chem 269: 27863–27868

Albelda SM (1991) Endothelial and epithelial cell adhesion molecules. Am J Respir Cell Mol Biol 4: 193–203

Albelda SM (1993) Role of integrins and other cell adhesion molecules in tumor progression and metastasis. Lab Invest 68: 4–17

Alby L, Auerbach R (1984) Differential adhesion of tumor cells to capillary endothelial cells in vitro. Proc Natl Acad Sci USA 81: 5739–5743

Auerbach R, Lu WC, Pardon E, Gumkowski F, Kaminska G, Kaminska M (1987) Specificity of adhesion between murine tumor cells and capillary endothelium: an in vitro correlate of preferential metastasis in vivo. Cancer Res 47: 1492–1496

Augustin-Voss HG, Johnson RC, Pauli BU (1991) Modulation of endothelial cell surface glycoconjugate expression by organ-derived biomatrices. Exp Cell Res 192: 346–351

Banks RE, Gearing AJ, Hemingway IK, Norfolk DR, Perren TJ, Selby PJ (1993) Circulating intercellular adhesion molecule-1 (ICAM-1), E-selectin and vascular cell adhesion molecule-1 (VCAM-1) in human malignancies. Br J Cancer 68: 122–124

Belloni PN, Nicolson GL (1988) Differential expression of cell surface glycoproteins on various organ-derived microvascular endothelia and endothelial cell cultures. J Cell Physiol 136: 398–410

Belloni PN, Tressler RJ (1990) Microvascular endothelial cell heterogeneity: interactions with leukocytes and tumor cells. Cancer Metastasis Rev 8: 353–389

Bevilacqua MP (1993) Endothelial-leukocyte adhesion molecules. Annu Rev Immunol 11: 767–894

Bevilacqua MP, Nelson RM (1993) Selectins. J Clin Invest 91: 379–387

Bevilacqua MP, Stengelin S, Gimbrone MA Jr, Seed B (1989) Endothelial leukocyte adhesion molecule 1: an inducible receptor for neutrophils related to complement regulatory proteins and lectins. Science 243: 1160–1165

Borsum T, Hagan T, Henrickson T, Hagan I, Henriksen T, Carlander B (1982) Alterations in the protein composition and surface structure of human endothelial cells during growth in primary culture. Atherosclerosis 44: 367–378

Butcher EC (1991) Leukocyte-endothelial cell recognition: three (or more) steps to specificity and diversity. Cell 67: 1033–1037

Butcher EC, Scollay RG, Weissman IL (1979) Lymphocyte adherence to high endothelial venules: characteristics of a modified in vitro assay and examination of the binding of syngeneic and allogeneic lymphocyte populations. J Immunol 123: 1996–2003

Chackal-Roy M, Niemeyer C, Moore M, Zetter BR (1989) Stimulation of human prostatic carcinoma cell growth by factors present in human bone marrow. J Clin Invest 84: 43–50

Couldwell WT, de Tribolet N, Antel JP, Gauthier T, Kuppner MC (1992) Adhesion molecules and malignant gliomas: implications for tumorigenesis. J Neurosurg 76: 782–791

Damjanovich L, Albelda SM, Mette SA, Buck CA (992) Distribution of Integrin cell adhesion receptors in normal and malignant lung tissue. Am J Respir Cell Mol Biol 6: 197–206

De Bono DT, Green C (1984) The adhesion of different cell types to vascular endothelium: effects of culture density and age. Br J Exp Pathol 65: 145–154

Dejana E, Martin-Padura I, Lauri D, Bernasconi S, Bani MR, Garofalo A, Giavazzi R, Magnani J, Mantovani A, Menard S (1992) Endothelial leukocyte adhesion molecule-1-dependent adhesion of colon carcinoma cells to vascular endothelium is inhibited by an antibody to Lewis fucosylated type I carbohydrate chain. Lab Invest 66: 324–330

Drickamer K (1987) Membrane receptors that mediate glycoprotein endocytosis: structure and biosynthesis. Kidney Int 32: S167–S180

Drickamer K (1988) Two distinct classes of carbohydrate-recognition domains in animal lectins. J Biol Chem 263: 9557–9560

Elble RC, Widom J, Cheng CF, Levine RA, Pauli BU (1993) A lung-specific endothelial cell adhesion molecule mediates lung metastasis of murine melanomas. Mol Biol Cell 4: 217a
El-Sabban ME (1993) The role of endothelial cell-tumor cell interaction in tumor metastasis. Ph D dissertation, Cornell University
El-Sabban ME, Pauli BU (1991) Cytoplasmic dye transfer between metastatic tumor cells and vascular endothelium. J Cell Biol 115: 1375–1382
Fearson ER, Vogelstein B (1990) A genetic model for colorectal tumorigenesis. Cell 61: 759–767
Fidler IJ (1990) Critical factors in the biology of human cancer metastasis: twenty-eighth G.H.A. Clowes memorial award lecture. Cancer Res 50: 6130–6138
Goetz DJ, El-Sabban ME, Hammer DA, Pauli BU (1995) Lu-EAM-1 mediated adhesion of melanoma cells to endothelium under conditions of flow. Int J Cancer (in press)
Goetz DJ, Hammer DA (1994) Dynamics of neutrophil rolling over stimulated endothelium in vitro. Biophys J 66: 2202–2209
Harning R, Ainolfi E, Bystryn J-C, Henn M, Merluzzi VJ, Rothlein R (1991) Serum levels of circulating intercellular adhesion molecule 1 in human malignant melanoma. Cancer Res 51: 5003–5005
Hart IR, Saini A (1992) Biology of tumor metastasis. Lancet 339: 1453–1457
Hartel S, Gossrau R, Hanski C, Reutter W (1988) Dipeptidyl peptidase IV (DPP IV) in rat organs: comparison of immunohistochemistry and activity histochemistry. Histochemistry 89: 151–162
Honn KV, Tang DG, Grossi I, Duniec ZM, Timar J, Renaud C, Leithauser M, Blair I, Diglio CA, Taylor JD, Marnett LJ (1993) Tumor cell-derived 12(S) HETE induces microvascular endothelial cell retraction. Cancer Res 53: 3421–3435
Horley KJ, Carpenito C, Baker B, Takei F (1989) Molecular cloning of murine intercellular adhesion molecule (ICAM-1). EMBO J 8: 2889–2896
Huhtala P, Humphries MJ, McCarthy JB, Tremble PM, Werb Z, Damsky CH (1995) Cooperative signaling by α5β1 and α4β1 integrins regulates metalloproteinase gene expression in fibroblast adhering to fibronectin. J Cell Biol 129: 867–879
Humphries MJ, Yamada KM, Olden K (1988) Investigation of the biological effects of anti-cell adhesive synthetic peptides that inhibit experimental metastasis of B16-F10 murine melanoma cells. J Clin Invest 81: 782–790
Hynes RO (1992) Integrins: versatility, modulation, and signaling in cell adhesion. Cell 69: 11–25
Iwai K, Ishikura H, Kaji M, Sugiura H, Ishizu A, Takahashi C, Kato H, Tanabe T, Yoshiki T (1993) Importance of E-selectin ELAM-1 and sialyl Lewis-A in adhesion of pancreatic carcinoma cells to activated endothelium. Int J Cancer 54: 972–977
Johnson RC, Augustin-Voss HG, Zhu D, Pauli BU (1991) Endothelial cell membrane vesicles in the study of organ preference of metastasis. Cancer Res 51: 394–399
Johnson RC, Zhu D, Augustin-Voss HG, Pauli BU (1993) Lung endothelial dipeptidyl peptidase IV is an adhesion molecule for lung-metastatic rat breast and prostate carcinoma cells. J Cell Biol 121: 1423–1432
Johnston GI, Cook RG, McEver RP (1989) Cloning of GMP-140, a granule membrane protein of platelets and endothelium: sequence similarity to proteins involved in cell adhesion and inflammation. Cell 56: 1033–1044
Kinsey DL (1960) An experimental study of preferential metastasis. Cancer 13: 674–676
Lafrenie RM, Buchanan MR, Orr FW (1993) Adhesion molecules and their role in cancer metastasis. Cell Biophysiol 23: 3–89
Larsen GR, Sako D, Ahern TJ, Shaffer M, Erban J, Sajer SA, Gibson RM, Wagner DD, Furie BC, Furie B (1992) P-selectin and E-selectin: distinct but overlapping leukocyte ligand specificities. J Biol Chem 267: 11104–11110
Lasky LA (1991) Lectin cell adhesion molecules (LEC-CAMs): a new family of cell adhesion proteins involved with inflammation. J Cell Biochem 45: 139–146
Lasky LA, Singer MS, Dowbenko D, Imal Y, Henzel WJ, Grimley C, Fennie C, Gillett H, Watson SR, Rosen SD (1992) An endothelial ligand for L-selectin is a novel mucin-like molecule. Cell 69: 927–938
Lauri D, Martin-Padura I, Biondelli T, Rossi G, Bernasconi S, Giavazzi R, Passerini F, van Hinbergh V, Dejana E (1991a) Role of beta-1 integrins in tumor cell adhesion to cultured human endothelial cells. Lab Invest 65: 525–531
Lauri D, Needham L, Martin-Padura I, Dejana E (1991b) Tumor cell adhesion to endothelial cells: endothelial-leukocyte adhesion molecule-1 as an inducible adhesive receptor specific for colon carcinoma cells. J Natl Cancer Inst 83: 1321–1324
Lawrence MB, Springer TA (1991) Leukocytes roll on a selectin at physiologic flow rates: distinction from and prerequisite for adhesion through integrins. Cell 65: 859–873

Matsushita Y, Cleary KR, Ota DM, Hoff SD, Irimura T (1990) Sialyl-dimeric Lewis X antigen expressed on mucin-like glycoproteins in colorectal cancer metastases. Lab Invest 63: 780–791
McCarthy SA, Ku I, Gatter KC, Bicknell R (1991) Heterogeneity of the endothelial cell and its role in organ preference of tumor metastasis. TIPS 12: 462–467
McEver RP, Beckstead JH, Moore KL, Marshall-Carlson L, Bainton DF (1989) GMP-140, a platelet alpha-granule membrane protein, is also synthesized by vascular endothelial cells and is localized in Weibel-Palade bodies. J Clin Invest 84: 92–99
Mege RM, Matsuzaki F, Gallin WJ, Goldberg JI, Cunningham BA, Edelman GM (1988) Construction of epitheloid sheets by transfection of mouse sarcoma cells with cDNA for chicken cell adhesion molecule. Proc Natl Acad Sci USA 85: 7274–7278
Musil LS, Cunningham BA, Edelman GM, Goodenough DA (1990) Differential phosphorylation of the gap junction protein connexin 43 in junctional communication-competent and deficient cell lines. J Cell Biol 111: 2077–2088
Naito S, Giavazzi R, Fidler IJ (1987) Correlation between the in vitro interactions of tumor cells with an organ environment and metastatic behavior in vivo. Invasion Metastasis 7: 126–129
Nicolson GL (1988) Cancer metastasis: tumor cell and host organ properties important in metastasis to specific secondary sites. Biochim Biophys Acta 948: 175–224
Nicolson GL, Belloni PN, Tressler RJ, Dulski K, Inoue T, Cavanough PG (1989) Adhesive, invasive and growth properties of selected metastatic variants of a murine large cell lymphoma. Invasion Metastasis 9: 102–116
Okahara H, Yagita H, Miyake K, Okumura K (1994) Involvement of very late activation antigen (VLA-4) and vascular cell adhesion molecule 1 (VCAM-1) in tumor necrosis factor α enhancement of experimental metastasis. Cancer Res 54: 3230–3236
Paget S (1889) The distribution of secondary growths in cancer of the breast. Lancet 1: 571–573
Pauli BU (1995) Consitutitve and inducible endothelial cell adhesion molecules in cancer metastasis. In: Bertino JR (ed) Molecular biology of cancer. Academic, New York (in press)
Pauli BU, Lee CL (1988) Organ preference of metastasis: the role of organ-specifically modulated endothelial cells. Lab Invest 58: 379–387
Pauli BU, Augustin-Voss HG, El-Sabban ME, Johnson RC, Hammer DA (1990) Organ preference of metastasis: the role of endothelial cell adhesion molecules. Cancer Metastasis Rev 9: 175–189
Pauli BU, Johnson RC, El-Sabban ME (1991) Organotypic endothelial cell surface molecules mediate organ preference of metastasis. In: Simionescu N, Siminionescu M (eds) Endothelial cell dysfunction. Plenum, New York, pp 439–454
Pauli BU, Johnson RC, Widom J, Cheng CF (1992) Endothelial cell adhesion molecules and their role in organ-preference of metastasis. TiGG 4: 405–414
Piazza GA, Callanan HM, Mowery J, Hixson DC (1989) Evidence for a role of dipeptidyl peptidase IV in fibronectin-mediated interactions of hepatocytes with extracellular matrix. Biochem J 262: 327–334
Pilewski JM, Albelda SM (1993) Adhesion molecules in the lung. Am Rev Respir Dis 148: 531–537
Qian F, Vaux DL, Weissman IL (1994) Expression of $\alpha_4\beta_1$ on melanoma cells can inhibit the invasive stage of metastasis formation. Cell 77: 335–347
Rossi MC, Zetter BR (1992) Selective stimulation of prostatic carcinoma cell proliferation by transferrin. Proc Natl Acad Sci USA 89: 6197–6201
Roskelley CD, Desprez PY, Bissell MJ (1994) Extracellular matrix-dependent tissue-specific gene expression in mammary epithelial cells requires both physical and biochemical signal transduction. Proc Natl Acad Sci USA 91: 12378–12382
Sargent NSE, Oestreicher M, Haidvogl H, Madnick HM, Burger MM (1988) Growth regulation of cancer metastasis by their host organ. Proc Natl Acad Sci USA 85: 7251–7255
Rustgi AK, Daniel K, Podolsky K (1992) The molecular basis of colon cancer. Annu Rev Med 43: 61–68
Springer TA (1990) Adhesion receptors of the immune system. Nature 346: 425–434
Stamper HB Jr, Woodruff JJ (1976) Lymphocyte homing into lymph nodes: in vitro demonstration of selective affinity of recirculating lymphocytes to high endothelial venules. J Exp Med 144: 828–833
Suter MM, Greenberger LJ, Wilkinson JE, Lewis RM (1990) Differential expression of cell surface antigens of canine keratinocytes defined by monoclonal antibodies. J Histochem Cytochem 38: 541–548
Takada A, Ohmori K, Yoneda T, Tsuyuoka K, Hasegawa A, Kiso M, Kannagi R (1993) Contribution of carbohydrate antigens sialyl Lewis A and sialyl Lewis X to adhesion of human cancer cells to vascular endothelium. Cancer Res 53: 354–361
Van Heijne G (1986) A new method for predicting signal sequence cleavage sites. Nucleic Acid Res 14: 4683–4690

Walz G, Aruffo A, Kolanus W, Bevilacqua M, Seed B (1990) Recognition by ELAM-1 of sialyl-Le^x determinant on myeloid and tumor cells. Science 250: 1132–1135
Weiss L, Orr FW, Honn KV (1988) Interactions of cancer cells with the microvasculature during metastasis. FASEB J 2: 12–21
Weterman MJ, van Muijen GNP, Bloemers HPJ, Ruiter DJ (1994) Molecular markers of melanocytic tumor progression. Lab Invest 70: 593–608
Zetter BR (1990) The cellular basis of site-specific tumor metastasis. N Engl J Med 322: 605–612
Zhu D, Pauli BU (1991) Generation of monoclonal antibodies directed against organ-specific endothelial cell surface determinants. J Histochem Cytochem 39: 1137–1142
Zhu D, Pauli BU (1993) Correlation between the lung distribution patterns of Lu-ECAM-1 and melanoma experimental metastases. Int J Cancer 53: 628–633
Zhu D, Cheng CF, Pauli BU (1991) Mediation of lung metastasis of murine melanomas by a lung-specific endothelial cell adhesion molecule. Proc Natl Acad Sci USA 88: 9568–9572
Zhu D, Cheng CF, Pauli BU (1992) Blocking of lung endothelial cell adhesion molecule-1 (Lu-ECAM-1) inhibits murine melanoma lung metastasis. J Clin Invest 89: 1718–1724

The E-cadherin/Catenin Complex in Invasion and Metastasis

M.E. BRACKE[1], F.M. VAN ROY[2], and M.M. MAREEL[1]

1 Does Cell Adhesion Promote or Prevent Cancer Invasion?

Cancer cells are genetically predisposed to contribute to the formation of a malignant tumor. As a population they do not respect tissue boundaries, but penetrate into the surrounding normal tissues. On their way to lymph or blood vessels they create a path of destruction. This invasive behavior eventually gives access to the circulation, and opens a gate to metastatic dissemination (NICOLSON et al. 1977; FIDLER and HART 1982; MAREEL et al. 1993).

[1] Laboratory of Experimental Cancerology, Department of Radiotheraphy, Nuclear Medicine and Experimental Cancerology, Poli 7, University Hospital, De Pintelaan 185, 9000 Gent, Belgium

[2] Laboratory of Molecular Biology, Section of Molecular Cell Biology, K.L. Ledeganckstraat 35, University of Gent, 9000 Gent, Belgium

During the last 25 years, research on the mechanisms of tumor invasion and metastasis has become a distinct discipline. Straightforward concepts on the role of tumor cell motility and extracellular protease activity have led to the discovery and characterisation of new families of motility factors and enzymes. It is easy to conceive that typical motility factors and their receptors on the plasma membrane of cancer cells play a role in invasion (STOKER and GHERARDI 1991). Examples of such ligands are autocrine motility factor (AMF) (LIOTTA et al. 1986) and scatter factor/ hepatocyte growth factor (SF/HGF) (STOKER et al. 1987; WEIDNER et al. 1990), which bind to their respective receptors, namely, gp78 (WATANABE et al. 1991) and the c-met tyrosine kinase receptor (NALDINI et al. 1991). It is also intellectually pleasing to attribute an important role to enzymes in the mechanisms of invasion. Plasminogen activators (DANØ et al. 1985), matrix metalloproteases (MURPHY et al. 1989; SATO et al. 1994), cathepsins (ROCHEFORT et al. 1988) and other cysteine proteases (SLOANE and HONN 1984), and heparinases (VLODAVSKY et al. 1991) all have been claimed to be implicated in tumor invasion. Their enzymatic activity is tuned by activators and inhibitors such as plasminogen activator inhibitors (PAIs) (MOTTONEN et al. 1992), tissue inhibitors of metalloproteases (TIMPs) (KHOKHA et al. 1989) and maspin (ZOU et al. 1994). Recently, receptors for some of these proteases have been described on cancer cells (receptors for urokinase and tissue-type plasminogen activators (ESTREICHER et al. 1990; PLOUG et al. 1991), plasmin (CORREC et al. 1992) and cathepsin D (KNUDSEN and WHEELOCK 1992), which localize the enzyme activity.

The role of cell-substrate adhesion molecules has been ambiguous, because theoretically they can act either as anchors to stabilize the cell's position or as grips for moving cancer cells. Thus, they can be interpreted both as invasion suppressor as well as invasion promoter molecules. Laminin, a glycoprotein of the basement membrane, is an example of this duality. The molecule can be considered as an anchor for normal epithelial cells, and, indeed, it partly disappears during the transition of carcinoma in situ to invasive carcinoma (ALBRECHTSEN et al. 1981) and during embryonic ingression of the primitive streak (BORTIER et al. 1989). In his three-step hypothesis of invasion, however, Liotta presents laminin as a temporary grip, which is subsequently broken down to give way to the invading cancer cells (LIOTTA 1986). The difficulty to assess the exact role of laminin in invasion is probably due to its multiple functions. Laminin does not only mediate cell adhesion, but also stimulates the secretion of type IV collagenase (TURPEENIEMI-HUJANEN et al. 1986), increases the motility of some cell types (FLIGIEL et al. 1985), and stops migration of others (COOPMAN et al. 1991). Moreover, the effect of laminin on cancer cells may depend on the number of integrin-type and other laminin receptors on their plasma membranes (MECHAM 1991; CASTRONOVO 1993).

Heterotypic cell-cell adhesion molecules regulate adhesive interactions between different types of cells. With these molecules a certain consensus has been obtained concerning their promoting role in invasion and metastasis. Intercellular adhesion molecule-1 (ICAM-1) on melanoma cells, for instance, makes these cells more invasive through an interaction with leukocytes (JOHNSON 1991; PANDOLFI et al. 1992). Adhesion of circulating tumor cells to the vascular endothelium, the initial step of extravasation, is thought to be governed by heterotypic cell-cell adhesion (DUSTIN and SPRINGER 1991). Here, a scenario similar to the one presented for leukocyte extravasation can be invoked (HYNES and LANDER 1992).

A challenge in cancer research is the recognition of efficient invasion suppressor molecules and the exploration of the mechanisms of their regulation. Homotypic cell-cell adhesion molecules—which hold cells of the same type together—are conceptually good candidates for invasion suppressors. These molecules can be divided into two main groups: (1) the immunoglobulin superfamily showing calcium-independent cell-cell adhesion and (2) the calcium-dependent cadherin superfamily (EDELMAN and CROSSIN 1991; GEIGER and AYALON 1992; ÖBRINK 1991; BIRCHMEIER and BEHRENS 1994). Members of the immunoglobulin superfamily that have been studied for their putative role in invasion, are N-CAM (neural cell adhesion molecule) (BRADY-KALNAY et al. 1993), DCC (deleted in colon cancer) (FEARON and VOGELSTEIN 1990), CEA (carcino-embryonic antigen) (WAGNER et al. 1992; HAUCK and STANNERS 1991) and MUC-18 (KUZU et al. 1993). It was, however, from the cadherin superfamily that a universal invasion suppressor molecule for epithelioid tissues emerged, namely, E-cadherin. In this review we will first describe the different molecular domains of E-cadherin, its interactions with the catenins and the cytoskeleton, its localisation on the plasma membrane and its organising activities. Next we will critically comment on the actual evidence for considering this molecule as an invasion suppressor, with emphasis on the regulation of its function. Finally, E-cadherin will be considered as a possible receptor, and its down-regulation in a number of noncancerous tissues will be discussed.

2 The E-cadherin/Catenin Complex

2.1 E-cadherin Is a Member of the Cadherin Family

The cadherins compose a superfamily of closely related cell surface molecules that require calcium and physiological temperatures in order to establish cell-cell adhesion. The members are usually indicated by a prefix letter that refers to the tissue or organ in which the molecule was found originally. The three classical cadherins are E (epithelial), P (placental) (NOSE and TAKEICHI 1986) and N (neural)-cadherin (HATTA and TAKEICHI 1986), but other members such as B (HERZBERG et al. 1991; NAPOLITANO et al. 1991), U (HERRENKNECHT et al. 1991), EP (GINSBERG et al. 1991), M (KAUPMANN et al. 1992), R (MIYATANI et al. 1992; MATSUNAMI et al. 1993), K(XIANG et al. 1994), C (BRIEHER and GUMBINER 1994), LI (BERNDORFF et al. 1994) and OB-cadherin (OKAZAKI et al. 1994), and cadherin-4 (TANIHARA et al. 1994) and -5 (SUZUKI et al. 1991) have been described. Another group of related molecules called desmosomal cadherins (desmocollins and desmogleins) falls beyond the scope of this review and will only be mentioned in relation to the classical cadherins. Up to now about 30 different cadherins are known.

Molecular cloning and amino acid analysis of the classical cadherins and of other cell-cell adhesion molecules have revealed strong homology or identity between surface molecules indicated as cadherins by one research group, or as CAMs (cell adhesion molecules) by another. This holds for N-cadherin vs A-CAM, and for E-cadherin vs L-CAM. Furthermore, E-cadherin was shown to be identical to uvomorulin in mouse and cell-CAM 120/80 in humans.

The classical cadherins seem to have evolved from a common precursor gene that has duplicated several times during evolution and has migrated to different chromosomal locations. While the P- and the E-cadherin genes are linked, both on mouse chromosome 8 (Hatta et al. 1991) and on human chromosome 16 (Bussemakers et al. 1994b), the N-cadherin gene is located on a different chromosome, i.e., on mouse and human chromosome 18 (Miyatani et al. 1992; Walsh et al. 1990). Although the cadherin gene products possess similar extracellular domains, they prefer homophilic association over interaction with other members of the cadherin family (Nose and Takeichi 1986).

The human E-cadherin gene, which has recently been cloned and characterized (Berx et al. 1995), is located on chromosome 16q22.1 (Mansouri et al. 1988; Natt et al. 1989) in a large conserved linkage group including the loci for chymotrypsinogen B, haptoglobin, lecithin:cholesterol acyltransferase, metallothionein-1,-2, and tyrosine aminotransferase (Scherer et al. 1989). Loci of serum esterase 1 (Eistetter et al. 1988) and zinc finger protein 1 (Chowdhury et al. 1989) have also been mapped in the immediate vicinity of E-cadherin. The E-cadherin promoter contains a palindromic sequence, called E-pal element, with influence on the epithelial-cell specific activity, an initiation site for transcription and a GC-rich sequence with a putative binding site for Sp1 (Behrens et al. 1991a; Ringwald et al. 1991; Sorkin et al. 1993). Interestingly, the activity of the promoter was reduced in dedifferentiated breast carcinoma cells, indicating that the identified elements are subject to negative regulation during tumor progression (Behrens et al. 1991a). In prostate cancer cells the activity of the human E-cadherin promoter was also suppressed, probably due to the binding of a repressor protein (Bussemakers et al. 1994a). One candidate as a modulator molecule of E-cadherin gene expression is the Slug gene product, which triggers epithelial to mesenchymal transitions in the embryo (Nieto et al. 1994).

Allelic loss of the E-cadherin gene with skipping of exon 8 or 9 in the remaining allele (Becker et al 1993, 1994), an in-frame deletion and a frame shift have been found in gastric carcinomas (Oda et al. 1994). Moreover, nonsense and missense mutations have been detected in the E-cadherin gene of carcinomas of the endometrium and the ovaries (Risinger et al. 1994). These phenomena indicate that down-regulation of E-cadherin in tumors can be the result of irreversible genomic alterations.

Remarkably, no splicing variants of the 4 kilobase (Kb) mRNA of E-cadherin have been published up to now (Gallin et al. 1985; Sorkin et al. 1988). This contrasts with the many examples of alternative splicing among cell-substrate and cell-cell adhesion molecules. E-cadherin is a transmembrane glycoprotein and its turnover is fast, with a half-life of 5 h in MDCK cells (SHORE and NELSON 1991). The expression of E-cadherin was reported to be up-regulated in some cell types by herbimycin A (REBUT-BONNETON et al. 1993), interferon-γ (FENYVES et al. 1993), estradiol (MACCALMAN et al. 1994b) and relaxin (SACCHI et al. 1994). Down-regulation was described by an autocrine motility factor-like molecule (ISHISAKI et al. 1994).

2.2 From E-cadherin to the Cytoskeleton

2.2.1 E-cadherin from N to C

The extracellular NH_2-terminal end of E-cadherin results from cytoplasmic enzymatic removal of part of the 135 kDa precursor to yield the mature 120 kDa molecule. The molecule is then routed mainly towards the basolateral surface of epithelial cells (LE BIVIC et al. 1990). The enzymatic trimming of the precursor is essential for cell-cell adhesion (OZAWA and KEMLER 1990).

Also, extracellularly, five homologous repeated domains (CAD repeats) are encountered, which presumably result from partial gene duplication (RINGWALD et al. 1987). The three-dimensional structures of these domains were elucidated recently (OVERDUIN et al. 1995; SHAPIRO et al. 1995). The domains contain two conserved regions, which are the putative calcium-binding sites of E-cadherin. Substitution of one amino acid (Asp by Lys or Ala) in one of these domains is sufficient to abrogate the calcium-binding property and hence the cell-cell adhesion function of E-cadherin (OZAWA et al. 1990a). It should be noted that isolated extracellular E-cadherin fragments do not associate in the presence of calcium, which indicates the E-cadherin-dependent cell-cell adhesion requires more elements than the calcium-binding domains only (POKUTTA et al. 1994). The outermost domain contains an HAV sequence, which appears to be essential for homophilic recognition. Although flanked by different sequences, this HAV sequence is also found in P- and N-cadherin, in superoxide dismutase (WILLEMS et al. 1993) and in the fibroblast growth factor receptor (BYERS et al. 1992; WILLIAMS et al. 1994). Upon calcium depletion, the five repeated domains change their organisation from a rod-like to a more globular assembly (POKUTTA et al. 1994).

The extracellular part of E-cadherin possesses a flexible hinge region (BECKER et al. 1989), and five potential asparagine glycosylation sites (GALLIN et al. 1987). Endoglycosidase treatments of chicken L-CAM have revealed one high mannose and three complex oligosaccharide chains (CUNNINGHAM et al. 1984). Glycosylation may be important for cell-cell adhesion, but does not affect processing and transport of the precursor towards the plasma membrane (SHORE and NELSON 1991).

An 80 kDa fragment, containing the NH_2-terminal and most of the calcium-binding sites of the molecule, can be cleaved off from the extracellular part of E-cadherin (CUNNINGHAM et al. 1984; HYAFIL et al. 1981, BEHRENS et al. 1985). A major cleavage site in the extracellular part of E-cadherin has recently been localised

(Pokutta et al. 1994). In the presence of calcium this fragment is resistant to trypsin, and it is recognized by the majority of polyclonal and monoclonal antibodies that can neutralize the cell-cell adhesion function of E-cadherin. Calcium, manganese and strontium can stabilize this fragment against enzymatic breakdown and sustain the function of E-cadherin, while lanthanium and cadmium have a negative effect (Hyafil et al. 1981; Behrens et al. 1985). This fragment, also coined UMt or Ft1, is shed into the circulation and has recently been proposed as a circulating tumor marker in blood (Katayama et al. 1994). This region also contains a cysteine cluster.

An extracellular region located close to the plasma membrane is also involved in cell-cell adhesion. Binding of the DECMA-1 antibody to an epitope in the vicinity of the cysteine cluster of this region inhibits adhesion (Vestweber and Kemler 1985). Reducing the disulfide bonds of the DECMA-1 region leads to an increased trypsin sensitivity and inhibits cell compaction (Ozawa et al. 1990b).

The extracellular part of E-cadherin is connected to its cytoplasmic part via a single hydrophobic membrane-spanning domain of 31 amino acids (Becker et al. 1989).

In the cytoplasmic part a highly conserved region common to all members of the cadherin family is present (Ozawa et al. 1989). This region can be used as a target to screen for the presence of cadherins immunologically. Two domains in this region, termed CH2 and CH3, were recently found to be characteristic for all classical cadherins, but absent or divergent in the more distantly related cadherins such as desmosomal cadherins, T-cadherin, fat gene product, and the human ret oncogene product (Rimm and Morrow 1994). A 20 kDa region of the cytoplasmic part contains a number of phosphorylation sites on serine and threonine moieties (Gallin et al. 1987; Stappert and Kemler 1994), and we will discuss further in this review the possible implications of phosphorylation on the function of E-cadherin.

Last but not least, the cytoplasmic part contains a 72 amino acid recognition site for the catenins, which forms the link with the cytoskeleton (Ozawa et al. 1990c). This region was recently narrowed down to 30 amino acid residues, comprising up to eight well conserved serine residues (Stappert and Kemler 1994). Within this site, strong homology with the cytoplasmic tail of desmocollin has been described.

2.2.2 Via the Catenins to the Cytoskeleton

Three types of catenins bind to E-cadherin in a noncovalent way, but additional molecules having the same property may show up in the future (Fig. 1). These molecules (or their mRNA) may also be present in the cytoplasm of cells that do not express E-cadherin, such as NIH 3T3, L cells and avian fibroblasts, and become associated with E-cadherin after transfection and expression of E-cadherin cDNA (Ozawa et al. 1989). The catenins are structurally related in different species Kemler and Ozawa 1989) and are implicated in the cell-cell adhesion function of E-cadherin (Nagafuchi and Takeichi 1990). Recently, however, a cadherin in the testis was determined to be responsible for the adhesion between Sertoli cells and germ cells in the absence of catenins (BYERS et al. 1994).

α-Catenin (102 kDa) is related to vinculin (HERRENKNECHT et al. 1991), and a variant, α-N-catenin, is found in nervous tissues (HIRANO et al. 1992). For both α-catenins, two splice variants were described (UCHIDA et al. 1994). The human α-catenin gene has been assigned to chromosome 5q31 (FURUKAWA et al. 1994; NOLLET et al. 1995). β-Catenin (88, 92 or 94 kDa) is homologous with plakoglobin and with the product of the segment polarity gene armadillo in Drosophila (McCREA et al. 1991; McCREA and GUMBINER 1991; PEIFER et al. 1992). The human β-catenin gene was localised at chromosome 3p21 (KRAUS et al. 1994; VAN HENGEL et al. 1995). γ-catenin (80, 83 or 86 kDa) is related to, and presumably identical to, plakoglobin (KNUDSEN and WHEELOCK 1992; PIEPENHAGEN and NELSON 1993). Plakoglobin is suggested to play a suppressive role in the tumorigenic behavior of carcinoma cells (NAVARRO et al. 1993). The assembly of α-, β-, and γ -catenin has been analyzed in vitro using recombinant proteins (ABERLE et al. 1994). There is evidence for the existence of two distinct E-cadherin/catenin complexes in the same cell: one complex is composed of E-cadherin, α - and β-catenin, the other of E-cadherin, α- and γ-catenin (plakoglobin) (BUTZ and KEMLER 1994; HINCK et al. 1994a). A fourth catenin-like molecule, $p120^{cas}$ (120 kDa), has recently been found to be part of the E-cadherin/catenin complex, and four isoforms have been detected up to now (REYNOLDS et al. 1994).

Several cytoplasmic molecules other than the catenins have been reported to associate with the E-cadherin complex, such as vinculin and talin (DUBAND and THIERY 1990), fodrin and ankyrin (NELSON et al. 1990), and Na^+, K^+-ATPase (NELSON et al. 1991). These associations are stressed by the fact that some of these molecules are endocytosed as complexes in common vesicles, when the cells are shifted to low calcium concentrations (KARTENBECK et al. 1991). The cytoplasmic APC gene product associates with the catenins, but not with E-cadherin (SU et al. 1993; RUBINFELD et al. 1993). The Wnt-1 gene product was found to stabilize the interaction between β-catenin and E-cadherin, and this effect was shown to promote cell-cell adhesion (HINCK et al. 1994b). The c-erbB-2 gene product is another molecule which was recently found to be part of the complex via interaction with β-catenin and plakoglobin (OCHIAI et al. 1994a).

Catenins are believed to form the link between E-cadherin and the actin cytoskeleton, and β-catenin (or plakoglobin) appears to be the first catenin to interact with E-cadherin when the latter molecule is being routed towards the plasma membrane (NAGAFUCHI and TAKEICHI 1990; HINCK et al. 1994a). The β-catenin domains responsible for binding to either E-cadherin or the APC gene product were characterized (HÜLSKEN et al. 1994; Fig. 1). Once α-catenin associates with a short region near to the NH_2-terminal of β-catenin (ABERLE et al. 1994) stable connections of the E-cadherin/catenin complex with the cytoskeleton are established. This is apparent after 10 min of cell-cell contact and reflected by the resistance of E-cadherin to detergent extraction (McNEILL et al. 1993; WHEELOCK and KNUDSEN 1991). Furthermore, disruption of the cytoskeleton affects the organisation of E-cadherin at the plasma membrane (PASDAR and LI 1993). Spectrin, which colocalises with the E-cadherin/catenin complex, binds to the

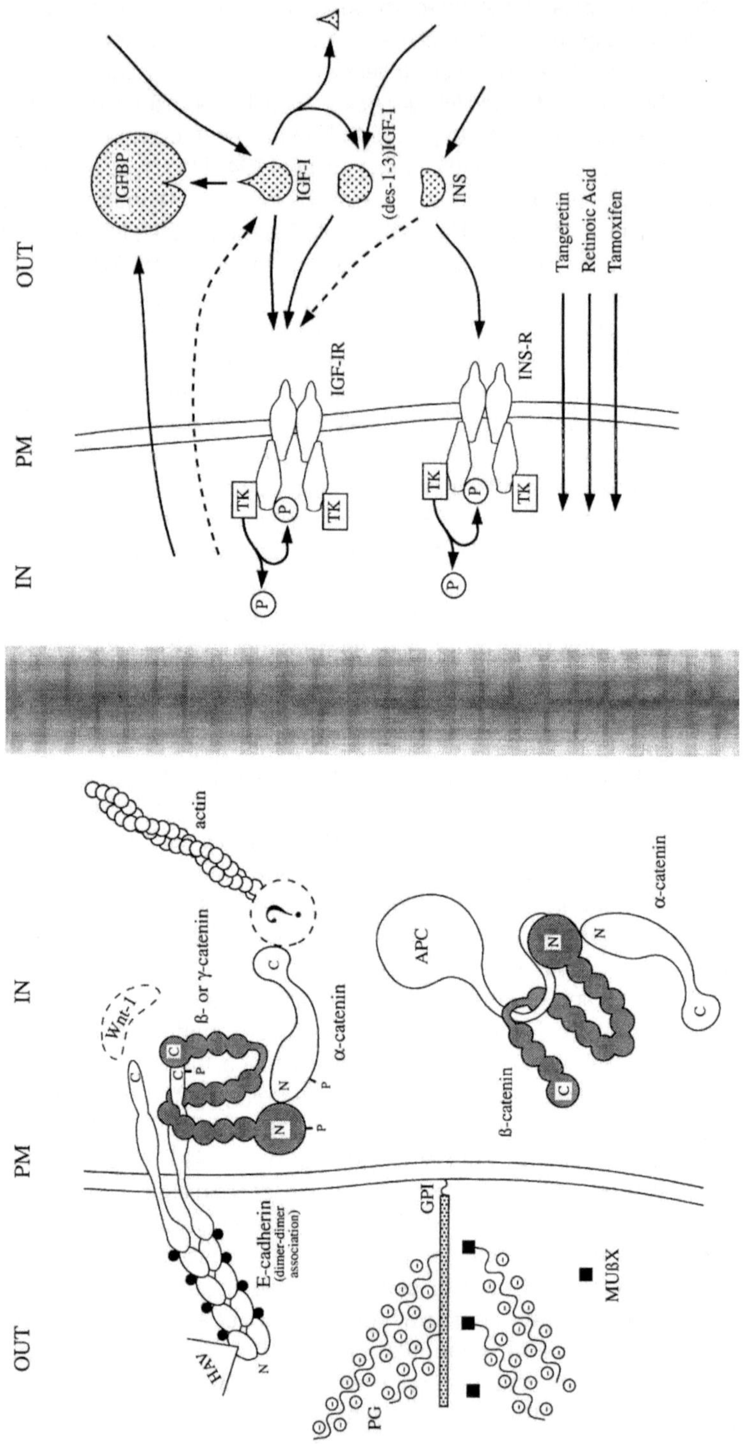
OUT
PM
IN
IGFBP
IGF-I
(des-1-3)IGF-I
INS
IGF-IR
INS-R
TK
P
Tangeretin
Retinoic Acid
Tamoxifen
actin
?
Wnt-1
β- or γ-catenin
α-catenin
E-cadherin
(dimer-dimer association)
HAV
GPI
PG
MUßX
APC
ß-catenin

NH_2-terminal region of α-catenin, a domain found to be responsible for the link to actin (LOMBARDO et al. 1994).

The formation of a link between E-cadherin and the cytoskeleton via the catenins does not exclude, however, that the catenins of the membrane complex can be exchanged by free α/β-catenin complexes from the cytoplasm, and that catenins can form complexes with molecules other than E-cadherin, both in the cytoplasm (e.g. with APC) and at the plasma membrane, e.g. with the epidermal growth factor receptor (EGFR) (HINCK et al. 1994a; NÄTHKE et al. 1994). The interaction between β-catenin and the EGFR has been proposed as a link between EGF-induced signals and E-cadherin function (HOSCHUETZKY et al. 1994).

3 Localisation of E-cadherin at Cell Surfaces

E-cadherin is expressed by many if not all epithelial cells. The molecule is even more widespread in the embryo, where it is transiently present in, e.g. the brain (SHIMAMURA and TAKEICHI 1992). In the adult, expression of E-cadherin is still found

Fig. 1. Functional regulation of the E-cadherin/catenin complex. *Left* The plasma membrane (*PM*)-associated complex between E-cadherin, β- and α-catenin is illustrated. *N* and *C*, respectively NH_2- and COOH-terminal residues; *HAV*, histidine-alanine-valine sequence, characteristic for the first extracellular domain of type 1 cadherins; *P*, phosphorylation site. *Filled circles* indicate the position of Ca^{2+} ions at the proximal end of cadherin protomers (in accordance with HÜLSKEN et al. 1994, OVERDUIN et al. 1995, SHAPIRO et al. 1995; ABERLE et al. 1994 and NÄTHKE et al. 1994). The binding between E-cadherin and β-catenin may be stabilized by Wnt-1, and α-catenin is associated with the actin cytoskeleton possibly via a hithertho unknown link molecule (*question mark*). Free cytoplasmic catenin complexes may be linked to other molecules than E-cadherin, for instance to the APC gene product. Functional inactivity of the E-cadherin/catenin complex can be due to several reasons, including the presence of GPI-anchored, membrane-associated extracellular proteoglycans (*PG*). By their large, negatively charged glycosaminoglycan side chains these molecules are thought to shield E-cadherin from homophilic interactions (VLEMINCKX et al. 1994). Restoration of the E-cadherin function can be achieved by 3-methylumbellipheryl-βD-xyloside (*MUβX*). The latter molecule acts as an acceptor for the attachment of glycosaminoglycan moieties, which eventually results in the synthesis of free glycosaminoglycans in the culture medium, and of naked proteoglycan core proteins at the plasma membrane.

Right A number of molecules are depicted that have been able to correct E-cadherin function and to inhibit invasion at least in vitro. One of these molecules is insulin-like growth factor I (*IGF-I*), which can activate the IGF-I receptor (*IGF-IR*) at the plasma membrane via an autocrine or a paracrine loop. The action of IGF-I can be tempered after binding to IGF binding proteins (*IGFBP*), which again can be secreted via autocrine or paracrine mechanisms. Cleaving off the first three amino acids from IGF-I yields (*des*1-3)IGF-I, and makes the molecule unable to bind to the IGFBP's, while increasing its potency to activate the IGF-IR. Insulin at supraphysiological concentrations can mimic the IGF-I effect via the insulin receptor (*INS-R*) or possibly via the IGF-IR. Triggering of both receptor types activates their intrinsic tyrosine kinase (*TK*) domains and leads to autophosphorylation and to phosphorylation of cytoplasmic substrates (*P*). The citrus flavonoid tangeretin, the vitamin A analogue retinoic acid and the anti-estrogen tamoxifen are able to correct the function of the E-cadherin/catenin complex through so far unknown signaling pathways. In between the left and the right part of the figures lies a gray shaded area, indicating our lack of knowledge concerning the signal transducing pathways in arachnoid villi (YAMASHIMA et al. 1992) and in a subset of sensory neurons (SHIMAMURA et al. 1992). It has been used to discriminate between insulinomas (E-cadherin positive) and glucagonomas (E-cadherin negative) of the pancreas (MOLLER et al. 1992).

Only traces of E-cadherin are detectable on the surface of unfertilised mouse oocytes, whereas between 6 and 11 h after activation detergent-resistant surface expression becomes evident. Embryos at subsequent stages up to the 8-cell stage exhibit a uniform distribution of E-cadherin at the plasma membrane. Upon compaction, however, E-cadherin accumulates at intercellular contacts, whilst free surface E-cadherin is reduced (CLAYTON et al. 1993).

When epithelioid cells, such as MDCK dog kidney cells, form cell-cell contacts in vitro, E-cadherin is detectable both on the apical and basolateral plasma membranes during the first 2 days. Later on, the molecule becomes restricted to the basolateral membrane (WOLLNER et al. 1992), and keeps this localisation during cell division in the polarized monolayer (REINSCH and KARSENTI 1994). In most epithelia E-cadherin is eventually associated with adherens junctions or with the adhesion belt at the basolateral site.

Tight junctions, which are located in a more apical position with respect to adherens junctions, form in epithelial sheets an occluding barrier for antibodies against E-cadherin. This explains why, in vitro, antibodies that can neutralize the E-cadherin function are unable to dissociate epithelioid cell islands on a tissue culture substratum if added to the apical cell side via the culture medium (BEHRENS et al. 1985). These antibodies can exert their activity only if added beneath a porous filter support of the cells and allowed in this way to diffuse to the basolateral region of the cells. Similar junctional organisations may be responsible for the fact that some anti-E-cadherin antibodies decompact cell aggregates but do not affect cell monolayers when added to the culture medium (MAILLET and BUC-CARON 1985).

When MDCK cells are cultured on tissue culture plastic for more than 5 days, the distribution of the catenins at the basolateral membrane becomes accentuated (NÄTHKE et al. 1994). From apical to basal one encounters: first, the apical junctional complex, combining the adherens and the tight junctions (high concentration of E-cadherin, β-catenin, actin and c-src and c-yes tyrosine kinases); second, the lateral plasma membrane with low concentration of E-cadherin/catenin (α and β) complexes; and third, the desmosomes, where desmoglein and desmocollins are associated with β-catenin and plakoglobin.

Recent reports show that the localisation of E-cadherin can be influenced by culture conditions. The polarity and E-cadherin localisation of MDCK cell cysts depend on whether they have been formed in suspension culture (apex towards the central lumen) or in collagen gel (apex towards the outside) (WANG et al.1990). Parathyroid hormone addition to the culture medium of rat osteosarcoma cells induces a diffuse distribution of E-cadherin (BABICH and FOTI 1994).

E-cadherin is a homophilic and homotypic cell-cell adhesion molecule, and type-specific sorting out occurs when cadherin-negative cells, transfected with different members of the cadherin family, are cultured together (NOSE et al. 1988;

FRIEDLANDER et al. 1989; JAFFE et al. 1990). Exceptionally heterophilic adhesion between E- and N-cadherin (VOLK et al. 1987) and heterotypic E-cadherin mediated adhesion of Langerhans cells (TANG et al. 1993) or melanocytes (TANG et al. 1994) to keratinocytes have been reported. Heterotypic E-cadherin mediated adhesion has also been held responsible for the interaction of thymocytes with thymic epithelial cells (LEE et al. 1994) and of T lymphocytes with intestinal epithelial cells (CEPEK et al. 1994).

4 E-cadherin as an Organizer

E-cadherin is not only involved in processes such as cell-cell recognition and adhesion, but also in the induction of cell polarity. Epithelial cell polarity implicates that the position of many cell molecules is restricted to certain regions of the plasma membrane, e.g. close to the intercellular junctions. Cell polarity is required for many epithelial functions, one of which is vectorial fluid transport.

When the E-cadherin gene is transfected into E-cadherin-negative cells, the expression of α-catenin is up-regulated by posttranscriptional mechanisms (NAGAFUCHI and TSUKITA 1994), and its distribution changes from diffuse cytoplasmic to plasma membrane-associated (OTTO et al. 1994). Recovery experiments from functional inhibition have shown that E-cadherin is an organizer of the junctional complex in epithelial cells (GUMBINER et al. 1988). The formation and functional integrity of not only the adherens junctions (through protein kinase C activity) (LEWIS et al. 1994), but also of tight junctions (GUMBINER and SIMONS 1986; AMAGAI et al. 1995), gap junctions (JONGEN et al. 1991) and desmosomes (AMAGAI 1995) are regulated by E-cadherin, whereby ZO-1, connexin 43 and desmoplakin seem to be major junctional targets involved in these various junctions.

As a master molecule E-cadherin is responsible for epithelial cell polarity MCNEILL et al. 1990), glandular differentiation (PIGNATELLI et al. 1992) and epidermal stratification (LEWIS et al. 1994). Within this context it is noteworthy that E-cadherin concentrates molecules such as urokinase-type plasminogen activator (JENSEN and WHEELOCK 1992) and epidermal growth factor receptor (SUAREZ-QUIAN and BYERS 1993) at cell-cell contact sites.

It should, however, not be overlooked that other molecules such as fodrin (ESKELINEN et al. 1992) have also been proposed as crucial organizers of cell polarity. Moreover, acidic fibroblast growth factor has been shown to counteract the organizing function of E-cadherin (BOYER et al. 1992), and sustained expression of E-cadherin does not seem to be sufficient under these circumstances to protect NBT-II rat bladder carcinoma cells from dispersing in vitro (BOYER et al. 1993). Also, formation of desmosomes and tight junctions was observed in early (disorganized) embryos of E-cadherin knock-out mice (LARUE et al. 1994; RIETHMACHER et al. 1995).

5 E-cadherin Is an Invasion Suppressor Molecule

The concept of E-cadherin as an invasion suppressor molecule stems from two types of studies. First, correlation studies have shown that the molecule is down-regulated in invasive tumors. Second, experimental up- or down-regulation of E-cadherin leads to, respectively, inhibition and induction of invasion.

5.1 Down-Regulation of E-cadherin in Invasive Tumors In Vivo

A number of inverse correlations has been published between the expression of E-cadherin and tumor malignancy. Homogeneous or heterogeneous down-regulation of the molecule was studied in histological sections of tumors examined via immunohistochemical reactions or in situ hybridizations. As a general rule, E-cadherin down-regulation in these studies was correlated with malignancy parameters, such as tumor progression, loss of differentiation, invasion, metastatic potential and poor prognosis. Furthermore, loss of heterozygosity for chromosome 16q predicts an adverse outcome in cancer patients (GRUNDY et al. 1994).

Down-regulation of E-cadherin was found in many carcinomas (Fig. 2). In human pathology these tumors were localized in the skin (squamous; NICOLOSON et al. 1991; CZECH et al. 1993 and basal cell carcinomas NICOLSON et al. 1991;

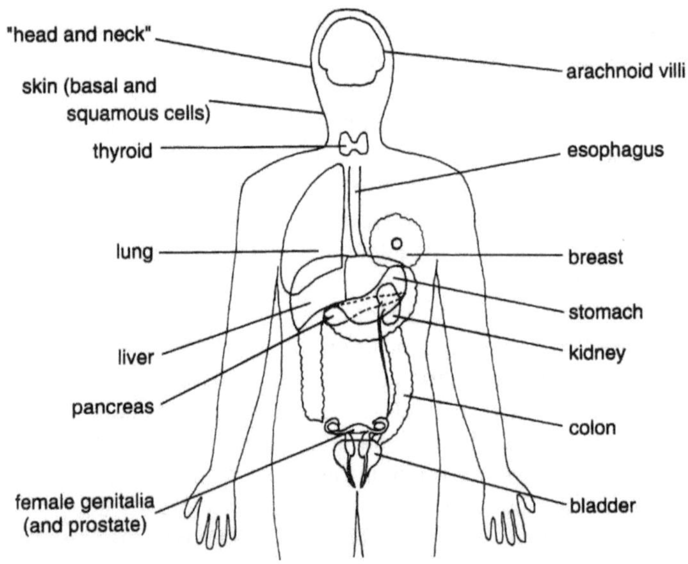

Fig. 2. Overview of different organs and tissues in the human body that can give rise to carcinomas in which down-regulation of E-cadherin has been documented by immunohistological techniques

Pizarro et al. 1994), the meninges (Tohma et al. 1992), the head and neck region (Schipper et al. 1991; Field 1992; Mattijssen et al. 1993; Bowie et al. 1993; McNeill et al. 1990; Sorscher et al. 1995), the esophagus (Miyata et al. 1990; Shiozaki et al. 1991; Kadowaki et al. 1994; Shiozaki et al. 1993), the thyroid (Brabant et al. 1993), the lungs (Böhm et al. 1994b), the female breast (Shiozaki et al. 1993; Böhm et al. 1994a; Rasbridge et al. 1993; Moll et al. 1993; Oka et al. 1992a; Lipponen et al. 1994), the stomach (Shiozaki et al. 1993; Oka et al. 1990; Shino et al. 1991; Shimoyama and Hirohashi 1991b; Matsuura et al. 1992; Oka et al. 1992b; Shino et al. 1992; Mayer et al. 1993; Cai et al. 1994), the liver (Shimoyama and Hirohashi 1991a), the pancreas (Moller et al. 1992), the kidney (Terpe et al. 1993; Katagiri et al. 1995), the colon (Van der Wurff et al. 1992; Nigam et al. 1993; Kinsella et al. 1993; Dorudi et al. 1992, 1993; Van Aken et al. 1993; Kinsella et al. 1994), the female genitalia (Honda 1992; Inoue et al. 1992; Mori et al. 1994; Sakuragi et al. 1994), the bladder (Bringuier et al. 1993; Vet et al. 1994; Otto et al. 1994) and the prostate (Bussemakers et al. 1992; Giroldi adn Schalken 1993; Otto et al. 1993; MacCalman et al. 1994a; Umbas et al. 1992, 1994). An example of differential down-regulation of E-cadherin can be found in some human adenocarcinomas: cells in the intravascular component show a higher E-cadherin expression than those in the extravascular component (Cowley and Smith 1995).

In experimentally transplanted invasive tumors in the mouse, down-regulation of E-cadherin was observed with human keratinocytes (Nicolson et al. 1991) and MDCK dog kidney cells (Mareel et al. 1991). The same holds for rat prostate cancer populations in syngeneic animals (Bussemakers et al. 1992).

The mechanisms behind this down-regulation in vivo have not yet been elucidated. Experiments with MO4 murine cells transfected with the dog E-cadherin cDNA under the control of a constitutively active promoter have shown the molecule to be sensitive to down-regulation in vivo. Tumors raised by these cells in syngeneic animals indeed showed down-regulation or mutation of the transfected dog E-cadherin at the transcriptional level, and were invasive (Gao 1993). The down-regulation of E-cadherin expression was rapidly reversible after ex vivo explantation of the tumor cells. The host factors responsible for the down-regulating effect are now a matter of further investigation in our laboratory. The down-regulation of E-cadherin by an autocrine motility-like factor in human oral squamous cell carcinoma cell line may be relevant in this respect (Ishisaki et al. 1994).

Exceptionally, E-cadherin was found to be expressed homogeneously in invasive human lung adenocarcinoma (Matsuura et al. 1992) and esophageal (Kadowaki et al. 1994) and gastric carcinoma (Shimoyama and Hirohashi 1991a). This phenomenon suggests that under certain circumstances, E-cadherin may not be sufficiently active as an invasion suppressor. One of the reasons for the disfunction of the E-cadherin-catenin complex may be the down-regulation of α-catenin, as shown in cancers of the esophagus, stomach, colon and breast (Shiozaki et al. 1994; Breen et al. 1993; Matsui et al. 1994; Ochiai et al. 1994b). Another reason may be a 107 amino acids deletion in the NH_2-terminal region of β-catenin, as described in a human stomach cancer cell line. This mutated β-catenin still interacts with E-cadherin, but not with α-catenin (OYAMA et al. 1994; KAWANISHI et al. 1995).

5.2 Experimental Evidence In Vitro for the Invasion Suppressor Role of E-cadherin

More direct evidence for the invasion suppressor role of E-cadherin could be gathered from experiments in vitro. Some of these experiments are biochemical equivalents of the immunohistological studies and confirmed, for instance, that highly and weakly metastatic cell sublines derived from the same murine ovarian tumor, showed, respectively, low and high E-cadherin expression, as evidenced by western and northern blots (HASHIMOTO et al. 1989). Human ascites cells from solid ovarian tumors expressed significantly less E-cadherin and were more invasive in vitro than their counterparts in the primary tumors (VEATCH et al. 1994). In other studies experimental up- and down-regulation correlated with cell-cell adhesion, cell aggregation and compaction without reference, however, to invasion (TANG et al. 1994; AO and ERICKSON 1992; FABRE and GARCIA DE HERREROS 1993; BRADLEY et al. 1993).

The most powerful evidence to consider E-cadherin as an invasion suppressor was derived from invasion experiments with cell lines differing in their expression of E-cadherin. The invasion assays used range from the formation of invasive and metastatic tumors in syngeneic or immunodeficient laboratory animals, over organotypic cultures made up of confrontations between aggregates of tumor cells and living embryonic heart fragments (confronting cultures) to penetration into nonliving substrates such as Matrigel and type I collagen gels. Although the presence of potentially up- and down-regulating factors of E-cadherin may differ substantially among these assays, the overall results appear to confirm an inverse correlation between E-cadherin expression and invasion.

Invasion was inhibited by transfection of the E-cadherin cDNA into constitutively E-cadherin-negative cell lines (CHEN and ÖBRINK 1991; BEHRENS et al. 1991b; FRIXEN et al. 1991; VLEMINCKX et al. 1991). Induction of invasion in constitutively E-cadherin-positive cell lines was achieved by antagonizing the expression of the molecule via the introduction of a plasmid encoding E-cadherin-specific antisense mRNA (VLEMINCKX et al. 1991), via ras and polyoma middle T oncogene transfection (HOFFMAN et al. 1993; CHASTRE et al. 1993), via Epstein-Barr virus transfection (FAHRAEUS et al. 1992), or via antibodies that neutralize the adhesion function of E-cadherin (CHEN and ÖBRINK 1991; BEHRENS et al. 1991b; VLEMINCKX et al. 1991; FRIXEN et al. 1991; DOKI et al. 1993. In mammary carcinoma cells, down-regulation of E-cadherin seems to be necessarily accompanied by expression of the intermediate filament component vimentin (typical for epithelial to mesenchymal transitions) for invasion to occur (SOMMERS et al. 1991; BAE et al. 1993). E-cadherin may, however, "override" the vimentin-associated invasiveness in this type of cells (THOMPSON et al. 1994). Overexpression of the erb-B2 receptor in human mammary carcinoma cells has been correlated with aggressiveness and was shown to inhibit the expression of E-cadherin in vitro (D'SOUZA and TAYLOR-PAPADIMITRIOU 1994). In a number of studies invasion could be switched on and off by down- and up-regulation of E-cadherin within the same population of cells (CHEN and ÖBRINK 1991; BEHRENS et al. 1991b; VLEMINCKX et al. 1991; FRIXEN and NAGAMINE 1993).

The concept of E-cadherin as an invasion suppressor was further confirmed by correlation studies, in which the invasiveness in vitro was tested for a number of E-cadherin-positive and-negative human cell lines. These studies included bladder, breast, lung, pancreas (FRIXEN et al. 1991), esophagus (DOKI et al. 1993), colon (KINSELLA et al. 1994) and endometrium (MORI et al. 1994) cell lines. Again invasion was correlated with absence of E-cadherin in these cell lines.

6 Functional Regulation of the E-cadherin/Catenin Complex

Up to now we have been concentrating on a large number of observations and experiments that led to the straightforward conclusion that E-cadherin can be considered as an invasion suppressor molecule in epithelial cells. However, a small number of exceptions to this rule, namely invasion of tumor cell populations showing homogeneous E-cadherin expression, have been reported. Cases of invasive human lung, esophagus and stomach cancers with an homogenous immunohistochemical staining pattern for E-cadherin were already mentioned (KADOWAKI et al. 1994; MATSUURA et al. 1992; SHIMOYAMA and HIROHASHI 1991a). In laboratory animals HaCaT keratinocytes can form invasive nodules (BOUKAMP and FUSENIG 1993) and human MCF-7/6 breast carcinoma cells can give rise to invading and metastatizing tumors (CORREC et al. 1990), both with homogenous expression of E-cadherin, as evident from immunohistochemistry. Also invasion in vitro was found with murine (NMuMG) (VLEMINCKX et al. 1994) and human (MCF-7/6) (BRACKE et al. 1991) mammary cells that show homogeneous E-cadherin expression. MDCK cells were also found to be invasive after passage through the peritoneal cavity, without down-regulation of their E-cadherin expression at the plasma membrane (VANDENBOSSCHE et al. 1994). The rat colon cancer cell line DHD/PROb is invasive both in vivo and in vitro, while it maintains homogeneous E-cadherin expression. For the latter cells, invasiveness depends on the collaboration of tumor-associated myofibroblasts (DIMANCHE-BOITREL et al. 1994b). Together, these data suggest that invasion can also be the result of the putative functional inactivity or inactivation of E-cadherin at the tumor cell surface.

6.1 For What Reasons May E-cadherin Be Functionally Inactive?

A number of discrete experimental mutations obtained with cells in vitro have allowed identification of critical regions for functional integrity of E-cadherin.

Impaired aggregation and/or compaction was the result of mutations in the domains that are involved in enzymatic trimming of the precursor (OZAWA and KEMLER 1990), calcium binding (OZAWA et al. 1990a) or complex formation with the catenins (KEMLER and OZAWA 1989; NAGAFUCHI and TAKEICHI 1988). Theoretically, similar mutations in vivo could also render E-cadherin inactive and abrogate its invasion suppressor function. However, mutations in the E-cadherin gene appear to be infrequent events, and, so far, functionally inactivating mutations have not been equivocally demonstrated in natural tumors, although candidates for such mutations were reported. A point mutation in exon 7 with possible functional impact (Asn to Ser) was recently detected in two cases of invasive mammary carcinoma (KANAI et al. 1994). In two human diffuse-type gastric carcinomas, point mutations in exons 8 and 10 result in, respectively, Asp to Ala and Val to Asp changes (BECKER et al. 1994). In endometrial carcinomas, point mutations in exons 12 and 13 result in Ala to Phe and Leu to Val changes, respectively, whereas exons 16 in an ovarian carcinoma bears a point mutation resulting in a Ser to Gly change (RISINGER et al. 1994). Except for the last mutation, all other mentioned changes affect the extracellular domain of E-cadherin. Experiments have suggested that lack of expression of α-catenin may be at the base of a functional defect of the E-cadherin-catenin complex. This is the case for both human PC-9 lung (SHIMOYAMA et al. 1992; ODA et al. 1993), PC-3 prostate (MORTON et al. 1993) and several colon (BREEN et al. 1993) carcinoma cells, apparently due to deletions in the α-catenin gene. Moreover, down-regulation of α-catenin was found to correlate with invasion and lymph node metastasis of human esophagal carcinomas (KADOWAKI et al. 1994). NAVARRO et al. suggested that a defect in γ-catenin or plakoglobin may be responsible for E-cadherin/catenin inactivity, although no direct evidence for this relation is available at the moment (NAVARRO et al. 1993). Low levels of plakoglobin were also expressed by invasive human breast cancer cells (SOMMERS et al. 1994). Another interesting finding is the deletion of the β-catenin gene in a human gastric cancer cell line (OYAMA et al. 1994; KAWANISHI et al. 1995).

Tyrosine phosphorylation of the E-cadherin complex appears to be a rapid mechanism to down-regulate the function of the E-cadherin/catenin complex. Transfection experiments with the v-src oncogene have shown that the tyrosine kinase activity of its gene product preferentially increases β-catenin tyrosine phosphorylation, and this leads to impaired cell compaction (MATSUYOSHI et al. 1992) and invasiveness in vitro (BEHRENS et al. 1993). Hepatocyte growth factor and epidermal growth factor, whose receptors both colocalize with E-cadherin (CREPALDI et al. 1994; SUAREZ-QUIAN and BYERS 1993), also enhance tyrosine phosphorylation of β-catenin and plakoglobin in human carcinoma cells with concomitant cell scattering in vitro (SHIBAMOTO et al. 1994; HOSCHUETZKY et al. 1994). Treatment with epidermal growth factor indeed counteracts E-cadherin mediated junctional assembly and induces a more invasive phenotype (SHIOZAKI et al. 1995). Furthermore, tyrosine phosphorylation of β-catenin can be obtained with tyrosine phosphatase inhibitors, such as pervanadate, and these treatments reduce cell compaction (MATSUYOSHI et al. 1992). The src gene product inhibitor herbimycin, by contrast, stimulates cell compaction (MATSUYOSHI et al. 1992).

Other observations concerning phosphorylation of the complex may be relevant to understand its function, but the link with cell-cell adhesion or invasion has not been shown yet. E-cadherin can, for instance, act as an initiator of gap junctional organisation and function via phosphorylation of connexin 43 (MUSIL et al. 1990); the role of this phenomenon in cancer invasion is not known. Tyrosine phosphorylation of plakoglobin is essential to prevent its interaction with the APC gene product and could in this way interfere with APC signals (SHIBATA et al. 1994). The complex formation of β-catenin and plakoglobin with the c-erbB-2 gene product suggests that this tyrosine kinase receptor may also participate in the regulation of the function of the E-cadherin/catenin complex (OCHIAI et al. 1994a). Finally, since $p120^{cas}$ is a tyrosine kinase substrate for growth factor receptors and for the src gene product, it may also play a role in signal transduction from ligand-triggered receptors towards the E-cadherin/catenin complex (REYNOLDS et al. 1994).

Little has been published on the role of glycosylation in E-cadherin function. Inhibition of N-linked glycosylation by tunicamycin had no effect on E-cadherin mediated cell aggregation (SHIRAYOSHI et al. 1986). A possible role of cell surface proteoglycans in the function of E-cadherin has recently been described (VLEMINCKX et al. 1994; VANDENBOSSCHE et al. 1994). Indeed, functional inactivity of the E-cadherin/catenin complex was shown to be due to the presence of GPI-anchored, membrane-associated extracellular proteoglycans (VLEMINCKX et al. 1994). By their large negatively charged glycosaminoglycan side chains these molecules are thought to shield E-cadherin from homophilic interactions. Restoration of the E-cadherin function could be achieved by addition of 3-methyl-umbellipheryl-β-D-xyloside. The latter molecule acts as an acceptor for the attachment of glycosaminoglycan moieties, which eventually results in the synthesis of free glycosaminoglycans in the culture medium and of naked proteoglycan core proteins at the cell surface.

6.2 Correction of the Function of the E-cadherin/Catenin Complex

From both a scientific and a therapeutic point of view, correction of the defective function of the E-cadherin/catenin complex appears to be an interesting goal. For this we examined a number of variants of the human MCF-7 cell line, which was originally derived from the pleural effusion of a metastasizing mammary adenocarcinoma (SOULE et al. 1973). Variants were obtained from different laboratories, and their MCF-7 origin was confirmed by specific immunoreactive markers (COOPMAN et al. 1991). All variants were confronted in vitro with embryonic chick heart fragments, cultured in suspension for 8 days, and their interaction with the chick heart fragments evaluated by histologic analysis (MAREEL et al. 1979). Some variants (e.g., MCF-7/6) invaded the heart tissue, while others (e.g. MCF-7/AZ) failed to do so (BRACKE et al. 1991). The presence of an apparently intact E-cadherin/catenin complex at the plasma membrane of both MCF-7/6 and MCF-7/AZ cells was demonstrated via multiple techniques (immunocytochemistry, immunohistochemistry, co-immunoprecipitation, flow cytometry and western blotting) (BRACKE et al. 1993). From these data we concluded that the E-cadherin/catenin complex was unable to exert its invasion suppressor function in MCF-7/6 cells.

Comparison between invasive MCF-7/6 and noninvasive MCF-7/AZ cells revealed differences in cell-cell adhesion in vitro. A fast aggregation assay, inspired by the work of TAKEICHI (1977) and KADMON et al. 1990), starts by detaching cells from their plastic tissue culture substrate under conditions that should leave the E-cadherin/catenin complex functionally intact. By Coulter counter measurements, the initial number of particles in suspension is compared with the number of particles after 30 min of aggregation (BRACKE et al. 1993). In this assay MCF-7/6 cells showed poor aggregation in contrast with their noninvasive MCF-7/AZ counterparts. MCF-7/AZ aggregation was specifically inhibited in the presence of monoclonal antibodies directed against a functional domain of E-cadherin (HECD-1, SHIMOYAMA et al. 1989; MB2, BRACKE et al. 1993), but not by other monoclonals binding to other glycoproteins of the plasma membrane (e.g. 5D10, PLESSERS et al. 1990). These results indicate that E-cadherin at the surface of MCF-7/6 cells is not functionally active. As mentioned before, our current research is aimed at finding agents that are able to correct the E-cadherin/catenin complex function in the invasive MCF-7/6 variant as far as cell-cell adhesion and invasion suppression are concerned.

6.2.1 Insulin-Like Growth Factor I Corrects the Defective Cell-Cell Adhesion and Invasion Suppressor Functions in Human Breast Carcinoma Cells

Insulin-like growth I (IGF-1) is a small 7kDa polypeptide hormone with multiple functions, as reflected by the many names that were given to this molecule in the past: sulphation factor, nonsuppressible insulin-like activity-soluble (NSILA-S), multiplication stimulating activity (MSA) and somatomedin C (DAUGHADAY 1992). Secreted by hepatocytes under the regulation of growth hormone, IGF-I is released into the blood and can act as an endocrine factor. More recently, however, short-range paracrine (JENNISCHE et al. 1992) and autocrine (SARA 1992) effects have been elucidated. We have shown that IGF-I can correct the defective cell-cell adhesion function of the E-cadherin/catenin complex in the invasive MCF-7/6 cell variant at near-physiological serum concentrations (BRACKE et al. 1993). Fast aggregation is stimulated by IGF-I within 30 min, and the effect is independent from de novo protein synthesis. This increased aggregation is mediated by E-cadherin, since it can be blocked by the monoclonal anti-E-cadherin antibodies HECD-1 and MB2. IGF-I interacts with the IGF-I receptor on the MCF-7 cell surface (DE LEON et al. 1988). This receptor is a heterotetramer of two α two β subunits. We have described this receptor on MCF-7/6 cells by Scatchard analysis, flow cytometry and immunoprecipitation with the anti-IGF-I-receptor monoclonal antibody α IR3 (Fig. 1). The α subunits contain a cysteine-rich region, while the β

subunits possess tyrosine kinase activity. Furthermore, the anti-IGF-I receptor monoclonal antibody inhibited completely the effect of IGF-I on MCF-7/6 aggregation. MCF-7 cells express in addition insulin receptors that show homology to the IGF-I receptor (MOUNTJOY et al. 1987), and insulin is indeed able to mimic the IGF-I effect on cell aggregation, albeit at supraphysiological concentrations. The insulin effect was not blocked by the monoclonal antibody αIR3, indicating that insulin does not act via the IGF-I receptor, but presumably via its own receptor. It is not excluded that our cells possess a hybrid of the IGF-I and the insulin receptor, which is less sensitive to IGF-I and insulin than the proper IGF-I or insulin receptors respectively (MOXHAM et al. 1989). Insulin-like growth factor II (IGF-II) has its own type of receptor and was not able to induce fast cell aggregation of MCF-7/6 cells.

Signal transduction from the triggered IGF-I receptor starts with switching on its intracellular tyrosine kinase activity, which leads to autophosphorylation of the receptor in our MCF-7/6 cells and to phosphorylation of (a) cytoplasmic substrate(s) (JACOBS et al. 1983; KADOWAKI et al. 1987). For the insulin receptor these phosphorylations have been shown to result in ras-activation through a cascade of a number of rapid molecular interactions (SKOLNIK et al. 1993; BALTENSPERGER et al. 1993). We have shown that activation of E-cadherin-mediated aggregation of MCF-7/6 cells can be blocked by a number of tyrosine kinase inhibitors:genistein (25 μM), Me-2,5-diOH cinnamate (50 μM) and 2-OH-5-(2,5 diOH benzyl) aminobenzoic acid (10 μM). The goal of our current research is to reveal possible effects of IGF-I on the phosphorylation of catenins and E-cadherin. As mentioned above, phosphorylation of β-catenin has recently been shown to modulate the function of the E-cadherin/catenin complex.

Extracellular regulation of IGF-I as a trigger for its receptor is mediated by the insulin-like growth factor binding proteins (IGFBPs). These proteins are present in blood (PALKA and PETERKOFSKY 1988), and are also secreted by MCF-7 cells in their culture medium. Six species of IGFBPs have been characterized, five of which may be secreted by MCF-7 cells (FIGUEROA and YEE 1992). Ligand blotting of MCF-7/6 conditioned media with ^{125}I-labeled IGF-I revealed at least three bands. While in a few cases binding of IGF-I with an IGFBP was reported to result in an increase of IGF-I potency, this usually prevents binding of the ligand to its receptor (FIGUEROA and YEE 1992).

The impact of IGFBPs on the correction of E-cadherin-mediated cell-cell adhesion function via IGF-I can be illustrated by the potency of (des1–3)IGF-I. The latter is a natural truncated variant of IGF-I which lacks the first three NH_2-terminal residues (SARA et al. 1986). This variant has lost most of its binding affinity for IGFBPs and is about 100 times more potent in activating E-cadherin mediated fast aggregation of MCF-7/6 cells, as compared with nontruncated IGF-I. A possible autocrine loop via this truncated IGF-I is not excluded in some types of MCF-7 cells, since (des 1–3)IGF-I has been shown to be produced by at least some MCF-7 cells (OGASAWARA et al. 1988). Preliminary data on the secretion of IGF-I into the MCF-7 culture medium indicate that with the noninvasive MCF-7/AZ variant higher levels are found than with the invasive MCF-7/6 variant (data in collaboration with J.-M. KAUFMAN). When added to the medium of confronting cultures of

MCF-7/6 cells with embryonic chick heart fragments, IGF-I exerted an anti-invasive effect which was reversible upon omission of the factor (BRACKE et al. 1993). Again, this effect appeared to be mediated by the IGF-I receptor, as it could be blocked by a monoclonal antibody against the IGF-I receptor. Our finding that IGF-I receptor triggering prevents invasion in vitro should be compared with studies on human mammary tissue samples indicating production of IGF-I by stromal cells of nonmalignant specimens and production of IGF-II by stromal cells of invasive carcinomas (CULLEN et al. 1991; PAIK 1992). Recently, it was also found that expression of high levels of IGF-I receptor in mammary tumors is an indicator of better prognosis (PEYRAT and BONNETERRE 1992; PAPA et al. 1993).

6.2.2 The Citrus Flavonoid Tangeretin Also Corrects the Defective Function of the E-cadherin/Catenin Complex via by a IGF-I Receptor-Independent Mechanism

Flavonoids are interesting tools to study the mechanisms of mammary tumor invasion. Examples are the weak estrogen and strong tyrosine kinase inhibitor genistein (MARTIN et al. 1978; AKIYAMA et al. 1987), the pp60^{src} inhibitor quercetin (GRAZIANI 1986), the laminin-binding anti-invasive flavanol (+)-catechin (BRACKE et al. 1987), the synthetic 3,7-dimethoxy flavone (PARMAR et al. 1994), and the biological response modifier flavone acetic acid (CHING and BAGULEY 1987). Tangeretin is a flavonoid extractable from citrus plants, which was retained in a screening program for potential anti-invasive molecules in the assay with embryonic chick heart (BRACKE et al. 1989). This molecule inhibits the invasion of virus-transformed MO4 mouse cells and of MCF-7/6 cells. As a polymethoxylated flavonoid, tangeretin is stable both in vitro and in vivo. After oral administration via the drinking water, it is found in the liver of laboratory mice (BRACKE et al. 1990).

Tangeretin corrects the defective function of the E-cadherin/catenin complex in MCF-7/6 cells in a way that resembles the effect of IGF-I. Tangeretin induces MCF-7 fast cell aggregation (BRACKE et al. 1994a). This phenomenon is mediated by E-cadherin, since it can be blocked by the anti-E-caherin monoclonal antibodies HECD-1 and MB2. It is not dependent on de novo protein synthesis, since cycloheximide does not inhibit the tangeretin effect. The flavonoid effect, however, differs from the IGF-I effect in that it does not seem to require triggering of the IGF-I receptor, as evidenced by the lack of effect of αIR3, the monoclonal antibody that inhibits the function of this receptor. Furthermore, tangeretin does not induce tyrosine autophosphorylation of the IGF-I receptor.

Experiments with estrogen-primed female nude mice are going on to explore whether tangeretin can prevent liver metastases from MCF-7 tumors raised in the spleen (KOPPER et al. 1982). Such experiments are also inspired by the retarding effect of flavone acetic acid on the development of spleen-derived artificial liver metastases (GIAVAZZI et al. 1988).

6.2.3 Retinoic Acid Induces Fast Aggregation of Human MCF-7/6 Breast Cancer Cells

All-trans-retinoic acid (RA) is a vitamin A derivative with opposite effects on invasion depending on the type of tumor cells under study. When added to the medium of confronting cultures with embryonic chick heart fragments, RA inhibits invasion of MCF-7/6 cells, but induces invasion in the constitutively noninvasive cell variant MCF-7/AZ (Bracke et al. 1991). RA affects multiple functions of MCF-7/6 cells: it decreases fast plasma membrane motility, it increases the enzymatic activity of tissue-type transglutaminases (Bracke et al. 1992) and in increases cell-cell adhesion. Thus, the anti-invasive effect of RA on MCF-7/6 can be explained by inducing a less motile, more rigid and more strongly adherent cell population.

Fast aggregation of MCF-7/6 cells was increased by addition of RA to the medium (Bracke et al. 1994c). This effect was about maximal after pretreatment of the cells for 4 h and was concentration-dependent. The anti-E-cadherin monoclonal antibodies HECD-1 and MB2 were able to block the effect of RA on cell aggregation, which indicates that RA specifically corrects the defective function of the E-cadherin/catenin complex in these cells. The RA effect was not sensitive to cycloheximide, which shows that de novo protein synthesis is not required.

The mechanism of action of RA on MCF-7/6 cells is not well understood. RA is considered to be a morphogen (Giguere et al. 1987; Eichele 1993) that penetrates the plasma membrane and is transported in the cytoplasm towards the nucleus via the cytoplasmic RA-binding protein (CRABP) present in MCF-7/6 cells (Bracke et al. 1992). Interaction with the nuclear RA receptors modulates gene transcription, and other molecules can interact with this zinc-finger mediated phenomenon (Mader et al. 1993). Cycloheximide insensitivity makes classical promotion of DNA transcription/translation of relevant genes by RA rather improbable. Up to now we have also gathered information leading us to believe that RA does not act via triggering of the IGF-I receptor, because neither blocking of this receptor by the monoclonal antibody αIR3 nor inhibition of its tyrosine kinase activity diminish the effect of RA on fast aggregation.

In another human mammary cancer cell line, SK-BR-3, RA has been shown to induce the expression of E-cadherin in vitro (Anzano et al. 1994). This observation indicates that RA can influence E-cadherin mediated cell-cell adhesion via multiple mechanisms.

6.2.4 Correction of E-cadherin Mediated Cell-Cell Adhesion by the Anti-estrogen Tamoxifen

Tamoxifen has proven to be useful in the adjuvant therapy of mammary carcinoma and has been found to act on several targets implicated in tumor invasion. Tamoxifen inhibits estradiol binding to its receptor (Mader et al. 1993), increases natural killer cell activity (Mandeville et al. 1984), increases the secretion of transforming growth factor (TGF)-β (Knabbe et al. 1987), reduces the secretion of

TGF-α (KOGA and SUTHERLAND 1987) and reduces cell-matrix adhesion (MILLON et al. 1989). Recently we have found that tamoxifen also restores the defective E-cadherin function in MCF-7/6 cells; it increases fast cell aggregation and inhibits invasion, in the chick heart assay (BRACKE et al. 1994b). This anti-invasive effect in vitro contrasts with the results of other laboratories, in which tamoxifen was found to stimulate MCF-7 invasion (THOMPSON et al. 1988). Apart from differences in the MCF-7 variants and in the assays for invasion used in both studies, the control cultures to which the treated ones are compared with are definitely not similar as far as the presence of estrogens is concerned. In our study, the control cultures are kept in culture medium supplemented with fetal bovine serum, and these cultures show invasion. In the other study, the controls are kept serum-free, and they show no invasion. So, depending on the presence of estrogens in the controls and on their invasive behavior, one can presumably observe an anti-invasive or an invasion promoting effect of tamoxifen.

Tamoxifen increases MCF-7/6 aggregation at 10^{-6} *M* or higher, but not at lower concentrations. The metabolites 4-OH-tamoxifen and *N*-desmethyl-tamoxifen also increased MCF-7/6 aggregation, but were less potent than tamoxifen itself.

The effect of tamoxifen on fast cell aggregation resembles the effect of tangeretin, as it is fast (observable within minutes) and mediated by E-cadherin (inhibited by the monoclonal anti-E-cadherin antibody HECD-1). It is not inhibited by the protein synthesis inhibitor cycloheximide nor by the monoclonal anti-IGF-I-receptor antibody αIR3.

It is tempting to speculate that the estrogen receptor on MCF-7/6 is involved in the tamoxifen effect, since estrogen receptor-negative human breast cell lines (SKBR-3 and HBL-100) do not respond to tamoxifen in the fast aggregation assay. One can argue that too high concentrations (10^{-6} *M* or higher) are required for the effects described above. Studies on intratumoral tamoxifen concentrations, however, have shown that 10^{-6} *M* is usually exceeded in human mammary carcinomas after daily oral intake of one tablet (20 mg) of tamoxifen by the patient (JOHNSTON et al. 1993). We therefore believe that the restoration of the E-cadherin function in human mammary carcinoma cells may contribute to the established therapeutic benefit of tamoxifen in breast cancer patients.

7 Can E-cadherin Be Considered as a Receptor?

In the above mentioned experiments E-cadherin has been considered as an effector molecule that can be activated by external triggers. Recent reports, however, have led to the E-cadherin/catenin complex to be considered as a receptor too. The high degree of sequence similarity between β-catenin and the armadillo gene product in Drosophila suggests that β-catenin is involved in developmental signaling. Microinjection of anti-β-catenin indeed perturbs

Xenopus development (GUMBINER 1993). This raises the possibility that β-catenin is involved in a similar (same?) signaling pathway as the Wnt factor. In Drosophila, armadillo distribution in the cell is indeed regulated by the Wingless signal, which is homologous to the vertebrate Wnt signal, and by Zeste-White 3 kinase (PEIFER et al. 1994). Furthermore, both E-cadherin (MATSUZAKI et al. 1990) and Wnt (OLSON et al. 1991) expression are able to stimulate gap junctional communication between adjacent cells, which may be considered as a reflection of changes in E-cadherin-mediated cell-cell adhesion (MOON et al. 1993). Together those data indicate that the E-cadherin/catenin complex may be part of a signaling pathway(s).

Other experiments have strengthened this view by showing that E-cadherin, upon homophilic binding, can activate protein kinase C, which leads to assembly of tight junctions (BALDA et al. 1993). A signaling role for E-cadherin was also found in the down-regulation of integrin expression during terminal differentiation of keratinocytes (HODIVALA and WATT 1994).

With relation to invasion, an inverse link has been established between the expression of E-cadherin and of urokinase-type plasminogen activator (u-PA). Immunological blocking of the E-cadherin function in human breast cancer cells leads to invasiveness into type I collagen, along with increased u-PA expression (FRIXEN and NAGAMINE 1993). This invasion appeared to be u-PA-dependent, since it could be stopped by applying neutralizing antibodies against the u-PA activity.

A few recent reports indicate that E-cadherin mediated cell-cell adhesion is correlated with an increase in the fraction of nonproliferating cells (SACCHI et al. 1994; WATABE et al. 1994; DIMANCHE-BOITREL et al. 1994a). Future research will be necessary to elucidate whether or not E-cadherin possesses a receptor function which controls cell growth in epithelioid cells.

8 Down-Regulation of E-cadherin Is Not a Unique Feature of Cancer Cell Populations

Down-regulation of E-cadherin in epithelia is a common phenomenon in embryonic morphogenesis. At the morula stage of the pig, cycles of compaction/decompaction are accompanied by alterations in E-cadherin distribution (REIMA et al. 1993). After gastrulation some epithelia modulate from an E-cadherin-positive towards an E-cadherin-negative state (CROSSIN et al. 1985; EDELMAN et al. 1987). One example is the process of neurulation, in which the developing neural tube increases its N-CAM expression and down-regulates L-CAM (the E-cadherin analogue in chicken) (EDELMAN et al. 1983; THIERY et al. 1984). Another example is found in the developing murine cochlea: E-cadherin in the reticular lamina is important in maintaining a barrier between fluid compartments with different composition (endolymph and perilymph). During formation of fluid spaces in the organ of Corti, E-cadherin disappears from the lateral cell membranes, which suggests that down-regulation or redistribution of E-cadherin facilitates the process of fluid space opening (WHITLON 1993).

Other physiological processes in which E-cadherin is down-regulated are trophoblast formation and liver regeneration. While the use of E-cadherin null mutant embryos has demonstrated that this molecule is necessary to form a trophectodermal epithelium (LARUE et al. 1994; RIETHMACHER et al. 1995), fusion of cytotrophoblast cells to syncytiotrophoblast later on coincides with down-regulation of E-cadherin at the transcriptional level (COUTIFARIS et al. 1991). The causal relation between these two phenomena was stressed by the demonstration that neutralizing antibodies against E-cadherin inhibit cell fusion and that fusion-incompetent choriocarcinoma cell lines do not down-regulate E-cadherin. During liver regeneration after partial hepatectomy, down-regulation at the transcriptional level was observed, when the regenerating tissue was compared with normal liver (GLUCK et al. 1992).

Down-regulation of E-cadherin has been observed in a few nonmalignant pathologies. Reduced staining of E-cadherin has been reported at the periphery of some acantholytic keratinocytes in Darier's disease and Hailey-Hailey disease (BURGE and SCHOMBERG 1992). Both skin diseases are characterized by reduced adhesion between keratinocytes. In murine polycystic kidney disease another example can be found of down-regulation of E-cadherin (ROCCO et al. 1992). The down-regulation is apparent in the epithelial cells of proximal and collecting tubules and is accompanied by low mRNA levels for E-cadherin. The gene responsible for human polycystic kidney disease has been localized on chromosome 16 (REEDERS et al. 1985).

9 Conclusions

E-cadherin is a transmembrane glycoprotein with both extracellular and intracellular interactions in epithelial cells. The extracellular binding is usually homophilic and homotypic, while cytoplasmic interaction with the cytoskeleton via catenins is imperative for functional integrity. These interactions attribute a cell-cell adhesive and an invasion-suppressive role to this molecule.

Down-regulation of E-cadherin is a common characteristic of invasive epithelial tumors. E-cadherin and the catenins are highly conserved molecules. Apart from gross deletions of chromosome 16 that involve the E-cadherin gene, mutations seem to be infrequent, except for particular tumor types. No natural splicing variants have been found so far. Factors from the host have been held responsible for frequent E-cadherin down-regulation in invasive tumors. Characterization of those putative factors and revealing their action targets is one of the goals of future research.

Impaired function in the presence of an intact E-cadherin/catenin complex points towards another level of modulation. Posttranslational modulations via proteolytic modification, glycosylation and/or phosphorylation are now under study. From a clinical point of view, up-regulating molecules of the E-cadherin function may be of great value in the future therapy of cancer. Molecules with such an up-regulating action so far include IGF-I, tangeretin, RA and tamoxifen.

Acknowledgments. Studies on tangeretin mentioned in this review were supported by the Department of Citrus of the State of Florida, the Belgian NFWO, the Vereniging voor Kankerebestrijding vzw, Brussels, the ASLK-VIVA, Brussels, the GOA of the RUG, the Effel Prize and the Sport Vereniging tegen Kanker, Brussels. Frans M. Van Roy is Research Director with the Belgian N.F.W.O. The authors thank Stefan Vermeulen, Krist'l Vennekens, Erik Bruyneel, Georges De Bruyne, Caroline Labit, Veerle Van Marck, Geert Berx, Annick Keirsebilck, Friedel Nollet, Jolanda van Hengel and Ann Vanlandschoot for helpful discussions.

References

Aberle H, Butz S, Stappert J, Weissig H, Kemler R, Hoschuetzky H (1994) Assembly of the cadherin-catenin complex in vitro with recombinant proteins. J Cell Sci 107: 3655–3663

Akiyama T, Ishida J, Nakagawa S, Ogawara H, Watanabe S, Itoh N, Shibuya M, Fukami Y (1987) Genistein, a specific inhibitor of tyrosine-specific protein kinases. J Biol Chem 262: 5592–5595

Albrechtsen R, Nielsen M, Wewer U, Engvall E, Ruoslahti E (1981) Basement membrane changes in breast cancer detected by immunohistochemical staining for laminin. Cancer Res 41: 5076–5081

Amagai M (1995) Adhesion molecules. I: Keratinocyte-keratinocyte interactions; cadherins and pemphigus. J Invest Dermatol 104: 146–152

Amagai M, Fujimori T, Masunaga T, Shimizu H, Nishikawa T, Shimizu N, Takeichi M, Hashimoto T (1995) Delayed assembly of desmosomes in keratinocytes with disrupted classic-cadherin-mediated cell adhesion by a dominant negative mutant. J Invest Dermatol 104: 27–34

Anzano MA, Byers SW, Smith JM, Peer CW, Mullen LT, Brown CC, Roberts AB, Sporn MB (1994) Prevention of breast cancer in the rat with 9-cis-retinoic acid as a single agent and in combination with tamoxifen. Cancer Res 54: 4614–4617

Ao A, Erickson RP (1992) Injection of antisense RNA specific for E-cadherin demonstrates that E-cadherin facilitates compaction, the first differentiative step of the mammalian embryo. Antisense Res Dev 2: 153–163

Babich M, Foti LR (1994) E-cadherins identified in osteoblastic cells: effects of parathyroid hormone and extracellular calcium on localization. Life Sci 54: PL201–208

Bae SN, Arand G, Azzam H, Pavasant P, Torri J, Frandsen TL, Thompson EW (1993) Molecular and cellular analysis of basement membrane invasion by human breast cancer cells in Matrigel-based in vitro assays. Breast Cancer Res Treat 24: 241–255

Balda MS, Gonzalez-Mariscal L, Matter K,Cereijido M, Anderson JM (1993) Assembly of the tight junction: the role of diacylglycerol. J Cell Biol 123: 293–302

Baltensperger K, Kozma LM, Cherniack AD, Klarlund JK, Chawla A, Banerjee U, Czech MP (1993) Binding of the ras activator son of sevenless to insulin receptor substrate-1 signaling complexes. Science 260: 1950–1952

Becker JW, Erickson HP, Hoffman S, Cunningham BA, Edelman GM (1989) Topology of cell adhesion molecules. Proc Natl Acad Sci USA 86: 1088–1092

Becker KF, Atkinson MJ, Reich U, Huang HH, Nekarda H, Siewert JR, Höfler H (1993) Exon skipping in the E-cadherin gene transcript in metastatic human gastric carcinomas. Hum Mol Genet 2: 803–804

Becker KF, Atkinson MJ, Reich U, Becker I, Nekarda H, Siewert JR, Höfler H (1994) E-cadherin gene mutations provide clues to diffuse type gastric carcinomas. Cancer Res 54: 3845–3852

Behrens J, Birchmeier W, Goodman SL, Imhof BA (1985) Dissociation of Madin-Darby canine kidney epithelial cells by the monoclonal antibody anti-arc-1: mechanistic aspects and identification of the antigen as a component related to uvomorulin. J Cell Biol 101: 1307–1315

Behrens J, Löwrick O, Klein-Hitpass L, Birchmeier W (1991a) The E-cadherin promoter:functional analysis of a G C-rich region and an epithelial cell-specific palindromic regulatory element. Proc Natl Acad Sci USA 88: 11495–11499

Behrens J, Weidner KM, Frixen UH, Schipper JH, Sachs M, Arakaki N, Daikuhara Y, Birchmeier W (1991b) The role of E-cadherin and scatter factor in tumor invasion and cell motility. EXS 59: 109–126

Behrens J, Vakaet L, Friis R, Winterhager E, Van Roy F, Mareel MM, Birchmeier W (1993) Loss of epithelial morphotype and gain of invasiveness correlates with tyrosine phosphorylation of the E-cadherin/β-catenin complex in cells transformed with a temperature-sensitive v-src gene. J Cell Biol 120: 757–766

Berndorff D, Gessner R, Kreft B, Schnoy N, Lajous-Petter AM, Loch N, Reutter W, Hortsch M, Tauber R (1994) Liver-intestine cadherin: molecular cloning and characterization of a novel Ca^{2+}-dependent cell adhesion molecule expressed in liver and intestine. J Cell Biol 125: 1353–1369

Berx G, Staes K, van Hengel J, Molemans F, Bussemakers MJG, van Bokhoven A, Van Roy F (1995) Cloning and characterization of the human invasion suppressor gene E-cadherin (CDH1). Genomics 26: 281–289

Birchmeier W, Behrens J (1994) Cadherin expression in carcinomas: role in the formation of cell junctions and the prevention of invasiveness. Biochim Biophys Acta 1198: 11–26

Böhm M, Totzeck B, Wieland I (1994a) Differences of E-cadherin expression levels and patterns in human lung cancer. Ann Hematol 68: 81–83

Böhm M, Totzeck B, Birchmeier W, Wieland I (1994b) Differences of E-cadherin expression levels and patterns in primary and metastatic human lung cancer. Clin Exp Metastasis 12: 55–62

Bortier H, De Bruyne G, Espeel M, Vakaet L (1989) Immunohistochemistry of laminin in early chicken and quail blastoderms. Anat Embryol 180: 65–69

Boukamp P, Fusenig NE (1993) "Trans-differentiation" from epidermal to mesenchymal/myogenic phenotype is associated with a drastic change in cell-cell and cell-matrix adhesion molecules. J Cell Biol 120: 981–993

Bowie GL, Caslin AW, Roland NJ, Field JK, Jones AS, Kinsella A (1993) Expression of the cell-cell adhesion molecule E-cadherin in squamous cell carcinoma of the head and neck. Clin Otolaryngol 18: 196–201

Boyer B, Dufour S, Thiery JP (1992) E-cadherin expression during the acidic FGF-induced dispersion of a rat bladder carcinoma cell line. Exp Cell Res 201: 347–357

Boyer B, Valles AM, Tucker GC, Delouvee A, Thiery JP (1993) Involvement of cell motility in tumor progression. Symp Soc Exp Biol 47: 183–195

Brabant G, Hoang-Vu C, Cetin Y, Dralle H, Scheumann G, Mölne J, Hansson G, Jansson S, Ericson LE, Nilsson M (1993) E-cadherin: a differentiation marker in thyroid malignancies. Cancer Res 53: 4987–4993

Bracke ME, Castronovo V, Van Cauwenberge RM-L, Vakaet L Jr, Strojny P, Foidart J-M, Mareel MM (1987) The anti-invasive flavonoid (+)-catechin binds to laminin and abrogates the effect of laminin on cell morphology and adhesion. Exp Cell Res 173: 193–205

Bracke ME, Vyncke BM, Van Larebeke NA, Bruyneel EA, De Bruyne GK, De Pestel GH, De Coster WJ, Espeel MF, Mareel MM (1989) The flavonoid tangeretin inhibits invasion of MO4 mouse cells into embryonic chick heart in vitro. Clin Exp Metatasis 7: 283–300

Bracke ME, Vyncke B, De Pestel G, Vakaet L Jr, Bourgois L, Van Larebeke N, Bortier H, Mareel MM (1990) Effect of flavonoids on cancer cell invasiveness. In: Das NP (ed) Plant flavonoids in biology and medicine. III. Current issues in flavonoid research. National University of Singapore Press, Singapore, pp 279–292

Bracke ME, Van Larebeke NA, Vyncke BM, Mareel MM (1991) Retinoic acid modulates both invasion and plasma membrane ruffling of MCF-7 human mammary carcinoma cells in vitro. Br J Cancer 63: 867–872

Bracke M, Romijn H, Vakaet L Jr, Vyncke B, De Mets M, Mareel M (1992) The use of spheroids in the study of invasion. In: Bjerkvig R (ed) Spheroid culture in cancer research. CRC Press, Boca Raton, pp 73–105

Bracke ME, Vyncke BM, Bruyneel EA, Vermeulen SJ, De Bruyne GK, Van Larebeke NA, Vleminckx K, Van Roy FM, Mareel MM (1993) Insulin-like growth factor I activates the invasion suppressor function of E-cadherin in MCF-7 human mammary carcinoma cells in vitro. Br J Cancer 68: 282–289

Bracke ME, Bruyneel EA, Vermeulen SJ, Vennekens K, Van Marck V, Mareel MM (1994a) Citrus flavonoid effect on tumor invasion and metastasis. Food Technol 48: 121–124

Bracke ME, Charlier C, Bruyneel EA, Labit C, Mareel MM, Castronovo V (1994b) Tamoxifen restores the E-cadherin function in human breast cancer MCF-7/6 cells and suppresses their invasive phenotype. Cancer Res 54: 4607–4609

Bracke ME, Vermeulen SJ, Bruyneel EA, Vennekens KM, De Bruyne GK, Van Roy FM, Mareel MM (1994c) The invasion suppressor function of E-cadherin in mammary epitheloid cells. In: Schmitt M,

Graeff H, Kindermann G (eds) Prospects in diagnosis and treatment of breast cancer. Elsevier Science, Amsterdam, pp 107–115 (Excerpta Medica international congress series no 1050)

Bradley RS, Cowin P, Brown AM (1993) Expression of Wnt-1 in PC12 cells results in modulation of plakoglobin and E-cadherin and increased cellular adhesion. J Cell Biol 123: 1857–1865

Brady-Kalnay SM, Boghaert ER, Zimmer S, Brackenbury R (1993) Increasing N-CAM-mediated cell-cell adhesion does not reduce invasion of RSV-transformed WC5 rat cerebellar cells. Clin Exp Metastasis 11: 313–324

Breen E, Clarke A, Steele G Jr, Mercurio AM (1993) Poorly differentiated colon carcinoma cell lines deficient in a catenin expression express high levels of surface E-cadherin but lack Ca^{2+}- dependent cell-cell adhesion. Cell Adhesion Commun 1: 239–250

Brieher WM, Gumbiner BM (1994) Regulation of C-cadherin function during activin induced morphogenesis of Xenopus animal caps. J Cell Biol 126: 519–527

Bringuier PP, Umbas R, Schaafsma HE, Karthaus HFM, Debruyne FMJ, Schalken JA (1993) Decreased E-cadherin immunoreactivity correlates with poor survival in patients with bladder tumors. Cancer Res 53: 3241–3245

Burge SM, Schomberg KH (1992) Adhesion molecules and related proteins in Darier's disease and Hailey-hailey disease. Br J Dermatol 127: 335–343

Bussemakers MJG, van Moorselaar RJA, Giroldi LA, Ichikawa T, Isaacs JT, Takeichi M, Debruyne FMJ, Schalken JA (1992) Decreased expression of E-cadherin in the progression of rat prostatic cancer. Cancer Res 52: 2916–2922

Bussemakers MJ, Giroldi LA, van Bokhoven A, Schalken JA (1994a) Transcriptional regulation of the human E-cadherin gene in human prostate cancer cell lines: characterization of the human E-cadherin gene promoter. Biochem Biophys Res Commun 203: 1284–1290

Bussemakers MJ, van Bokhoven A, Voller M, Smit FP, Schalken JA (1994b) The genes for the calcium-dependent cell adhesion molecules P- and E-cadherin are tandemly arranged in the human genome. Biochem Biophys Res Commun 203: 1291–1294

Butz S, Kemler R (1994) Distinct cadherin-catenin complexes in CA (2+)- dependent cell-cell adhesion. FEBS Lett 355: 195–200

Byers S, Amaya E, Munro S, Blaschuk O (1992) Fibroblast growth factor receptors contain a conserved HAV region common to cadherins and influenza strain A hemagglutinins: a role in protein-protein interactions? Dev Biol 152: 411–414

Byers SW, Sujarit S, Jegou B, Butz S, Hoschutzky H, Herrenknecht K, MacCalman C, Blaschuk OW (1994) Cadherins and cadherin-associated molecules in the developing and maturing rat testis. Endocrinology 134: 630–639

Cai JC, Jiang SJ, Chen ZP (1994) Relationship between E-cadherin and growth pattern of gastric cancer and degree of cell differentiation. Chung Hua Ping Li Hsueh Tsa Chih 23: 132–134

Castronovo V (1993) Laminin receptors and laminin-binding proteins during tumor invasion and metastasis. Invasion Metastasis 13: 1–30

Cepek KL, Shaw SK, Parker CM, Russell GJ, Morrow JS, Rimm DL, Brenner MB (1994) Adhesion between epithelial cells and T lymphocytes mediated by E-cadherin and the alphaEbeta 7 integrin. Nature 372: 190–193

Chastre E, Empereur S, Di Gioia Y, El Mahdani N, Mareel M, Vleminckx K, Van Roy F, Bex V, Emami S, Spandidos DA, Gespach C (1993) Neoplastic progression of human and rat intestinal cell lines after transfer of the ras and polyoma middle T oncogenes. Gastroenterology 105: 1776–1789

Chen W, Öbrink B (1991) Cell-cell contacts mediated by E-cadherin (Uvomorulin) restrict invasive behavior of L-cells. J Cell Biol 114: 319–327

Ching LM, Baguley BC (1987) Induction of natural killer cell activity by the antitumor compound flavone acetic acid (NSC 347512). Eur J Cancer Clin Oncol 23: 1047–1050

Chowdhury K, Dietrich S, Balling R, Guenet JL, Gruss P (1989) Structure, expression and chromosomal localization of zfp-1, a murine zinc finger protein gene. Nucleic Acids Res 17: 10427–10438

Clayton L, Stinchcombe SV, Johnson MH (1993) Cell surface localisation and stability of uvomorulin during early mouse development. Zygote 1: 333–344

Coopman PJ, Bracke ME, Lissitzky JC, De Bruyne GK, Van Roy FM, Foidart J-M, Mareel MM (1991) Influence of basement membrane molecules on directional migration of human breast cell lines in vitro. J Cell Sci 98: 395–401

Correct P, Fondanèche M-C, Bracke M, Burtin P (1990) The presence of plasmin receptors on three mammary carcinoma MCF-7 sublines. Int J Cancer 46: 745–750

Correc P, Zhang S, Komano O, Laurent M, Burtin P (1992) Visualization of the plasmin receptor on carcinoma cells. Int J Cancer 50: 767–771

Coutifaris C, Kao L-C, Sehdev HM, Chin U, Babalola GO, Blaschuk OW, Strauss JF III (1991) E-cadherin expression during the differentiation of human trophoblasts. Development 113: 767–777
Cowley GP, Smith MEF (1995) Modulation of E-cadherin expression and morphological phenotype in the intravascular component of adenocarcinomas. Int J Cancer 60: 325–329
Crepaldi T, Pollack AL, Prat M, Zborek A, Mostov K, Comoglio PM (1994) Targeting of the SF/HGF receptor to the basolateral domain of polarized epithelial cells. J Cell Biol 125: 313–320
Crossin KL, Chuong CM, Edelman GM (1985) Expression sequences of cell adhesion molecules. Proc Natl Acad Sci USA 82: 6942–6946
Cullen KJ, Smith HS, Hill S, Rosen N, Lippman ME (1991) Growth factor messenger RNA expression by human breast fibroblasts from benign and malignant lesions. Cancer Res 51: 4978–4985
Cunningham BA, Leutzinger Y, Gallin WJ, Sorkin BC, Edelman GM (1984) Linear organization of the liver cell adhesion molecule L-CAM. Proc Natl Acad Sci USA 81: 5787–5791
Czech W, Krutmann J, Herrenknecht K, Schopf E, Kapp A (1993) Human cell adhesion molecule uvomorulin is differentially expressed in various skin tumors. J Cutan Pathol 20: 168–172
Danø K, Andreasen PA, Grondahl-Hansen G, Kristensen P, Nielsen LS, Skriver L (1985) Plasminogen activators, tissue degradation and cancer. Adv Cancer Res 44: 139–266
Daughaday WH (1992) The historical evolution of concepts of the role of the insulin-like growth factors. In: Schofield PN (ed) The insulin-like growth factors. Oxford University Press, Oxford, pp 5–11
De Leon DD , Bakker B, Wilson DM, Hintz RL, Rosenfeld RG (1988) Demonstration of insulin-like growth factor (IGF-I and -II) receptors and binding protein in human breast cancer cell lines. Biochem Biophys Res Commun 152: 390–405
Dimanche-Boitrel MT, Genne P, Duchamp O, Chauffert B (1994a) Confluence dependent resistance (CDR) to doxorubicin and E-cadherin expression in murine mammary cells. Cancer Lett 85: 171–176
Dimanche-Boitrel MT, Vakaet L Jr, Pujuguet P, Chauffert B, Martin MS, Hammann A, Van Roy F, Mareel M, Martin F (1994b) In vivo and in vitro invasiveness of a rat colon cancer cell line maintaining E-cadherin expression. An enhancing role of tumor-associated myofibroblasts. Int J Cancer 56: 512–521
Doki Y, Shiozaki H, Tahara H, Inoue M, Oka H, Iihara K, Kadowaki T, Takeichi M, Mori T (1993) Correlation between E-cadherin expression and invasiveness in vitro in a human esophageal cancer cell line. Cancer Res 53: 3421-3426
Dorudi S, Sheffield JP, Poulsom R, Northover JMA, Hart I (1992) E-cadherin expression in colorectal cancer. Clin Exp Metastasis 10 [Suppl 1]: 39
Dorudi S, Sheffield JP, Poulsom R, Northover JMA, Hart IR (1993) E-cadherin expression in colorectal cancer. Am J Pathol 142: 981–986
D'Souza B, Taylor-Papadimitriou J (1994) Overexpression of ERBB2 in human mammary epithelial cells signals inhibition of transcription of the E-cadherin gene. Proc Natl Acad Sci USA 91: 7202–7206
Duband JL, Thiery JP (1990) Spatio-temporal distribution of the adherens junction-associated molecules vinculin and talin in the early avian embryo. Cell Differ Dev 30: 55–76
Dustin ML, Springer TA (1991) Role of lymphocyte adhesion receptors in transient interctions and cell locomotion. Annu Rev Immunol 9: 27–66
Edelman GM (1987) Epigenetic rules for expression of cell adhesion molecules during morphogenesis. Ciba Found Symp 125: 192–216
Edelman GM, Crossin KL (1991) Cell adhesion molecules: implications for a molecular histology. Annu Rev Biochem 60: 155–190
Edelman GM, Gallin WJ, Delouvee A, Cunningham BA, Thiery J-P (1983) Early epochal maps of two different cell adhesion molecules. Proc Natl Acad Sci USA 80: 4384–4388
Eichele G (1993) Retinoids in embryonic development. Ann NY Acad Sci 678: 22–36
Eistetter HR, Adolph S, Ringwald M, Simon-Chazottes D, Schuh R, Guenet JL, Kemler R (1988) Chromosomal mapping of the structural gene coding for the mouse cell adhesion molecule uvomorulin. Proc Natl Acad Sci USA 85: 3489–3493
Eskelinen S, Huotari V, Sormunen R, Palovuori R, Kok JW, Lehto VP (1992) Low intracellular pH induces redistribution of fodrin and instabilization of lateral walls in MDCK cells. J Cell Physiol 150: 122–133
Estreicher A, Mühlhauser J, Carpentier J-L, Orci L, Vassalli J-D (1990) The receptor for urokinase type plasminogen activator polarizes expression of the protease to the leading edge of migrating monocytes and promotes degradation of enzyme inhibitor complexes. J Cell Biol 111: 783–792
Fabre M, Garcia de Herreros A (1993) Phorbol ester-induced scattering of HT-29 human intestinal cancer cells is associated with down-modulation of E-cadherin. J Cell Sci 106: 513–522

Fahraeus R, Chen W, Trivedi P, Klein G, Öbrink B (1992) Decreased expression of E-cadherin and increased invasive capacity in EBV-LMP-transfected human epithelial and murine adenocarcinoma cells. Int J Cancer 52: 834–838

Fearon ER, Vogelstein B (1990) A genetic model for colorectal tumorigenesis. Cell 61: 759–767

Fenyves AM, Behrens J, Spanel-Borowski K (1993) Cultured microvascular endothelial cells (MVEC) differ in cytoskeleton, expression of cadherins and fibronectin matrix. A study under the influence of interferon-gamma. J Cell Sci 106: 879–890

Fidler I, Hart I (1982) Biological diversity in metastatic neoplasms: origins and implications. Science 217: 998–1003

Field JK (1992) Oncogenes and tumour-suppressor genes in squamous cell carcinoma of the head and neck. Eur J Cancer 28B: 67–76

Figueroa JA, Yee D (1992) The insulin-like growth factor binding proteins (IGFBPs) in human breast cancer. Breast Cancer Res Treat 22: 81–90

Fligiel SEG, Rodriguez AF, Knibbs RN, McCoy JP Jr, Varani J (1985) Characterization of laminin-stimulated adherence and motility in tumor cells. Oncology 42: 265–271

Friedlander DR, Mège RM, Cunningham BA, Edelman GM (1989) Cell sorting-out is modulated by both the specificity and amount of different cell adhesion molecules (CAMs) expressed on cell surfaces. Proc Natl Acad Sci USA 86: 7043–7047

Frixen UH, Nagamine Y (1993) Stimulation of urokinase-type plasminogen activator expression by blockage of E-cadherin-dependent cell-cell adhesion. Cancer Res 53: 3618–3623

Frixen UH, Behrens J, Sachs M, Eberle G, Voss B, Warda A, Löchner D, Birchmeier W (1991) E-cadherin-mediated cell-cell adhesion prevents invasiveness of human carcinoma cells. J Cell Biol 113: 173–185

Furukawa Y, Nakatsuru S, Nagafuchi A, Tsukita S, Muto T, Nakamura Y, Horii A (1994) Structure, expression and chromosome assignment of the human catenin (cadherin-associated protein) alpha 1 gene (CTNNA 1). Cytogenet Cell Genet 65: 74–78

Gallin WJ, Prediger EA, Edelman GM, Cunningham BA (1985) Isolation of a cDNA clone for the liver cell adhesion molecule (L-CAM). Proc Natl Acad Sci USA 82: 2809–2813

Gallin WJ, Sorkin BC, Edelman GM, Cunningham BA (1987) Sequence analysis of a cDNA clone encoding the liver cell adhesion molecule, L-CAM. Proc Natl Acad Sci USA 84: 2808–2812

Gao Y (1993) Downregulation of E-cadherin during experimental tumor formation in syngeneic mice and in human colorectal tumors. PhD thesis, Universiteit Gent

Geiger B, Ayalon O (1992) Cadherins. Annu Rev Cell Biol 8: 307–332

Giavazzi R, Garofalo A, Damia G, Garattini S, D'Incalci M (1988) Response to flavone acetic acid (NSC 347512) of primary and metastatic human colorectal carcinoma xenografts. Br J Cancer 57: 277–280

Giguere V, Ong ES, Segui P, Evans RM (1987) Identification of a receptor for the morphogen retinoic acid. Nature 330: 624–629

Ginsberg D, DeSimone D, Geiger B (1991) Expression of a novel cadherin (EP-cadherin) in unfertilized eggs and early Xenopus embryos. Development 111: 315–325

Giroldi LA, Schalken JA (1993) Decreased expression of the intercellular adhesion molecule E-cadherin in prostate cancer: biological significance and clinical implications. Cancer Metastasis Rev 12: 29–37

Gluck U, Rodriguez-Fernandez JL, Pankov R, Ben-Ze'ev A (1992) Regulation of adherens junction protein expression in growth-activated 3T3 cells and in regenerating liver. Exp Cell Res 202: 477–486

Graziani Y (1986) The effect of quercetin on $pp60^{V-Src}$ kinase activities. In: Cody V, Middleton E Jr, Harborne JB (eds) Plant flavonoids in biology and medicine: biochemical, pharmacological and structure-activity relationships. Liss, New York, pp 301–313

Grundy PE, Telzerow PE, Breslow N, Moksness J, Huff V, Paterson MC (1994) Loss of heterozygosity for chromosomes 16q and 1p in Wilms' tumors predicts an adverse outcome. Cancer Res 54: 2331–2333

Gumbiner BM (1993) Proteins associated with the cytoplasmic surface of adhesion molecules. Neuron 11: 551–564

Gumbiner B, Simons K (1986) A functional assay for proteins involved in establishing an epithelial occluding barrier: identification of a Uvomorulin-like polypeptide. J Cell Biol 102: 457–468

Gumbiner B, Stevenson B, Grimaldi A (1988) The role of the cell adhesion molecule uvomorulin in the formation and the maintenance of the epithelial junctional complex. J Cell Biol 107: 1575–1587

Hashimoto M, Niwa O, Nitta Y, Takeichi M, Yokoro K (1989) Unstable expression of E-cadherin adhesion molecules in metastatic ovarian tumor cells. Jpn J Cancer Res 80: 459–463

Hatta K, Takeichi M (1986) Expression of N-Cadherin adhesion molecules associated with early morphogenetic events in chick development. Nature 320: 447–449
Hatta M, Miyatani S, Copeland NG, Gilbert DJ, Jenkins NA, Takeichi M (1991) Genomic organization and chromosomal mapping of the mouse P-cadherin gene. Nucleic Acids Res 19: 4437–4441
Hauck W, Stanners CP (1991) Control of carcinoembryonic antigen gene family expression in a differentiating colon carcinoma cell line, Caco-2. Cancer Res 51: 3526–3533
Herrenknecht K, Ozawa M, Eckerskorn C, Lottspeich F, Lenter M, Kemler R (1991) The uvomorulin-anchorage protein alpha catenin is a vinculin homologue. Proc Natl Acad Sci USA 88: 9156–9160
Herzberg F, Wildermuth V, Wedlich D (1991) Expression of XBcad, a novel cadherin, during oogenesis and early development of Xenopus. Mech Dev 35: 33–42
Hinck L, Näthke IS, Papkoff J, Nelson WJ (1994a) Dynamics of cadherin/catenin complex formation: novel protein interactions and pathways of complex assembly. J Cell Biol 125: 1327–1340
Hinck L, Nelson WJ, Papkoff J (1994b) Wnt-1 modulates cell-cell adhesion in mammalian cells by stabilizing β-catenin binding to the cell adhesion protein cadherin. J Cell Biol 124: 729–741
Hirano S, Kimoto N, Shimoyama y, Hirohashi S, Takeichi M (1992) Identification of a Neural alpha-catenin as a key regulator of cadherin function and multicellular organisation. Cell 70: 293–301
Hodivala KJ, Watt FM (1994) Evidence that cadherins play a role in the down-regulation of integrin expression that occurs during keratinocyte terminal differentiation. J Cell Biol 124: 589–600
Hoffman AG, Burghardt RC, Tilley R, Auersperg N (1993) An in vitro model of ovarian epithelial carcinogenesis: changes in cell-cell communication and adhesion occurring during neoplastic progression. Int J Cancer 54: 828–838
Honda S (1992) Immunohistochemical study on the expression of E-cadherin in normal tissues and squamous cell carcinomas of the uterine cervix. Nippon Sanka Fujinka Gakkai Zasshi 44: 517–523
Hoschuetzky H, Aberle H, Kemler R (1994) β-catenin mediates the interaction of the cadherin-catenin complex with epidermal growth factor receptor. J Cell Biol 127: 1375–1380
Hülsken J, Birchmeier W, Behrens J (1994) E-cadherin and APC compete for the interaction with β-catenin and the cytoskeleton. J Cell Biol 127: 2061–2069
Hyafil F, Babinet C, Jacob F (1981) Cell-cell interactions in early embryogenesis: a molecular approach to the role of calcium. Cell 26: 447–454
Hynes RO, Lander AD (1992) Contact and adhesive specificities in the associations, migrations, and targeting of cells and axons. Cell 68: 303–322
Inoue M, Ogawa H, Miyata M, Shiozaki H, Tanizawa O (1992) Expression of E-cadherin in normal, benign, and malignant tissues of female genital organs. Am J Clin Pathol 98: 76–80
Ishisaki A, Oida S, Momose F, Amagasa T, Rikimaru K, Ichijo H, Sasaki S (1994) Identification and characterization of autocrine-motility-factor-like activity in oral squamous-cell-carcinoma cells. Int J Cancer 59: 783–788
Jacobs J, Kull FC Jr, Earp HS, Svoboda ME, Van Wyck JJ, Cuatrecasas P (1983) Somatomedine-C stimulates the phosphorylation of the β subunit of its own receptor. J Biol Chem 258: 9581–9584
Jaffe SH, Friedlander DR, Matsuzaki F, Crossin KL, Cunningham BA, Edelman GM (1990) Differential effects of the cytoplasmic domains of cell adhesion molecules on cell aggregation and sorting-out. Proc Natl Acad Sci USA 87: 3589–3593
Jennische E, Isgaard J, Isaksson OGP (1992) Local expression of insulin-like growth factors during tissue growth and regeneration. In: Schofield PN (ed) The insulin-like growth factors. Oxford University Press, Oxford, pp 221–239
Jensen PJ, Wheelock MJ (1992) Regulation of urokinase plasminogen activator localization in keratinocytes by calcium ion and E-cadherin. Exp Cell Res 202: 190–198
Johnson JP (1991) Cell adhesion molecules of the immunoglobulin supergene family and their role in malignant transformation and progression to metastatic disease. Cancer Metastasis Rev 10: 11–22
Johnston SRD, Haynes BP, Smith IE, Jarman M, Sacks NPM, Ebbs SR, Dowsett M (1993) Acquired tamoxifen resistance in human breast cancer and reduced intra-tumoral drug concentration. Lancet 342: 1521–1522
Jongen WMF, Fitzgerald DJ, Asamoto M, Piccoli C, Slaga TJ, Gros D, Takeichi M, Yamasaki H (1991) Regulation of connexin 43-mediated gap junctional intercellular communication by CA2+ in mouse epidermal cells is controlled by E-cadherin. J Cell Biol 114: 545–555
Kadmon G, Korvitz A, Altevogt P, Schachner M (1990) The neural cell adhesion molecule N-CAM enhances L1-dependent cell-cell interactions. J Cell Biol 110: 193–208
Kadowaki T, Koyasu S, Nishida E, Tobe K, Izumi T, Takaku F, Sakai H, Yahara I, Kasuga M (1987) Tyrosine phosphorylation of common and specific sets of cellular proteins rapidly induced by insulin, insulin-like growth factor I, and epidermal growth factor in an intact cell. J Biol Chem 262: 7342–7350

Kadowaki T, Shiozaki H, Inoue M, Tamura S, Oka H, Doki Y, Iihara K, Matsui S, Iwazawa T, Nagafuchi A, Tsukita S, Mori T (1994) E-cadherin and a-catenin expression in human esophageal cancer. Cancer Res 54: 291–296

Kanai Y, Oda T, Tsuda H, Ochiai A, Hirohashi S (1994) Point mutation of the E-cadherin gene in invasive lobular carcinoma of the breast. Jpn J Cancer Res 85: 1035–1039

Kartenbeck J, Schmelz M, Franke WW, Geiger B (1991) Endocytosis of junctional cadherins in bovine kidney epithelial (MDBK) cells cultured in low Ca^{2+} ion medium. J Cell Biol 113: 881–892

Katagiri A, Watanabe R, Tomita Y (1995) E-cadherin expression in renal cancer and its significance in metastasis and survival. Br J Cancer 71: 376–379

Katayama M, Hirai S, Kamihagi K, Nakagawa K, Yasumoto M, Kato I (1994) Soluble E-cadherin fragments increased in circulation of cancer patients. Br J Cancer 69: 580–585

Kaupmann K, Becker-Follmann J, Scherer G, Jockusch H, Starzinski-Powitz A (1992) The gene for the cell adhesion molecule M-cadherin maps to mouse chromosome 8 and human chromosome 16p24.1-qter and is near the E-cadherin (uvomorulin) locus in both species. Genomics 14: 488–490

Kawanishi J, Kato J, Sasaki K, Fujii S, Watanabe N, Niitsu Y (1995) Loss of E-cadherin-dependent cell-cell adhesion due to mutation of the β-catenin gene in a human cancer cell line, HSC-39. Mol Cell Biol 15: 1175–1181

Kemler R, Ozawa M (1989) Uvomorulin-catenin complex: cytoplasmic anchorage of a Ca^{2+}-dependent cell adhesion molecule. Bioessays 11: 88–91

Khokha R, Waterhouse P, Yagel S, Lala PK, Overall CM, Norton G, Denhardt DT (1989) Antisense RNA induced reduction in murine TIMP levels confers oncogenicity on Swiss 3T3 cells. Science 243: 947–950

Kinsella AR, Green B, Lepts GC, Hill CL, Bowie G, Taylor BA (1993) The role of the cell-cell adhesion molecule E-cadherin in large bowel tumour cell invasion and metastasis. Br J Cancer 67: 904–999

Kinsella AR, Lepts GC, Hill CL, Jones M (1994) Reduced E-cadherin expression correlates with increased invasiveness in colorectal carcinoma cell lines. Clin Exp Metastasis 12: 335–342

Knabbe C, Lippman ME, Wakefield LM, Flanders KC, Kasid A, Derynck R, Dickson RB (1987) Evidence that transforming growth factor-β is a hormonally regulated negative growth factor in human breast cancer cells. Cell 48: 417–428

Knudsen AK, Wheelock MJ (1992) Plakoglobin, or an 83-kD homologue distinct from beta-catenin, interacts with E-cadherin and N-cadherin. J Cell Biol 118: 671–679

Koga M, Sutherland RL (1987) Epidermal growth factor partially reverses the inhibitory effects of antiestrogens on T47D human breast cancer cell growth. Biochem Biophys Res Commun 146: 739–745

Kopper L, Van Hahn T, Lapis K (1982) Experimental model for liver metastasis formating using Lewis lung tumor. J Cancer Res Clin Oncol 103: 31–38

Kraus C, Liehr T, Hülsken J, Behrens J, Birchmeier W, Grzeschik KH, Ballhausen WG (1994) Localization of the human β-catenin gene (CTNNB1) to 3p21: a region implicated in tumor development. Genomics 23: 272–274

Kuzu I, Bicknell R, Fletcher CD, Gatter KC (1993) Expression of adhesion molecules on the endothelium of normal tissue vessels and vascular tumors. Lab Invest 69: 322–328

Larue L, Ohsugi M, Hirchenhain J, Kemler R (1994) E-cadherin null mutant embryos fail to form a trophectoderm epithelium. Proc Natl Acad Sci USA 91: 8263–8267

Le Bivic A, Sambuy Y, Mostov K, Rodriguez-Boulan E (1990) Vectorial targeting of an endogenous apical membrane sialoglycoprotein and uvomorulin in MDCK cells. J Cell Biol 110: 1533–1539

Lee MG, Sharrow SO, Farr AG, Singer A, Udey MC (1994) Expression of the homotypic adhesion molecule E-cadherin by immature murine thymocytes and thymic epithelial cells. J Immunol 152: 5653–5659

Lewis JE, Jensen PJ, Wheelock MJ (1994) Cadherin function is required for human keratinocytes to assemble desmosomes and stratify in response to calcium. J Invest Dermatol 102: 870–877

Liotta LA (1986) Tumor invasion and metastases-role of the extracellular matrix: Rhoads memorial award lecture. Cancer Res 46: 1–7

Liotta LA, Mandler R, Murano G, Katz DA, Gordon RK, Chiang PK, Schiffmann E (1986) Tumor cell autocrine motility factor. Proc Natl Acad Sci USA 83: 3302–3306

Lipponen P, Saarelainen E, Ji H, Aaltomaa S, Syrjanen K (1994) Expression of E-cadherin (E-CD) as related to other prognostic factors and survival in breast cancer. J Pathol 174: 101–109

Lombardo CR, Rimm DL, Koslov E, Morrow JS (1994) Human recombinant alpha-catenin binds to spectrin. Mol Biol Cell 5 [Suppl]: 47a

MacCalman CD, Brodt P, Doublet JD, Jednak R, Elhilali MM, Bazinet M, Blaschuk OW (1994a) The loss of E-cadherin mRNA transcripts in rat prostatic tumors is accompanied by increased expression of mRNA transcripts encoding fibronectin and its receptor. Clin Exp Metastasis 12: 101–107

MacCalman CD, Farookhi R, Blaschuk OW (1994b) Estradiol regulates E-cadherin mRNA levels in the surface epithelium of the mouse ovary. Clin Exp Metastasis 12: 276–282
Mader S, Leroy P, Chen JY, Chambon P (1993) Multiple parameters control the selectivity of nuclear receptors for their response elements. Selectivity and promiscuity in response element recognition by retinoic acid receptors and retinoic X receptors. J Biol Chem 268: 591–600
Maillet L, Buc-Caron MH (1985) Modulation of specific protein expression in teratocarcinoma cell aggregates by antibodies affecting cell-cell interactions. Dev Biol 111: 1–7
Mandeville R, Ghali SS, Chausseau JP (1984) In vitro stimulation of human NK activity by an estrogen antagonist (tamoxifen). Eur J Cancer Clin Oncol 20: 983–985
Mansouri A, Spurr N, Goodfellow PN, Kemler R (1988) Characterization and chromosomal localization of the gene encoding the human cell adhesion molecule uvomorulin. Differentiation 38: 67–71
Mareel M, Kint J, Meyvisch C (1979) Methods of study of the invasion of malignant C3H mouse fibroblasts into embryonic chick heart in vitro. Virchows Arch [B] Cell Pathol 30: 95–111
Mareel MM, Behrens J, Birchmeier W, De Bruyne GK, Vleminckx K, Hoogewijs A, Fiers WC Van Roy FM (1991) Down-regulation of E-cadherin expression in Madin Darby canine kidney (MDCK) cells inside nude mice tumors. Int J Cancer 47: 922–928
Mareel MM, Van Roy FM, Bracke ME (1993) How and when do tumor cells metastasize? Crit Rev Oncogenesis 4: 559–594
Martin PM, Horwitz KB, Ryan DS, McGuire WL (1978) Phytoestrogen interaction with estrogen receptors in human breast cancer cells. Endocrinology 103: 1860–1867
Matsui S, Shiozaki H, Inoue M, Tamura S, Doki Y, Kadowaki T, Iwazawa T, Shimaya K, Nagafuchi A, Tsukita S et al. (1994) Immunohistochemical evaluation of alpha-catenin expression in human gastric cancer. Virchows Arch 424: 375–381
Matsunami H, Miyatani S, Inoue T, Copeland NG, Gilbert DJ, Jenkins NA, Takeichi M (1993) Cell binding specificity of mouse R-cadherin and chromosomal mapping of the gene. J Cell Sci 106: 401–409
Matsuura K, Kawanishi J, Fujii S, Imamura M, Hirano S, Takeichi M, Niitsu Y (1992) Altered expression of E-cadherin in gastric cancer tissues and carcinomatous fluid. Br J Cancer 66: 1122–1130
Matsuyoshi N, Hamaguchi M, Tanigushi S, Nagafuchi A, Tsukita S, Takeichi M (1992) Cadherin-mediated cell-cell adhesion is perturbed by v-src tyrosine phosphorylation in metastatic fibroblasts. J Cell Biol 118: 703–714
Matsuzaki F, Mège RM, Jaffe SH, Friedlander DR, Gallin WJ, Goldberg JI, Cunningham BA, Edelman GM (1990) cDNAs of cell adhesion molecules of different specificity induce changes in cell shape and border formation in cultured S180 cells. J Cell Biol 110: 1239–1252
Mattijssen V, Peters HM, Schalkwijk L, Manni JJ, van't Hof Grootenboer B, de Mulder PH, Ruiter DJ (1993) E-cadherin expression in head and neck squamous cell carcinoma is associated with clinical outcome. Int J Cancer 55: 580–585
Mayer B, Johnson JP, Leitl F, Jauch KW, Heiss MM, Schildberg FW, Birchmeier W, Funke I (1993) E-cadherin expression in primary and metastatic gastric cancer: down-regulation correlates with cellular dedifferentiation and glandular disintegration. Cancer Res 53: 1690–1695
McCrea PD, Gumbiner BM (1991) Purification of a 92-kDa cytoplasmic protein tightly associated with the cell-cell adhesion molecule E-cadherin (uvomorulin) Characterization and extractability of the protein complex from the cell cytostructure. J Biol Chem 266: 4514–4520
McCrea PD, Turck CW, Gumbiner B (1991) A homolog of the armadillo protein in Drosophila (plakoglobin) associated with E-cadherin. Science 254: 1359–1361
McNeill H, Ozawa M, Kemler R, Nelson WJ (1990) Novel function of the cell adhesion molecule uvomorulin as an inducer of cell surface polarity. Cell 62: 309–316
McNeill H, Ryan TA, Smith SJ, Nelson WJ (1993) Spatial and temporal dissection of immediate and early events following cadherin-mediated epithelial cell adhesion. J Cell Biol 120: 1217–1226
Mecham RP (1991) Laminin receptors. Annu Rev Cell Biol 7: 71–91
Millon R, Nicora F, Muller D, Eber M, Klein-Soyer C, Abecassis J (1989) Modulation of human breast cancer cell adhesion by estrogens and antiestrogens. Clin Exp Metastasis 7: 405–415
Miyata M, Shiozaki H, Kobayashi K, Yano H, Tamura S, Tahara H, Mori T (1990) Correlation between expression of E-cadherin and metastases in human esophageal cancer: preliminary report. Nippon Geka Gakkai Zasshi 91: 1761
Miyatani S, Copeland NG, Gilbert DJ, Jenkins NA, Takeichi M (1992) Genomic structure and chromosomal mapping of the mouse N-cadherin gene. Proc Natl Acad Sci USA 89: 8443–8447
Moll R, Mitze M, Frixen UH, Brichmeier W (1993) Differential loss of E-cadherin expression in infiltrating ductal and lobular breast carcinomas. Am J Pathol 143: 1731–1742

Moller CJ, Christgau S, Williamson MR, Madsen OD, Niu ZP, Bock E, Baekkeskov S (1992) Differential expression of neural cell adhesion molecule and cadherins in pancreatic islets, glucagonomas, and insulinomas. Mol Endocrinol 6: 1332–1342
Moon RT, DeMarais A, Olson DJ (1993) Responses to Wnt signals in vertebrate embroys may involve changes in cell adhesion and cell movement. J Cell Sci [Suppl] 17: 183–188
Mori Y, Mizuuchi H, Sato K, Okamura N, Kudo R (1994) The factors involved in invasive ability of endometrial carcinoma cells. Nippon Sanka Fujinka Gakkai Zasshi 46: 509–516
Morton RA, Ewing CM, Nagafuchi A, Tsukita S, Isaacs WB (1993) Reduction of E-cadherin levels and deletion of the alpha-catenin gene in human prostate cancer cells. Cancer Res 53: 3585–3590
Mottonen J, Strand A, Symersky J, Sweet RM, Danley DE, Geoghegan KF, Gerard RD, Goldsmith EJ (1992) Structural basis of latency in plasminogen activator inhibitor-1. Nature 355: 270–273
Mountjoy KG, Finlay GJ, Holdaway IM (1987) Abnormal insulin-receptor down regulation and dissociation of down regulation from insulin biological action in cultured human tumor cells. Cancer Res 47: 6500–6504
Moxham CP, Duronio V, Jacob S (1989) Insulin-like growth factor I receptor beta-subunit heterogeneity. Evidence for hybrid tetramers composed of insulin-like growth factor I and insulin receptor heterodimers. J Biol Chem 264: 13238–13244
Murphy G, Reynolds JJ, Hembry RM (1989) Metalloproteinases and cancer invasion and metastasis. Int J Cancer 44: 757–760
Musil LS, Cunningham BA, Edelman GM, Goodenough DA (1990) Differential phosphorylation of the gap junction protein connexin43 in junctional communicattion-competent and -deficient cell lines. J Cell Biol 111: 2077–2088
Nagafuchi A, Takeichi M (1988) Cell binding function of E-cadherin is regulated by the cytoplasmic domain. EMBO J 7: 3679–3684
Nagafuchi A, Takeichi M (1990) Transmembrane control of cadherin-mediated cell adhesion: a94 kDa protein functionally associated with a specific region of the cytoplasmic region of E-cadherin. Cell Regul 1: 37–44
Nagafuchi A, Tsukita S (1994) The loss of the expression of alpha catenin, the 102 kD cadherin associated protein, in central nervous tissues during development. Dev Growth Differ 36: 59–71
Naldini L, Weidner KM, Vigna E, Gaudino G, Bardelli A, Ponzetta C, Narsimhan RP, Hartmann G, Zarnegar R, Michalopoulos GK, Birchmeier W, Comoglio PM (1991) Scatter factor and hepatocyte growth factor are indistinguishable ligands for the MET receptor. EMBO J 10: 2867–2878
Napolitano EW, Venstrom K, Wheeler EF, Reichardt LF (1991) Molecular cloning and characterization of B-cadherin, a novel chick cadherin. J Cell Biol 113: 893–905
Näthke IS, Hinck L, Swedlow JR, Papkoff J, Nelson WJ (1994) Defining interactions and distributions of cadherin and catenin complexes in polarized epithelial cells. J Cell Biol 125: 1341–1352
Natt E, Magenis RE, Zimmer J, Mansouri A, Scherer G (1989) Regional assignment of the human loci for uvomorulin (UVO) and chymotrypsinogen B (CTRB) with the help of two overlapping deletions on the long arm of chromosome 16. Cytogenet Cell Genet 50: 145–148
Navarro P, Lozano E, Cano A (1993) Expression of E- or P-cadherin is not sufficient to modify the morphology and the tumorigenic behavior of murine spindle carcinoma cells. Possible involvement of plakoglobin. J Cell Sci 105: 923–934
Nelson WJ, Shore EM, Wang AZ, Hammerton RW (1990) Identification of a membrane-cytoskeletal complex containing the cell adhesion molecule uvomorulin (E-cadherin), ankyrin, and fodrin in Madin-Darby Canine Kidney epithelial cells. J Cell Biol 110: 349–357
Nelson WJ, Hammerton RW, McNeill H (1991) Role of the membrane-cytoskeleton in the spatial organization of the Na, K-ATPase in polarized epithelial cells. Soc Gen Physiol Ser 46: 77–87
Nicolson GL, Birdwell CR, Brunson KW, Robbins JC, Beattie G, Fidler IJ (1977) Cell interactions in the metastatic process: some cell surface properties associated with successful blood-borne tumor spread. In: Lash JW, Burger MM (eds) Cell and tissue interactions. Raven, New York, pp 225–241
Nicholson LJ, Pei XF, Watt FM (1991) Expression of E-cadherin, P-cadherin and involucrin by normal and neoplastic keratinocytes in culture. Carcinogenesis 12: 1345–1349
Nieto MA, Sargent MG, Wilkinson DG, Cooke J (1994) control of cell behavior during vertebrate development by Slug, a zinc finger gene. Science 264: 835–839
Nigam AK, Savage FJ, Boulos PB, Stamp GW, Liu D, Pignatelli M (1993) Loss of cell-cell and cell-matrix adhesion molecules in colorectal cancer. Br J Cancer 68: 507–514
Nollet F, Van Hengel J, Berx G, Molemans F, Van Roy F (1995) Isolation and characterization of a human pseudogene (CTNNAP1) for alphaE-catenin (CTNNA1): assignment of the pseudogene to 5q22 and the alphaE-catenin gene to 5q31. Genomics 26: 410–413

Nose A, Takeichi M (1986) A novel cadherin cell adhesion molecule: its expression patterns associated with implantation and organogenesis of mouse embroys. J Cell Biol 103: 2649–2658
Nose A, Nagafuchi A, Takeichi M (1988) Expressed recombinant cadherins mediate cell sorting in model systems. Cell 54: 993–1001
Öbrink B (1991) C-CAM (Cell-CAM 105) – a member of the growing immunoglobulin superfamily of cell adhesion proteins. Bioessays 13: 227–234
Ochiai A, Akimoto S, Kanai Y, Shibata T, Oyama T, Hirohashi S (1994a) c-erbB-2 gene product associates with catenins in human cancer cells. Biochem Biophys Res Commun 205: 73–78
Ochiai A, Akimoto S, Shimoyama Y, Nagafuchi A, Shoichiro T, Hirohashi S (1994b) Frequent loss of alpha catenin expression in scirrhous carcinomas with scattered cell growth. Jpn J Cancer Res 85: 266–273
Oda T, Kanai Y, Shimoyama Y, Nagafuchi A, Tsukita S, Hirohashi S (1993) Cloning of the human alpha-catenin cDNA and its aberrant mRNA in a human cancer cell line. Biochem Biophys Res Commu 193: 897–904
Oda T, Kanai Y, Oyama T, Yoshiura K, Shimoyama Y, Birchmeier W, Sugimura T, Hirohashi S (1994) E-cadherin gene mutations in human gastric carcinoma cell lines. Proc Natl Acad Sci USA 91: 1858–1862
Ogasawara M, Karey KP, Sirbasku DA (1988) Insulin-like growth factor I: relationship between plasma and tissue sources, biological potency and N-alpha truncation. Proc AACR 29: A207
Oka H, Shiozoaki H, Kobayashi K, Tahara H, Tamura S, Yamazaki K, Mori T (1990) Immunoreactive expression of E-cadherin in human gastric cancer: preliminary report. Nippon Geka Gakkai Zasshi 91: 1814
Oka H, Shiozaki H, Kobayashi K, Tahara H, Kobayashi t, Takatsuka Y, Mori T (1992a) Correlation between E-cadherin expression and metastasis in human breast cancer: preliminary report. Nippon Geka Gakkai Zasshi 93: 105
Oka H, Shiozaki H, Kobayashi K, Tahara H, Tamura S, Miyata M, Doki Y, Iihara K, Matsuyoshi N, Hirano S, Takeichi M, Mori T (1992b) Immunohistochemical evaluation of E-cadherin adhesion molecule expression in human gastric cancer. Virchows Arch [A] Pathol Anat 421: 149–156
Okazaki M, Takeshita S, Kawai S, Kikuno R, Tsujimura A, Kudo A, Amann E (1994) Molecular cloning and characterization of OB-cadherin, a new member of cadherin family expressed in osteoblasts. J Biol Chem 269: 12092–12098
Olson DJ, Christian JL, Moon RT (1991) Effect of Wnt-1 and related proteins in gap junctional communication in Xenopus embroys. Science 252: 1173–1176
Otto T, Rembrink K, Goepel M, Meyer-Schwickerath M, Rubben H (1993) E-cadherin: a marker for differentiation and invasiveness in prostatic carcinoma. Urol Res 21: 359–362
Otto T, Birchmeier W, Schmidt U, Hinke A, Schipper J, Rübben H, Raz A (1994) Inverse relation of E-cadherin and autocrine motility factor receptor expression as a prognostic factor in patients with bladder carcinomas. Cancer Res 54: 3120–3123
Overduin M, Harvey TS, Bagby S, Tong KI, Yau P, Takeichi M, Ikura M (1995) Solution structure of the epithelial cadherin domain responsible for selective cell adhesion. Science 267: 386–389
Oyama T, Kanai Y, Ochiai A, Akimoto S, Oda T, Yanagihara K, Nagafuchi A, Tsukita S, Shibamoto S, Ito F, Takeichi M, Matsuda H, Hirohashi S (1994) A truncated beta-catenin disrupts the interaction between E-cadherin and alpha-catenin: a cause of loss of intercellular adhesiveness in human cancer cell lines. Cancer Res 54: 6282–6287
Ozawa M, Kemler R (1990) Correct proteolytic cleavage is required for the cell adhesive function of uvomorulin. J Cell Biol 111: 1645–1650
Ozawa M, Baribault H, Kemler R (1989) The cytoplasmic domain of the cell adhesion molecule uvomorulin associates with three independent proteins structurally related in different species. EMBO J8: 1711–1717
Ozawa M, Engel J, Kemler R (1990a) Single amino acid substitutions in one Ca^{2+} binding site of uvomorulin abolish the adhesive function. Cell 63: 1033–1038
Ozawa M, Hoschutzky H, Herrenknecht K, Kemler R (1990b) A possible new adhesive site in the cell-adhesion molecule uvomorulin. Mech Dev 33: 49–56
Ozawa M, Ringwald M, Kemler R (1990c) Uvomorulin-catenin complex formation is regulated by a specific domain in the cytoplasmic region of the cell adhesion molecule. Proc Natl Acad Sci USA 87: 4246–4250
Paik S (1992) Expression of IGF-I and IGF-II mRNA in breast tissue. Breast Cancer Res Treat 22: 31–38
Palka J, Peterkofsky B (1988) Salt stimulation of serum insulin-like growth factor binding protein activity. Anal Biochem 175: 442–449

Pandolfi F, Trentin L, Boyle LA, Stamenkovic I, Byers HR, Colvin RB, Kurnick JT (1992) Expression of cell adhesion molecules in human melanoma cell lines and their role in cytotoxicity mediated by tumor-infiltrating lymphocytes. Cancer 69: 1165–1173

Papa V, Gliozzo B, Clark GM, McGuire WL, Moore D, Fujita-Yamaguchi Y, Vigneri R, Goldfine ID, Pezzino V (1993) Insulin-like growth factor-I factors are overexpressed and predict a low risk in human breast cancer. Cancer Res 53: 3736–3740

Parmer VS, Jain R, Sharma SK, Vardhan A, Taeja P, Singh S, Vyncke BM, Bracke ME, Mareel MM (1994) Anti-invasive activity of 3,7-dimethoxyflavone in vitro. J Pharm Sci 83: 1217–1221

Pasdar M, Li Z (1993) Disorganization of microfilaments and intermediate filaments interfers with the assembly and stability of desmosomes in MDCK epithelial cells. Cell Motil Cytoskeleton 26: 163–180

Peifer M, McCrea PD, Green KJ, Wieschaus E, Gumbiner B (1992) The vertebrate adhesive junction proteins β-catenin and plakoglobin and the Drosophila segment polarity gene armadillo form a multigene family with similar properties. J Cell Biol 118: 681–691

Peifer M, Sweeton D, Casey M, Wieschaus E (1994) Wingless signal and Zeste-white 3 kinase trigger opposing changes in the intracellular distribution of Armadillo. Development 120: 369–380

Peyrat JP, Bonneterre J (1992) Type I IGF receptor in human breast diseases. Breast Cancer Res Treat 22: 59–67

Piepenhagen PA, Nelson WJ (!993) Defining E-cadherin-associated protein complexes in epithelial cells: plakoglobin, beta- and gamma-catenin are distinct components. J Cell Sci 104: 751–762

Pignatelli M, Liu D, Nasim MM, Stamp GWH, Hirano S, Takeichi M (1992) Morphoregulatory activities of E-cadherin and beta-1 integrins in colorectal tumour cells. Br J Cancer 66: 629–634

Pizarro A, Benito N, Navarro P, Palacios J, Cano A, Quintanilla M, Contreras F, Gamallo C (1994) E-cadherin expression in basal cell carcinoma. Br J Cancer 69: 157–162

Plessers L, Bosmans E, Cox A, Op de Beek L, Vandepitte J, Vanvuchelen J, Raus J (1990) Production of immunohistochemical reacticity of mouse anti-epithelial monoclonal antibodies raised against human breast cancer cells. Anticancer Res 10: 271–277

Ploug M, Rønne ER, Behrendt N, Jensen AL, Blasi F, Danø K (1991) Cellular receptor for urokinase plasminogen activator Carboxyl-terminal processing and membrane anchoring by glycosyl-phosphatidylinositol. J Biol Chem 266: 1926–1933

Pokutta S, Herrenknecht K, Kemler R, Engel J (1994) Conformational changes of the recombinant extracellular domain of E-cadherin upon calcium binding. Eur J Biochem 223: 1019–1026

Rasbridge SA, Gillett CE, Sampson SA, Walsh FS, Millis RR (1993) Epithelial (E-) and placental (P-) cadherin adhesion molecule expression in breast carcinoma. J Pathol 169: 245–250

Rebut-Bonneton C, Boutemy-Roulier S, Evain-Brion D (1993) Modulation of $pp60^{c\text{-}src}$ activity and cellular localization during differentiation of human trophoblast cells in culture. J Cell Sci 105: 629–636

Reeders ST, Breuning MH, Davies KE, Nicholls RD, Jarman AP, Higgs DR, Pearson PL, Weatherall DJ (1985) a highly polymorphic DNA marker linked to adult polycystic kidney disease on chromosome 16. Nature 317: 542–544

Reima I, Lehtonen E, Virtanen I, Flechon JE (1993) The cytoskeleton and associated proteins during cleavage, compaction and blastocyst differentiation in the pig. Differentiation 54: 35–45

Reinsch S, Karsenti E (1994) Orientation of spindle axis and distribution of plasma membrane proteins during cell division in polarized MDCKII cells. J Cell Biol 126: 1509–1526

Reynolds AB, Daniel J, McCrea PD, Wheelock MJ, Wu J, Zhang Z (1994) Identification of a new catenin: the tyrosine kinase substrate p120cas associates with E-cadherin complexes. Mol Cell Biol 14: 8333–8342

Riethmacher D, Brinkmann V, Birchmeier C (1995) A targeted mutation in the mouse E-cadherin gene results in defective preimplantation development. Proc Natl Acad Sci USA 92: 855–859

Rimm DL, Morrow JS (1994) Molecular cloning of human E-cadherin suggests a novel subdivision of the cadherin superfamily. Biochem Biophys Res Commun 200: 1754–1761

Ringwald M, Schuh R, Vestweber D, Eistetter H, Lottspeich F, Engel J, Dolz R, Jahnig F, Epplen J, Mayer S et al. (1987) The structure of cell adhesion molecule uvomorulin. Insights into the molecular mechanism of ca2+-dependent cell adhesion. EMBO J 6: 3647–3653

Ringwald M, Baribault H, Schmidt C, Kemler R (1991) The structure of the gene coding for the mouse cell adhesion molecule uvomorulin. Nucleic Acids Res 19: 6533–6539

Risinger JI, Berchuck A, Kohler MF, Boyd J (1994) Mutations of the E-cadherin gene in human gynecologic cancers. Nature Genet 7: 98–102

Rocco MV, Neilson EG, Hoyer JR, Ziyadeh FH (1992) Attenuated expression of epithelial cell adhesion molecules in murine polycystic kidney disease. Am J Physiol 262: F679–F686

Rochefort H, Augereau P, Briozzo P, Capony F, Cavailles V, Freiss G, Garcia M, Maudelonde T, Morisset M, Vignon F (1988) Structure, function, regulation and clinical significance of the 52K pro-cathepsin D secreted by breast cancer cells. Biochimie 70: 943–949

Rubinfeld B, Souza B, Albert I, Muller O, Chamberlain SH, Masiarz FR, Munemitsu S, Polakis P (1993) Association of the APC gene product with beta-catenin. Science 262: 1731–1734

Sacchi TB, Bani D, Brandi ML, Falchetti A, Bigazzi M (1994) Relaxin influences growth, differentiation and cell-cell adhesion of human breast-cancer cells in culture. Int J Cancer 57: 129–134

Sakuragi N, Nishiya M, Ikeda K, Ohkouch T, Furth EE, Hareyama H, Satoh C, Fujimoto S (1994) Decreased E-cadherin expression in endometrial carcinoma is associated with tumor dedifferentiation and deep myometrial invasion. Gynecol Oncol 53: 183–189

Sara VR (1992) The role of the insulin-like growth factors in the nervous system. In: Schofield PN (ed) The insulin-like growth factors. Oxford University Press, Oxford, pp 258–279

Sara VR, Carlsson-Skwirut C, Andersson C, Hall E, Sjogren B, Holmgren A, Jornvall H (1986) Characterization of somatomedins from human fetal brain: identification of a variant form of insulin-like growth factor I. Proc Natl Acad Sci USA 83: 4904–4907

Sato H, Takino T, Okada Y, Cao J, Shingawa A, Yamamoto E, Seiki M (1994) A matrix metalloproteinase expressed on the surface of invasive tumor cells. Nature 370: 61–65

Scherer G, Bausch E, Gaa A, von Deimling O (1989) Gene mapping on mouse chromosome 8 by interspecific crosses: new data on a linkage group conserved on human chromosome 16q. Genomics 5: 275–282

Schipper JH, Frixen UH, Behrens J, Unger A, Jahnke K, Birchmeier W (1991) E-cadherin expression in squamous cell carcinomas of head and neck: inverse correlation with tumor dedifferentiation and lymph node metastasis. Cancer Res 51: 6328–6337

Shapiro L, Fannon AM, Kwong PD, Thompson A, Lehmann MS, Grübel G, Legrand JF, Als-Nielsen J, Colman DR, Hendrickson WA (1995) Structural basis of cell-cell adhesion by cadherins. Nature 374: 327–337

Shibamoto S, Hayakawa M, Takeuchi K, Hori T, Oku N, Miyazawa K, Kitamura N, Takeichi M, I to F (1994) Tyrosine phosphorylation of β-catenin and plakoglobin enhanced by hepatocyte growth factor and epidermal growth factor in human carcinoma cells. Cell Adhesion Commun 1: 295–305

Shibata T, Gotoh M, Ochiai A, Hirohashi S (1994) Association of plakoglobin with APC, a tumor suppressor gene product, and its regulation by tyrosine phosphorylation. Biochem Biophys Res Commun 203: 519–522

Shimamura K, Takeichi M (1992) Local and transient expression of E-cadherin involved in mouse embryonic brain morphogenesis. Development 116: 1011–1019

Shimamura K, Takahashi T, Takeichi M (1992) E-cadherin expression in a particular subset of sensory neurons. Dev Biol 152: 242–254

Shimoyama Y, Hirohashi S (1991a) Cadherin intercellular adhesion molecule in hepatocellular carcinomas: loss of E-cadherin expression in an undifferentiated carcinoma. Cancer Lett 57: 131–135

Shimoyama Y, Hirohashi S (1991b) Expression of E- and P-cadherin in gastric carcinomas. Cancer Res 51: 2185–2192

Shimoyama Y, Hirohashi S, Hirano S, Nogushi M, Shimosato Y, Takeichi M, Abe O (1989) Cadherin cell-adhesion molecules in human epithelial tissues and carcinomas. Cancer Res 49: 2128–2133

Shimoyama Y, Nagafuchi A, Fujita S, Gotoh M, Takeichi M, Tsukita S, Hirohashi S (1992) Cadherin dysfunction in a human cancer cell line: possible involvement of loss of alpha-catenin expression in reduced cell-cell adhesiveness. Cancer Res 52: 5770–5774

Shino Y, Nakatani K, Watanabe A, Sawada H, Okumura T, Yamada Y, Nakano H (1992) Expression of E-cadherin in human gastric carcinomas. Gastroenterol Jpn 26: 231

Shino Y, Nakatani K, Watanabe A, Okumura T, Yamada Y, Yamada Y, Yano T, Nakano H (1991) Expression of E-cadherin in poorly differentiated carcinomas of human stomach: preliminary report. Nippon Geka Gakkai Zasshi 93: 217

Shiozaki H, Tahara H, Oka H, Miyata M, Kobayashi K, Tamura S, Iihara K, Doki Y, Hirano S, Takeichi M, Mori T (1991) Expression of immunoreactive E-cadherin adhesion molecules in human cancers. Am J Pathol 139: 17–23

Shiozaki H, Doki Y, Oka H, Iihara K, Miyata M, Kadowaki T, Matsui S, Tamura S, Inoue M, Mori T (1993) E-cadherin expression and cancer invasion and metastasis. Hum Cell 6: 94–99

Shiozaki H, Iihara K, Oka H, Kadowaki T, Matsui S, Gofuku J, Inoue M, Nagafuchi A, Tsukita S, Mori T (1994) Immunohistochemical detection of alpha-catenin expression in human cancers. Am J Pathol 144: 667–674

Shiozaki H, Kadowaki T, Doki Y, Inoue M, Tamura S, Oka H, Iwazawa T, Matsui S, Shimaya K, Takeichi M, Mori T (1995) Effect of epidermal growth factor on cadherin-mediated adhesion in a human oesophageal cancer cell line. Br J Cancer 71: 250–258

Shirayoshi Y, Nose A, Iwasaki K, Takeichi M (1986) N–linked oligosaccharides are not involved in the function of a cell-cell binding glycoprotein E-cadherin. Cell Struct Funct 11: 245–252

Shore EM, Nelson WJ (1991) Biosynthesis of the cell adhesion molecule uvomorulin (E-cadherin) in Madin-Darby Canine Kidney epithelial cells. J Biol Chem 266: 19672–19680

Skolnik EY, Batzer A, Li N, Lee C-H, Lowenstein E, Mohammadi M, Margolis B, Schlessinger J (1993) The function of GRB2 in linking the insulin receptor to ras signaling pathways. Science 260: 1953–1955

Sloane BF, Honn KV (1984) Cysteine proteinases and metastasis. Cancer Metastasis Rev 3: 249–263

Sommers CL, Thompson EW, Torri JA, Kemler R, Gelmann EP, Beyers SW (1991) Cell adhesion molecule uvomorulin expression in human breast cancer cell lines: relationship to morphology and invasive capacities. Cell Growth Differ 2: 365–372

Sommers CL, Gelmann EP, Kemler R, Cowin P, Byers SW (1994) Alterations in β-catenin phosphorylation and plakoglobin expression in human breast cancer cells. Cancer Res 54: 3544–3552

Sorkin BC, Hemperly JJ, Edelman GM, Cunningham BA (1988) Structure of the gene for the liver cell adhesion molecule, L-CAM. Proc Natl Acad Sci USA 85: 7617–7621

Sorkin BC, Jones FS, Cunningham BA, Edelman GM (1993) Identification of the promoter and a transcriptional enhancer of the gene encoding L-CAM, a calcium-dependent cell adhesion molecule. Proc Natl Acad Sci USA 90: 11356–11360

Sorscher SM, Russack V, Cagle M, Faramisco JR, Green MR (1995) Immunolocalization of E-cadherin in human head and neck cancer. Arch Pathol Lab Med 119: 82–84

Soule HD, Vazquez J, Long A, Albert S, Brennan M (1973) A human cell line from a pleural effusion derived from a breast carcinoma. J Natl Cancer Inst 51: 1409–1416

Stappert J, Kemler R (1994) A short core region of E-cadherin is essential for catenin binding and is highly phosphorylated. Cell Adhesion Commun (in press)

Stoker M, Gherardi E (1991) Regulation of cell movement: the motogenic cytokines. Biochim Biophys Acta 1072: 81–102

Stoker M, Gherardi E, Perryman M, Gray J (1987) Scatter factor is a fibroblast-derived modulator of epithelial cell mobility. Nature 327: 239–242

Su L-K, Vogelstein B, Kinzler KW (1993) Association of the APC tumor suppressor protein with catenins. Science 262: 1734–1737

Suarez-Quian CA, Byers SW (1993) Redistribution of epidermal growth factor receptor as a function of cell density, cell-cell adhesion and calcium in human (A-431) cells. Tissue Cell 25: 1–17

Suzuki S, Sano K, Tanihara H (1991) Diversity of the cadherin family: evidence for eight new cadherins in nervous tissue. Cell Regul 2: 261–270

Takeichi M (1977) Functional correlation between cell adhesive properties and some cell surface proteins. J Cell Biol 75: 464–474

Tang A, Amagai M, Granger LG, Stanley JR, Udey MC (1993) Adhesion of epidermal Langerhans cell to keratinocytes mediated by E-cadherin. Nature 361: 82–85

Tang A, Eller MS, Hara M, Yaar M, Hirohashi S, Gilchrest BA (1994) E-cadherin is the major mediator of human melanocyte adhesion to keratinocytes in vitro. J Cell Sci 107: 983–992

Tanihara H, Kido M, Obata S, Heimark RL, Davidson M, St John T, Suzuki S (1994) Characterization of cadherin-4 and cadherin-5 reveals new aspects of cadherins. J Cell Sci 107: 1697–1704

Terpe HJ, Tajrobehkar K, Gunthert U, Almannsberger M (1993) Expression of cell adhesion molecules alpha-2, alpha-5 and alpha-6 integrin, E-cadherin, N-CAM and CD-44 in renal cell carcinomas. An immunohistochemical study. Virchows Arch [A] Pathol Anat Histopathol 422: 219–224

Thiery P, Delouvee A, Gallin WJ, Cunningham BA, Edelman GM (1984) Ontogenetic expression of cell adhesion molecules: L-CAM is found in epithelia derived from the three primary germ layers. Dev Biol 102: 61–78

Thompson EW, Reich R, Shima TB, Albini A, Graf J, Martin GR, Dickson RB, Lippman ME (1988) Differential regulation of growth and invasiveness of MCF-7 breast cancer cells by antiestrogens. Cancer Res 48: 6764–6768

Thompson EW, Torri J, Sabol M, Sommers CL, Byers S, Valverius EM, Martin GR, Lippman ME, Stampfer MR, Dickson RB (1994) Oncogene-induced basement membrane invasiveness in human mammary epithelial cells. Clin Exp Metastasis 12: 181–194

Tohma Y, Yamashima T, Yamashita J (1992) Immunohistochemical localization of cell adhesion molecule epithelial cadherin in human arachnoid villi and meningiomas. Cancer Res 52: 1981–1987

Turpeenniemi-Hujanen T, Thorgeirsson UP, Rao CN, Liotta LA (1986) Laminin increases the release of type IV collagenase from malignant cells. J Biol Chem 261: 1883–1889

Uchida N, Shimamura K, Miyatani S, Copeland NG, Gilbert DJ, Jenkins NA, Takeichi M (1994) Mouse alphaN-catenin: two isoforms, specific expression in the nervous system, and chromosomal localization of the gene. Dev Biol 163: 75–85

Umbas R, Schalken , Aalders TW, Carter BS, Karthaus HFM, Schaafsma HE, Debruyne FMJ, Isaacs WB (1992) Expression of the cellular adhesion molecule E-cadherin is reduced or absent in high-grade prostate cancer. Cancer Res 52: 5104–5109

Umbas R, Isaacs WB, Bringuier PP, Schaafsma HE, Karthaus HFM, Oosterhof GON, Debruyne FMJ, Schalken JA (1994) Decreased E-cadherin expression is associated with poor prognosis in patients with prostate cancer. Cancer Res 54: 3929–3933

Van Aken J, Cuvelier CA, De Wever N, Gao Y, Mareel MM, Roels HJ (1993) Immunohistochemical analysis of E-cadherin expression in human colorectal tumours. Pathol Res Pract 189: 975–978

Van der Wurff AAM, Ten Kate J, Van der Linden EPM, Dinjens WNM, Arends J-W, Bosman FT (1992) L-Cam expression in normal, premalignant, and malignant colon mucosa. J Pathol 168: 287–291

Vandenbossche GMR, De Bruyne GK, Bruyneel EA, Clemminck G, Vleminckx K, Van Roy FM, Mareel MM (1994) Microencapsulation of MDCK-ras-e cells prevents loss of the E-cadherin invasion suppressor function in vivo. Int J Cancer 57: 73–80

Van Hengel J, Nollet F, Berx G, Van Roy N, Speleman F, Van Roy F (1995) Assignment of the human β-catenin gene (CTNNB1) to 3p22—> p21.3 by fluorescence in situ hybridization. cytogenet Cell Genet (in press)

Veatch AL, Carson LF, Ramakrishnan S (1994) Differential expression of the cell-cell adhesion molecule E-cadherin in ascites and solid human ovarian tumor cell. Int J Cancer 58: 393–399

Vestweber D, Kemler R (1985) Identification of a putatiave cell adhesion domain of uvomorulin. EMBO J 4: 3393–3398

Vet JA, Debruyne FM, Schalken JA (1994) Molecular prognostic factor in bladder cancer. World J Urol 12: 84–88

Vleminckx K, Vakaet L Jr, Mareel M, Fiers W, Van Roy F (1991) Genetic manipulation of E-cadherin expression by epithelial tumor cells reveals an invasion suppressor role. Cell 66: 107–119

Vleminckx KL, Deman JJ, Bruyneel EA, Vandenbossche GMR, Keirseblick AA, Mareel MM, Van Roy FM (1994) Enlarged cell-associated proteoglycans abolish E-cadherin functionally in invasive tumor cells. Cancer Res 54: 873–877

Vlodavsky I, Svahn CM, Mohsen M, Ishai-Michaeli R, Eldor A, Fuks Z (1991) Heparan sulfate degradation in tumor cell invasion and neovascularization. TIGG 3: 82–90

Volk T, Cohen O, Geiger B (1987) Formation of heterotypic adherens-type junctions between L-CAM-containing liver cells and A-CAM containing lens cells. Cell 50: 987–994

Wagner HE, Toth CA, Steele GD Jr, Thomas P (1992) Metastatic potential of human colon cancer cell lines: relationship to cellular differentiation and carcinoembryonic antigen production. Clin Exp Metastasis 10: 25–31

Walsh FS, Barton CH, Putt W, Moore SE, Kelsell D, Spurr N, Goodfellow PN (1990) N-cadherin gene maps to human chromosome 18 and is not linked to the E-cadherin gene. J Neurochem 55: 805–812

Wang AZ, Ojakian GK, Nelson WJ (1990) Steps in the morphogenesis of a polarized epithelium. I. Uncoupling the roles of cell-cell and cell-substratum contact in establishing plasma membrane polarity in multicellular epithelial (MDCK) cysts. J Cell Sci 95: 137–151

Watabe M, Nagafuchi A, Tsukita S, Takeichi M (1994) Induction of polarized cell-cell association and retardation of growth by activation of the E-cadherin-catenin adhesion system in a dispersed carcinoma line. J Cell Biol 127: 247–256

Watanabe H, Nabi IR, Raz A (1991) The relationship between motility factor receptor internalization and the lung colonization capacity of murine melanoma cells. Cancer Res 51: 2699–2705

Weidner KM, Behrens J, Vandekerckhove J, Birchmeier W (1990) Scatter factor: molecular characteristics and effect on the invasiveness of epithelial cells. J Cell Biol 111: 2097–2108

Wheelock MJ, Knudsen (1991) Cadherins and associated proteins. In Vivo 5: 505–514

Whitlon DS (1993) E-cadherin in the mature and developing organ of Corti of the mouse. J Neurocytol 22: 1030–1038

Willems J, Zwijsen A, Slegers H, Nicolaï S, Bettadapura J, Raymackers J, Scarcez T (1993) Purification and sequence of rat extracellular superoxide dismutase B secretede by C6 glioma. J Biol Chem 268: 24614–24621

Williams E, Furness J, Walsh FS, Doherty P (1994) Activation of the FGF receptor underlies neurite outgrowth stimulated by L1, NCAM and N-cadherin. Neuron 13: 583–594

Wollner DA, Krzeminiski KA, Nelson WJ (1992) Remodeling the cell surface distribution of membrane proteins during the development of epithelial cell polarity. J Cell Biol 116: 889–899
Xiang Y-Y, Tanaka M, Suzuki M, Igarashi H, Kiyokawa E, Naito Y, Ohtawara Y, Shen Q, Sugimura H, Kino I (1994) Isolation of complementary DNA encoding K-cadherin, a novel rat cadherin preferentially expressed in fetal kidney and kidney carcinoma. Cancer Res 54: 3034–3041
Yamashima T, Tohma Y, Yamashita J (1992) Expression of cell adhesion molecule E-cadherin in human arachnoid villi. J Neurosurg 77: 749–756
Zou Z, Anisowicz A, Hendrix MJ, Thor A, Neveu M, Sheng S, Rafidi K, Seftor E, Sager R (1994) Maspin, a serpin with tumor-suppressing activity in human mammary epithelial cells. Science 263: 526–529

Biological Characteristics of Micrometastatic Carcinoma Cells in Bone Marrow

S. BRAUN[1] and K. PANTEL[2]

1 Introduction

Individual epithelial tumor cells disseminated to mesenchymal organs early during primary tumor development can now be identified with monoclonal antibodies directed to cytokeratins as obligatory constituents of the epithelial cytoskeleton (for review see PANTEL and RIETHMÜLLER, this volume) Among the various organs investigated bone marrow appears to play a prominent role as indicator organ of such micrometastatic spread. Bone marrow is an easily accessible organ which normally lacks any epithelial cells. Thus, consecutive marrow samples can be taken from individual patients and "foreign" epithelial cells can be easily detected by use of anti-cytokeratin monoclonal antibodies. Moreover, bone or bone marrow is one of the major sites of skeletal metastasis in several types of epithelial tumors, including cancer of the breast, prostate or lung.

In contrast to the well-documented prognostic significance of "isolated tumor cells disseminated to bone marrow (ITC-BM)" (SCHLIMOK et al. 1991, 1992; LINDEMANN et al. 1992), their biological characteristics remain poorly understood. This ignorance is particularly disturbing in diseases like colorectal cancer, in which

[1] I. Frauenklinik der Ludwig-Maximilians-Universität München, Maistrasse 11, 80337 Munich, Germany
[2] Institut für Immunologie der Ludwig-Maximilians-Universität München, Goethestrasse 31, 80336 Munich, Germany

the presence of ITC-BM was revealed as a strong and independent predictor for reduced overall metastatic relapse, while manifest skeletal metastasis occurred rarely (LINDEMANN et al. 1992). Thus, tumor cells detected in bone marrow at the time of surgery of the primary tumor may not necessarily have the potential to form clinically detectable metastases in this particular environment, but may rather remain dormant for years. Support for the concept of dormancy is also derived from the clinical observation that distant metastases can manifest themselves as late as 10 years after the excision of a primary tumor.

Despite a prevalence of minimal residual cancer of about 30% throughout different types of carcinomas, such advances are rendered more difficult by the extremely low frequency of ITC-BM of 10^{-5}–10^{-6}. In a first attempt, we therefore designed and applied immunocytochemical double marker assays to phenotype ITC-BM derived from different types of epithelial malignancies. These assays provide new insights into the biology of bone marrow micrometastasis. The present review focuses on biological characteristics of micrometastatic carcinoma cells in bone marrow, including molecules that mediate cell adhesion, immune responsiveness and proliferation at the secondary site.

2 Histogenetic and Tumor-Associated Markers

Critics of reports describing the assessment of isolated carcinoma cells in bone marrow questioned the epithelial origin of these cells. Double marker and fluorescence in situ hybridization analyses, however, helped to resolve these concerns. Aberrant cytokeratin expression in single bone marrow cells could be excluded by labeling of bone marrow preparations with monoclonal antibodies CK2 to cytokeratin-18 and either T29/33 to CD45 (SCHLIMOK et al. 1987) or V9 to vimentin (PANTEL et al. 1995), both of which are reliable markers for hematopoietic cells (GATTER et al. 1985). While all of the hematopoietic cells were either CD45-positive or vimentin-positive, cytokeratin-positive epithelial cells consistently lacked detectable expression of these mesenchymal marker proteins (Fig. 1A,B).

In patients with prostatic cancer, we recently performed double marker assays employing monoclonal antibodies specific for prostate-specific antigen (PSA; RIESENBERG et al. 1993). Among the first 13 bone marrow samples admitted to this study, coexpression of PSA on cytokeratin-positive cells was seen in five (38%) samples, thereby revealing the prostatic origin of ITC-BM (Fig. 1C). This incidence is consistent with the rate of PSA expression in primary carcinomas and the LNCap cell line (GALLEE et al. 1986). The malignant nature of cytokeratin-positive cells in bone marrow of prostate cancer patients was supported by the recent work of Pallavicini and colleagues (PALLAVICINI et al. 1995), showing that these cells are cytogenetically aberrant. Taken together, these findings suggest that cytokeratin-positive cells detectable in bone marrow aspirates from patients with prostatic cancer are descendants of the primary tumor.

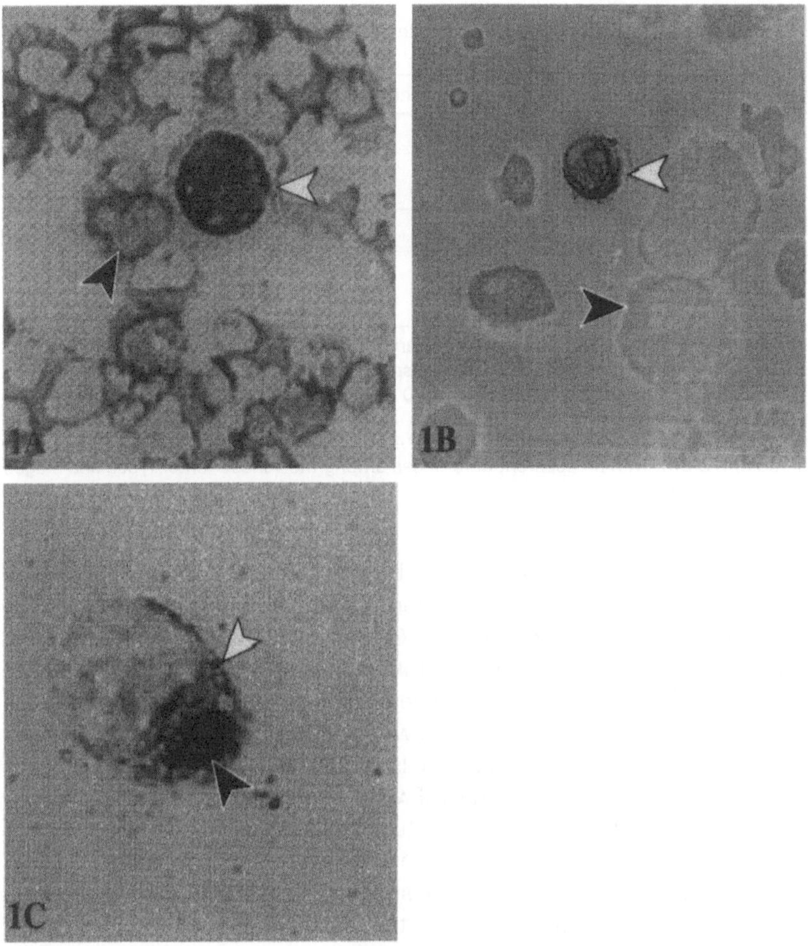

Fig. 1A–C. Histogenetic antigens on cytokeratin-positive micrometastatic carcinoma cells (*white arrows*) assessed by double labeling assays. **A** Monoclonal antibody T29/33 to CD45 (*black arrow*); x1600. **B** Monoclonal antibody V9 to vimentin; silver reflection of immunogold-silver staining (*black arrow*); x800. **C** Monoclonal antibody ER-PR8 to PSA (*black arrow*); x2800

3 Proliferation-Associated Antigens

Manifest skeletal metastasis is frequent in breast, lung and prostate cancer, less frequent in gastric cancer and rare in colorectal cancer. Yet, the incidences of ITC-BM in patients without overt metastasis (M_0) appear to be similar in these types of carcinomas amounting to somewhat around 25%–35% (Pantel et al. 1995). These data suggest that the capacities of disseminated epithelial tumor cells to home in bone marrow might be similar, whereas their potential to outgrow in this new environment might differ considerably. Applying our double labeling techniques, we observed that the proliferative fraction of ITC-BM obtained at the time

Table 1. Coexpression of proliferation-associated markers on isolated carcinoma cells disseminated to bone marrow

Tumor type	Number of patients per total number of patients		
	POX-AP double labeling[a]	IG-AP double labeling[a]	
	Ki-67-positive ITC-BM	Ki-67-positive ITC-BM	p120-positive ITC-BM
Breast cancer	1/9	0/3	1/11
Gastric cancer	0/7	0/1	4/13
Colorectal cancer	0/7	0/6	5/12
Total	1/23	0/10	10/36

ITC-BM, isolated tumor cells disseminated to bone marrow;
[a] A total of 4.0×10^5 nucleated cells per bone marrow aspirate were stained with monoclonal antibody CK2 and monoclonal antibodies to either ki-67 or p120 (FB2), respectively, using either an immunoperoxidase (POX)-alkaline phosphatase (AP) double labeling method or an immunogold(IG)-AP double labeling technique; the number of ITC-BM per 4.0×10^5 mononuclear cells ranged from 1 to 307 (mean: 38.5) ITC-BM with no apparent differences among the various tumor types and stages

of surgery of the primary tumor appears to be small in all subgroups of patients analyzed (Table 1). Surprisingly, Ki-67 antigen, known to be present in all phases of the cell cycle except G_0 and early G_1 (Gerdes et al 1983, 1984), was only detectable in one of 33 patients with cytokeratin-positive ITC-BM. As internal positive controls, Ki-67-positive hematopoietic marrow cells were present in all specimens (Fig. 2A). False negative results were further excluded by the fact that a sufficient number of micrometastatic cells (on average about 40 ITC-BM per patient) has been evaluated in this study (Pantel et al. 1993c).

A possible underestimation of the actual proliferative fraction might have occurred in our approach because the native Ki-67 is known to be quite unstable. We therefore re-examined an additional subgroup of 36 marrow specimens employing the monoclonal antibody FB2 to the nucleolar antigen p120, of which detectable levels appear throughout early G_1 and and peak in S phase (Freeman et al. 1988). Although ITC-BM expressed p120 in ten (28%) out of the 36 cases (Fig. 2B), the fraction of double-positive cells per specimen only amounted to 10% (Table 1). Since p120 is known to be undetectable in nonmalignant hematopoietic bone marrow cells (Freeman et al. 1988), the consistent absence of any p120-positive cells in control aspirates supported the specificity of our results.

The reduced proliferative reactivity of ITC-BM at this early stage of dissemination hampers therapeutic approaches with S phase specific chemotherapeutic agents and is consistent with the well-known phenomenon of tumor cell dormancy (Pantel et al. 1993 a–c). This phenomenon may be explained by experimental data showing that the acquisition of at least some characteristics of metastatic behavior, such as mutations of proto-oncogenes confering invasive potential or changing the adhesive interactions of individual cells, can occur prior to

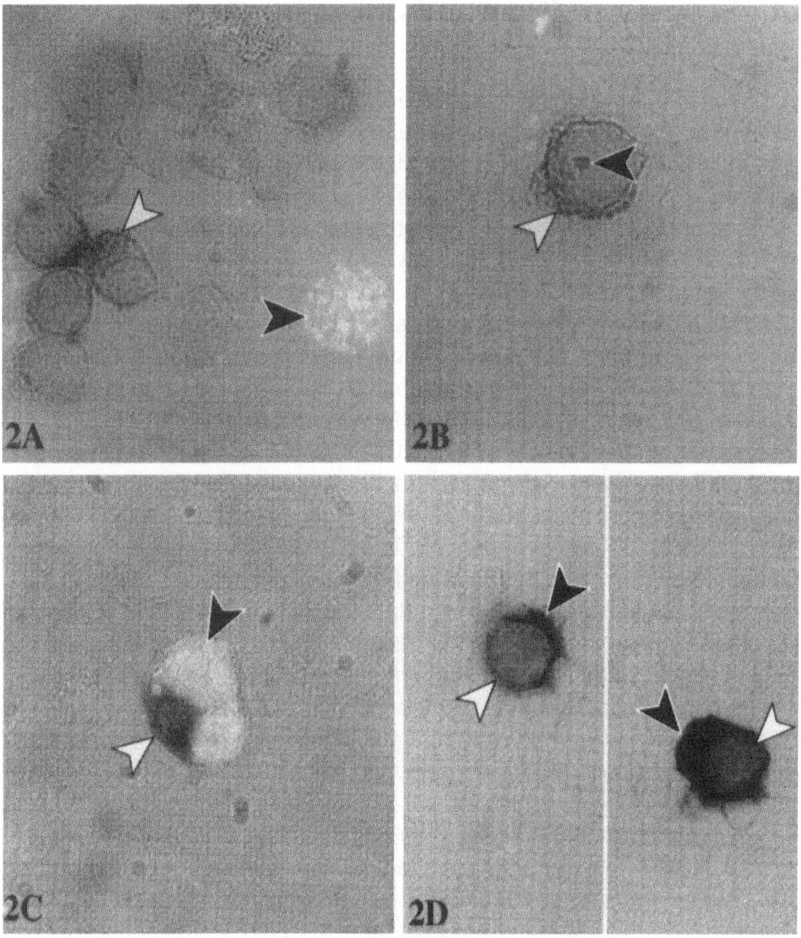

Fig. 2A–D. Expression of proliferation-associated antigens, tumor suppressor genes and oncogenes on cytokeratin-positive micrometastatic carcinoma cells (*white arrows*). **A** Monoclonal antibody MIB-1 to Ki-67; silver reflection of immunogold-silver staining (*black arrow*). Patient with invasive lobular breast cancer (pT2 G3 pN1 M1); x1200. **B** Monoclonal antibody FB2 to p120 (*black arrow*). Patient with adenocarcinoma of the colon (pT2 G2 pN0 M1); x1600. **C** Polyclonal antiserum CM1 to TP53; silver reflection of immunogold-silver staining (*black arrow*). Patient with adenocarcinoma of the colon (pT2 G2 pN1 M0); x1600. **D** Monoclonal antibody 9G6 to p185^{neu}(*black arrows*). Patient with invasive ductal breast carcinoma (pT2 G2 pN0 M0); x1200

unrestrained growth (Fidler and Hart 1982; Price et al. 1989).Thus, cells which have undertaken the first steps in the metastatic cascade (i.e. leaving the primary tumor) may develop their full growth potential only years later. It is therefore important to define markers on ITC-BM that predict their transition from a dormant into a proliferative state.

4 Tumor Suppressor Gene and Oncogene Products

By delaying the transition of cells from a quiescent state into S phase, wild-type p53 protein allows DNA repair to take place, thereby exerting a control function sometimes also addressed as "guardian of the genome" (LANE and BENCHIMOL 1990). Loss of this function may therefore result in the destabilization of the genome by an increased number of mutations that are carried over to the next generation of cells. Since many mutations in the p53 gene lead to the expression of a stabilized mutant protein (IGGO et al. 1990), accumulation of this protein has been claimed as a marker of malignant disease in diagnostic cytopathology (HALL et al. 1991). It was therefore obvious to evaluate whether such accumulation can be detected in ITC-BM by double marker analysis. Counter to our expectations, coexpression of p53 protein (TP53) on ITC-BM was only found in four (6%) of the 63 specimens examined in this series (Fig. 2C) and was entirely absent in another series applying further epithelial markers for the detection of ITC-BM, such as monoclonal antibodies to cytokeratin-19 and to an epitope shared by various cytokeratin molecules (PANTEL et al. 1993d). Thus, in our hands immunodetection of TP53 appeared of little value for the identification of individual micrometastatic carcinoma cells in bone marrow. The rare detection of increased TP53 levels in these cells argues against the assumption that protein-stabilizing mutations in the p53 gene provide a selective advantage for early tumor cell dissemination. Although this hypothesis needs to be tested in molecular analyses, indirect support for the presence of wild-type p53 gene derives from our observations that most cytokeratin-positive cells in marrow appear to be in G_0 or early G_1 phase (PANTEL et al. 1993a–c).

Previous studies have shown that gene amplification and overexpression of c-neu are associated with a more aggressive growth in human breast and ovarian cancer (BERGER et al. 1988; BIANCHI et al. 1993; SLAMON et al. 1989; TIWARI et al. 1992; VARLEY et al. 1987). In this context, immunodetection of $p185^{neu}$ by antibodies has been shown to be correlated with overexpression of the oncoprotein (VENTER et al. 1987). To assess this important growth factor receptor associated with the malignant transformation of epithelial cells, cytokeratin-positive aspirates from carcinoma patients were stained for $p185^{neu}$ overexpression (Fig. 2D). In contrast to proliferation-associated markers, we frequently observed oncogene overexpression on ITC-BM. As demonstrated in Table 2, patients with breast cancer exhibited distinctly higher incidences of c-neu expression on ITC-BM than did patients with gastrointestinal cancer at corresponding tumor stages. However, in both groups of patients, the incidence of cells coexpressing $p185^{neu}$ was positively correlated with tumor progression (Table 2). All of the analyzed breast carcinoma patients with manifest distant metastases (M_1) exhibited $p185^{neu}$ on ITC-BM as compared to about 50% of the patients with regional disease (M_0). Interestingly, the majority of the latter patients had been classified as node-negative by routine histopathological examination, which indicates that cancer cells expressing relevant growth factor receptors are released from the primary tumor in early phases of tumor progression.

Table 2. Incidence of c-neu overexpression on isolated carcinoma cells disseminated to bone marrow

Tumor type	Number of patients per total number of patients with p185^{neu}-positive ITC-BM (%)[a]
Breast cancer	
Total	48/71 (67.6)[b]
M_1	23/23 (100.0)[c]
M_0	25/48 (52.1)
Gastrointestinal cancer	
Total	14/50 (28.0)
M_1	9/21 (42.9)[d]
M_0	5/29 (17.2)
Gastric cancer	6/22 (27.3)
Colorectal cancer	8/28 (28.6)

ITC-BM, isolated tumor cells disseminated to bone marrow
[a]A total of 4.0×10^5 nucleated cells per bone marrow aspirate were stained with monoclonal antibody CK2 and monoclonal antibody 9G6 to the extracellular domain of p185^{neu}, using either an immunoperoxidase-alkaline phosphatase double labeling method (n=80) or an immunogold-alkaline phosphatase double labeling technique (n=41). Number of cytokeratin-positive ITC-BM per 4.0×10^5 mononuclear cells ranged from 1 to 265 (mean: 18.1 cells) with no apparent differences among the various tumor types and stages
[b]p=0.0001 as compared to gastrointestinal cancers (χ^2 test)
[c]p=0.0004 as compared to overt metastasis (χ^2 test)
[d]p=0.05 as compared to overt metastasis (χ^2 test)

Recent in vitro data reported that expression of p185^{neu} in human cells can induce a change in the homotypic epithelial adhesion interactions via down-regulation of E-cadherin expression (Taylor-Papadimitriou 1994). It is therefore conceivable that p185^{neu}-positive micrometastatic cells have been positively selected from a small metastatic subpopulation within the primary tumor by suppressing the metastasis suppressor function of E-cadherin. This conclusion, though tentative, points to c-neu expression as a marker of dissemination. In addition, the increased incidence of c-neu expression on metastatic tumor cells found in advanced stage patients suggests that such expression may also support the survival and/or outgrowth of these cells in the marrow environment. It may be therefore interesting to know whether proteins claimed to act as natural ligands for the c-neu receptor (Hoffman 1992; Lupu et al. 1992; Peles et al. 1992) are expressed in ITC-BM or bone marrow cells.

5 Proteins Relevant to the Immunological Anti-tumor Defense

Although isolated tumor cells present in an accessible environment, such as bone marrow, should be preferential targets for immune effector cells, it appears that ITC-BM can survive for extended periods of time without being killed by the immune system (Riethmüller and Johnson 1992). In principle, this discrepancy might be due to an inability of immune effector cells to recognize these cells and/or to an anergic state of the effector cells. In a first attempt to address the problem of how ITC-BM escape recognition by immune effector cells, we applied our double marker assay to phenotype ITC-BM for the expression of HLA class I molecules (Fig. 3A), which present tumor-specific peptides via the T cell receptor to T lymphocytes (Hämmerling et al. 1989; Pantel et al. 1991). In total, 25 (46%) of 54 patients yielded cytokeratin-positive ITC-BM that lacked a detectable expression of HLA class I molecules. In patients with breast cancer, the percentage

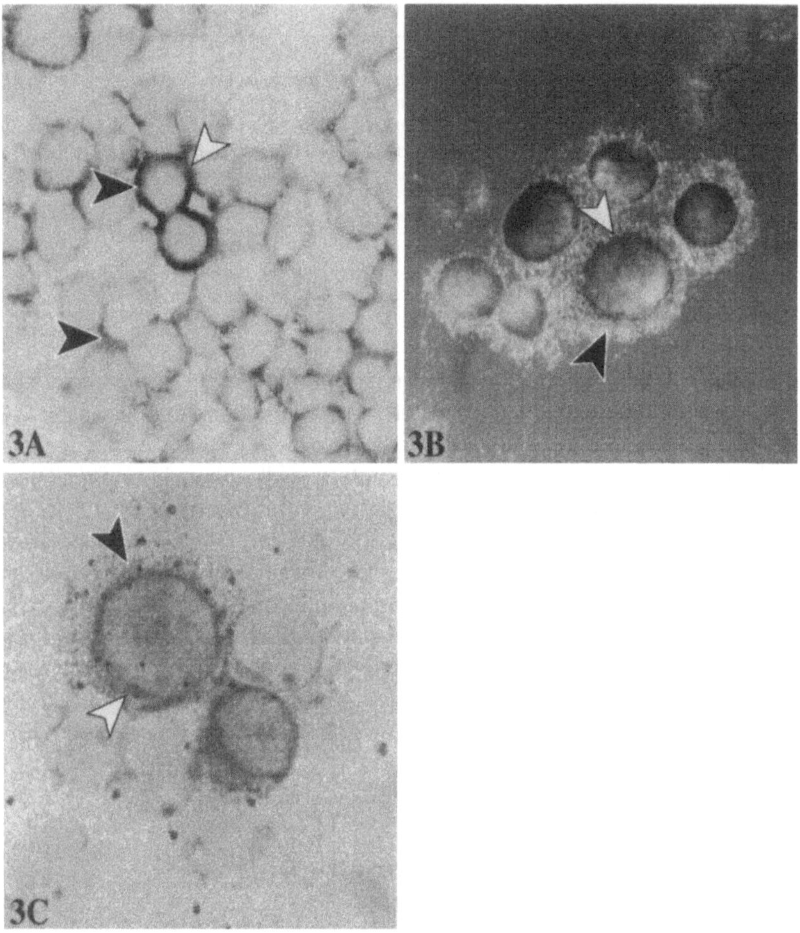

Fig. 3A–C. Expression of HLA class I antigens and adhesion molecules on cytokeratin-positive micrometastatic carcinoma cells (*white arrows*). **A** Monoclonal antibody W6/32 to HLA class I antigens (*black arrows*). Patient with invasive ductal breast carcinoma (pT2 G2 pN1 M0); x 800. **B** Monoclonal antibody PG5.1 to plakoglobin; silver reflection of immunogold-silver staining (*black arrow*). Patient with squamous lung carcinoma (pT2 G2 pN0 M0); x1000. **C** Monoclonal antibody 17-1A to EGP-40; silver reflection of immunogold-silver staining (*black arrow*). Patient with invasive ductal breast carcinoma (pT1 G2 pN0 M0); x1600

increased to 65% HLA class I-negative carcinoma cells in bone marrow as compared to only 27%–29% in patients with gastrointestinal carcinomas (Table 3). This observation matches well with the high incidence of metastatic disease of the bone or the bone marrow occurring in breast cancer patients. It is therefore conceivable that down-regulation of HLA class I molecules is an effective mechanism to escape from the anti-tumoral immune defense mediated by cytotoxic T lymphocytes. Moreover, the overall incidence of HLA-negative tumor cells in the bone marrow is higher than that reported for the respective primary tumors (HÄMMERLING et al. 1989), an observation which further supports the latter assumption that down-regulation of HLA class I molecules may confer a selective survival advantage to ITC-BM.

Several families of adhesion molecules participate in an effective immune response by mediating the tight binding of leukocytes to their target cells (HOGG and LANDIS 1993; SPRINGER 1990). One of these molecules, the intercellular adhesion molecule-1 (ICAM-1), mediates leukocyte binding through its interaction with the integrins $\alpha_L\beta_2$ (LFA-1) and $\alpha_M\beta_2$ (Mac-1). ICAM-1 is normally not expressed by epithelial cells, though it can be induced by a variety of stimuli and is sometimes expressed by carcinoma cells in vivo (JOHNSON et al. 1989; PASSLICK et al. 1994). Expression of ICAM-1 on cytokeratin-positive cells in the bone marrow was found in 13 (42%) of 31 patients with completely resectable non-small cell lung carcinoma and appeared to be of prognostic significance. Those patients with ICAM-1-positive micrometastatic cells had a reduced rate of tumor relapses and cancer-related deaths as compared to those patients with ICAM-1-negative tumor cells. Local and distant recurrence of lung cancer, or death directly related to the disease, occurred predominantly in patients with ICAM-1-negative ITC-BM, whereas only a few patients with ICAM-1-positive ITC-BM succumbed to distant metastatic disease. Even though the number of patients both examined immunocytochemically for ICAM-1 expression ($n = 31$) and evaluated prospectively for the course of the disease ($n = 19$) is considerably low thus far, our data suggest that ICAM-1-mediated binding of leukocytes may facilitate an effective immune response. This observation is consistent with studies on renal cell carcinoma in which the expression of ICAM-1 by the primary tumor was correlated with well differentiated tumors having a good prognosis (TOMITA et al. 1990). A recent study

Table 3. Deficient expression of HLA class I molecules on isolated carcinoma cells disseminated to bone marrow

Tumor type	Number of patients per total number of patients with HLA class I-negative ITC-BM (%)[a]	
Breast cancer	17/26	(65.4)
Colorectal cancer	5/17	(29.4)
Gastric cancer	3/11	(27.3)

[a]A total of 4.0×10^5 nucleated cells per bone marrow aspirate were stained with monoclonal antibody CK2 and monoclonal antibody W6/36 to an extracellular of HLA class I domain using immunoperoxidase-alkaline phosphatase double labeling

unraveled the induction of ICAM-1 by protein kinase C after stimulation with neu differentiation factor (NDF), also called heregulin (Bacus et al. 1993). NDF is a 44 kDa glycoprotein that stimulates tyrosine phosphorylation of the $p185^{neu}$ receptor and induces phenotypic differentiation of certain breast carcinoma cell lines to growth-arrested and milk-producing cells. Thus, expression of $p185^{neu}$ may exert a dual role for the malignant potential of ITC-BM, depending on the ligand present in the local environment.

An alternative approach to unraveling structures involved in the onset of micrometastatic spread is to evaluate possible correlations between the phenotype of primary breast carcinomas and the presence of ITC-BM. Expression of ICAM-1, known to facilitate metastatic spread of melanoma cells (Johnson et al. 1989), had no apparent influence on the metastatic behavior of primary non-small cell lung cancer cells (Passlick et al. 1994). By contrast, Ménard and colleagues revealed that the expression of the 67 kDa laminin receptor on primary breast cancer cells may support tumor cell dissemination into lymph nodes and bone marrow (Martignone et al. 1993; Ménard et al. 1994).

The frequent down-regulation of HLA class I molecules as well as the modulation of ICAM-1 expression appear to influence the recognition by immune effector cells and might thereby exert a significant effect on the survival of ITC-BM. Moreover, down-regulation of HLA class I antigens on residual tumor cells has also severe implications for future therapeutic approaches using genetically modified cancer vaccines (Pardoll 1993). The potential effectiveness of such vaccines has been proven in animal models and the first clinical trials have been initiated to evaluate their safety and efficacy in patients with advanced solid tumors (Nabel and Felgner 1993; Pardoll 1993). However, this active immunization approach works only if the residual tumor cells are recognized as foreign by T lymphocytes. It is therefore important to investigate whether the deficient expression of HLA class I molecules on ITC-BM can be up-regulated by systemic administration of class I-inducing cytokines such as interferon-γ.

6 Epithelial Cell Adhesion Molecules

The flexibility in the adhesive interactions of metastatic cells with different cell types plays an important role in the initiation of tumor cell emigration from the primary location to secondary sites (Behrens et al. 1992). In carcinomas, loss of homotypic adhesion is one of the first steps required for the successful dissemination of tumor cells (Hart et al. 1989). in the case of epithelial organs, a network of intercellular adhesive junctions is responsible for the tight integration of an individual cell within the tissue (Schwarz et al. 1990). The adherens junction complex is organized around the transmembrane E-cadherin protein that organizes a complex of cytoplasmatic proteins, including α-catenin, β-catenin and plakoglobin, a β-catenin relative found in desmosomes (Schwarz et al. 1990). The adherin-catenin complex mediates adhesion, cytoskeletal anchoring, and signaling. Catenins can also form a complex with the product of the tumor suppressor gene APC, which may mediate transmission of a growth regulatory signal (PFEIFER 1993). Employing our double marker assay, we were able to study the expression of plakoglobin on ITC-BM. Among the first 25 bone marrow samples admitted to this study, we found that in 17 (68%) samples expression of plakoglobin appears to be absent or at least below the immunocytochemical detection level in ITC-BM. In contrast, microaggregates of carcinoma cells present in bone marrow again appear to express plakoglobin (Fig. 3B). Therefore, it is conceivable that down-regulation of plakoglobin expression participates in mechanisms that determine the disseminative capacity of an individual carcinoma cell, whereas the up-regulation of expression might be necessary for solid metastasis formation at the secondary site.

Another interesting homophilic cell-cell adhesion molecule is the epithelial glycoprotein 40 (EGP-40, also known as GA733-2, ESA, KSA, and the 17-1A antigen) encoded by the GA-733-2 gene, which is expressed on the basolateral cell surface in most human simple epithelia (GÖTTLINGER et al. 1986b). The protein is also found in the vast majority of carcinomas and has attracted attention as a tumor marker (GÖTTLINGER et al. 1986a). Recently, LITVINOV and colleagues (LITVINOV et al. 1994) provided the first evidence that EGP-40 is an epithelium-specific intercellular adhesion molecule which acts in a calcium-independent manner. In its biological behavior EGP-40 resembles members of the immunoglobulin superfamily that comprises cell adhesion molecules. Therefore the name Ep-CAM was proposed. Our present double marker analysis demonstrated a presumably modulated, differential expression of EGP-40 on ITC-BM (Fig. 3C). Minimal residual breast and colorectal carcinoma cells remained EGP-40-negative in nine of 16 and two of six cases, respectively. Similar to the desmosomal adhesion molecule plakoglobin, down-regulation of EPG-40 expression might permit tumor cells to be released from contact-mediated controls within the primary tumor.

Since the expression of EGP-40 on ITC-BM is less frequently down-regulated than plakoglobin, it might serve as a suitable target for antibody therapy. This hypothesis has been recently sustained in colorectal cancer patients without clinical signs of overt distant metastases (stage Dukes C) who were treated in a randomized clinical trial with the murine monoclonal antibody CO17-1A specific for EGP-40 (RIETHMÜLLER et al. 1994). Randomized administration of this antibody was found to reduce the overall death rate by 30% and decrease the recurrence rate by 27%.

7 Concluding Remarks

To obtain further insights into the biology behind the phenomenon of minimal residual cancer, we designed and applied immunocytochemical double marker assays, which allow the detection and characterization of individual microdisseminated carcinoma cells in bone marrow. Our present data support the view that down-regulated expression of epithelial cell adhesion molecules on ITC-BM might be important to initiate early tumor cell dissemination. The survival of ITC-BM and thereby the influence they exert on the prognosis of individual cancer patients appears to be affected by the presence of molecules mediating an effective immune response, such as HLA class I molecules or ICAM-1. The acquisition of a selective survival advantage for ITC-BM may not necessarily comprise their ability for immediate autonomous proliferation in the bone marrow environment. Concordant with the phenomenon of dormancy, cells which have gained the potential to leave the primary tumor might develop their full metastatic potential only years later. This assumption is also supported by recent experimential observations that metastatic behavior and growth characteristics of individual cells are essentially controlled by two different sets of genes (FIDLER and RADINSKY 1990; LIOTTA et al. 1991). At present, it is unclear which changes in ITC-BM lead to the transition from dormancy to active proliferation and formation of overt metastases. Among many potential factors, the expression of growth factor receptors might be essential. The frequent overexpression of $p185^{neu}$ on ITC-BM could be the result of early selection within the primary tumor which might be explained by the observation that this expression results in down-regulation of the desmosomal adhesion molecule E-cadherin. After tumor cells have settled in bone marrow, $p185^{neu}$ may mediate at least two distinct pathways: (1) the growth factor activity of the tyrosine kinase receptor can support extended survival and outgrowth of solid metastases, or (2) binding of NDF/heregulin may lead to increased elimination of tumor cells by leukocytes due to induction of ICAM-1 expression of ITC-BM.

In addition to the investigation of biological features of ITC-BM, the application of double marker assays may contribute to the identification of cell surface target antigens suitable for passive antibody therapy. Support for the efficacy of antibody therapy is derived from the adjuvant trial on patients with Dukes C colorectal cancer, recently reported by RIETHMÜLLER and the German Cancer Aid 17-1A study Group (RIETHMÜLLER et al. 1994). While it is clear that a major consideration for the successful application of antibody therapy is the choice of the appropriate disease stage in which the tumor cells are accessible for intravenously administered immunoglobulins (RIETHMÜLLER and JOHNSON 1992), other aspects need to be taken into account. To cope with antigen heterogeneity a combination of antibodies directed to independently expressed antigens should be more efficient than a single agent (RIETHMÜLLER and JOHNSON 1992). In this context, $p185^{neu}$ expression frequently observed on ITC-BM appears to be another suitable target.

In conclusion, the present review demonstrates that, beyond the mere presence of micrometastatic cells in bone marrow, additional information can be obtained by double marker analyses. The phenotyping of ITC-BM may give new insights into the so often unpredictable and erratic course of an individual tumor. A cautionary note as to too far-reaching conclusions must be added since the number of carcinoma cells detectable per patient is rather small. With the development of new techniques like single-cell polymerase chain reaction and the in vitro expansion and isolation of micrometastatic cells, it may also be possible to determine the characteristic genotypic features of those cells. Careful follow-up is needed to prove conclusively that the clinical significance of our approach is superior to the current procedures of risk assessment. However, expression of proliferation-associated markers by disseminated tumor cells may indicate both that viable tumor cells are still present and that they pose a particular risk for later clinical relapse.

References

Bacus SS, Gudkov AV, Zelnick CR, Chin D, Stern R, Stancovski I, Peles E, Ben-Baruch N, Farbstein H, Lupu R (1993) Neu differentiation factor (heregulin) induces expression of intercellular adhesion molecule 1: implications for mammary tumors. Cancer Res 53: 5251–5261

Behrens J, Frixen U, Schipper J, Weidner M, Birchmeier W (1992) Cell adhesion in invasion and metastasis. Semin Cell Biol 3: 169–178

Berger MS, Locher GW, Saurer S, Gullick WJ, Waterfield MD, Groner B, Hynes NE (1988) Correlation of c-erbB-2 gene amplification and protein expression in human breast carcinoma with nodal status and nuclear grading. Cancer Res 48: 1238–1243

Bianchi S, Paglierani M, Zampi G, Cardona G, Cataliotti L, Bonardi R, Ciatto S (1993) Prognostic significance of c-erbB-2 expression in node negative breast cancer. Br J Cancer 67: 625–629

Fidler IJ, Hart IR (1982) Biological diversity in metastatic neoplasms: origin and implications. Science 217: 998–1003

Fidler IJ, Radinsky R (1990) Genetic control of cancer metastasis. J Natl Cancer Inst 82: 166–168

Freeman JW, Busch RK, Gyorkey F, Gyorkey P, Ross BE, Busch H (1988) Identification and characterization of a human proliferation-associated nucleolar antigen with a molecular weight of 120,000 expressed in early G1 phase. Cancer Res 48: 1244–1251

Gallee MPW, van Vroonhoven CCJ, van der Korput HAGM, van der Kwast TH, ten Kate FJW, Romijn JC, Trapman J (1986) Characterization of monoclonal antibodies raised against prostatic cell line PC-82. Prostate 9: 33–45

Gatter KC, Alcock C, Heryet A, Mason DY (1985) Clinical importance of analysing malignant tumours of uncertain origin with immunohistological techniques. Lancet 1: 1302–1305

Gerdes J, Schwab U, Lemke H, Stein H (1983) Production of a mouse monoclonal antibody reactive with a human nuclear antigen associated with cell proliferation. Int J Cancer 31: 13–20

Gerdes J, Lemke H, Baisch H, Wacker HH, Schwab U, Stein H (1984) Cell cycle analysis of a cell proliferation-associated human nuclear antigen defined by the monoclonal antibody *ki*67. J Immunol 133: 1710–1715

Göttlinger HG, Funke I, Johnson JP, Gokel JM, Riethmüller G (1986a) The epithelial cell surface antigen 17-1A, a target for antibody-mediated tumor therapy: its biochemical nature, tissue distribution and recognition by different monoclonal antibodies. Int J Cancer 38: 47–53

Göttlinger HG, Johnson JP, Riethmüller G (1986b) Biochemical and epitope analysis of the 17-1A membrane antigen. Hybridoma 5: 29–37

Hall PA, Ray A, Lemoine NR, Midley CA, Krausz T, Lane DP (1991) p53 immunostaining as a marker of malignant disease in diagnostic cytopathology. Lancet 338: 531

Hämmerling G, Maschek V, Sturmhöfel K, Momburg F (1989) Regulation and functional role of MHC expression on tumors. In: Melchers F (ed) Progress in immunology. Springer, Berlin Heidelberg New York, pp 1071–1078

Hart IR, Goode NT, Wilson RE (1989) Molecular aspects of the metastatic cascade. Biochim Biophys Acta 989: 65–84

Hoffman M (1992) New clue found to oncogene's role in breast cancer. Science 256: 1129

Hogg N, Landis RC (1993) Adhesion molecules in cell interaction. Curr Opin Immunol 5: 383–390

Iggo R, Gatter K, Bartek J, Lane D, Harris AL (1990) Increased expression of mutant forms of p53 oncogene in primary lung cancer. Lancet 335: 675–679

Johnson JP, Stade BG, Holzmann B, Schwäble W, Riethmüller G (1989) De novo expression of intercellular-adhesion molecule 1 in melanoma correlates with increased risk of metastasis. Proc Natl Acad Sci USA 86: 641–644

Lane DP, Benchimol S (1990) p53: oncogene or anti-oncogene? Genes Dev 4: 1–8

Lindemann F, Schlimok G, Dirschedl P, Witte J, Riethmüller G (1992) Prognostic significance of micrometastatic tumour cells in bone marrow of colorectal cancer patients. Lancet 340: 685–689

Liotta LA, Steeg PS, Stetler-Stevenson WG (1991) Cancer metastasis and angiogenesis: an imbalance of positive and negative regulation. Cell 64: 327–336

Litvinov SV, Velders MP, Bakker HAM, Fleuren GJ, Warnaar SO (1994) Ep-CAM: a human epithelial antigen is a homophilic cell-cell adhesion molecule. J Cell Biol 125: 437–446

Lupu R, Colomer R, Kannan B, Lippman ME (1992) Characterization of a growth factor that binds exclusively to the erbB2 receptor and induces cellular responses. Proc Natl Acad Sci USA 89: 2287–2291

Martignone S, Ménard S, Bufalino R, Cascinelli N, Pellegrini R, Tagliabue E, Andreola S, Rilke F, Colnaghi MI (1993) Prognostic significance of the 67-kilodalton laminin receptor expression in human breast carcinomas. J Natl Cancer Inst 85: 398–402

Ménard S, Squicciarini P, Luini A, Sacchini V, Rovini D, Tagliabue E, Veronesi U, Colnaghi MI (1994) Immunodetection of bone marrow micrometastases in breast carcinoma patients and its correlation with tumor prognostic features. Br J Cancer 69: 1126–1129

Nabel GJ, Felgner PL (1993) Direct gene transfer for immunotherapy and immunization. In: Robinson C, Watson F (eds) Trends in biotechnology: gene therapy–therapeutic strategies and commercial prospects. Bousfield, Cambridge, pp 211–214

Pallavicini MG, Cher ML, Bowers EE, Wessman M, Balouch M, Presti J, Carroll PR (1995) Chromosomal aneusomies in prostate cancer marrow micrometastases (abstract). Proc Am Assoc Cancer Res 36: 69

Pantel K, Schlimok G, Kutter D, Schaller G, Genz T, Wiebecke B, Backmann R, Funke I, Riethmüller G (1991) Frequent down-regulation of major histocompatibility class I antigen expression on individual micrometastatic carcinoma cells. Cancer Res 51: 4712–4715

Pantel K, Braun S, Schlimok G, Riethmüller G (1993a) Micrometastatic tumour cells in bone marrow in colorectal cancer. Lancet 341: 501

Pantel K, Izbicki JR, Angstwurm M, Braun S, Passlick B, Karg O, Thetter O, Riethmüller G (1993b) Immunocytological detection of bone marrow micrometastasis in operable non-small cell lung cancer. Cancer Res 53: 1027–1031

Pantel K, Schlimok G, Braun S, Kutter D, Schaller G, Funke I, Izbicki J, Riethmüller G (1993c) Differential expression of proliferation-associated molecules in individual micrometastatic carcinoma cells. J Natl Cancer Inst 85: 1419–1424

Pantel K, Schmaus W, Schlimok G, Riethmüller G (1993d) p53 immunostaining of megakaryocytes is preferentially observed in patients with epithelial malignancies. Blood 82: 610a

Pantel K, Schlimok G, Angstwurm M, Weckermann C, Schmaus W, Gath H, Passlick B, Izbicki J, Riethmüller G (1994) Methodological analysis of immunocytochemical screening for disseminated epithelial tumor cells in bone marrow. J Hematother 3: 165–173

Pantel K, Braun S, Passlick B, Schlimok G (1995) Progress in histochemistry and cytochemis try – Minimal residual epithelial cancer: diagnostic approaches and prognostic relevance. Fischer, Stuttgart

Pardoll DM (1993) Cancer vaccines. Immunol Today 14: 310–316

Passlick B, Izbicki JR, Simmel S, Kubuschok B, Karg O, Habekost M, Thetter O, Schweiberer L, Pantel K (1994) Expression of major histocompatibility class I and class II antigens and intercellular adhesion molecule-1 on operable non-small cell lung carcinomas. Eur J Cancer 30: 376–381

Peles E, Bacus SS, Koski RA, Lu HS, Wen D, Ogden SG, Levy RB, Yarden Y (1992) Isolation of the neu/HER-2 stimulatory ligand: a 44 kD glycoprotein that induces differentiation of mammary tumor cells. Cell 69: 205–216

Pfeifer M (1993) Cancer, catenins and cuticle pattern: a complex connection. Science 262: 667

Price JE, Aukerman SL, Ananthaswamy N, McIntyre BW, Schackert G, Fidler IJ (1989) Metastatic potential of cloned murine melanoma cells transfected with activated c-Ha-ras. Cancer Res 49: 4274–4281

Riesenberg R, Oberneder R, Kriegmair M, Epp M, Bitzer U, Hofstetter A, Braun S, Riethmüller G, Pantel K (1993) Immunocytochemical double staining of cytokeratin and prostate specific antigen individual prostatic tumor cells. Histochemistry 99: 61–66

Riethmüller G, Johnson JP (1992) Monoclonal antibodies in the detection and therapy of micrometastatic epithelial cancer. Curr Opin Immunol 4: 647–655

Riethmüller G, Schneider-Gädicke E, Schlimok G, Schmiegel W, Raab R, Höffken K, Gruber R, Pichlmaier H, Hirche H, Pichlmayr R, Buggisch P, Witte J. The German Cancer Aid 17-1A Study Group (1994) Randomised trial of monoclonal antibody for adjuvant therapy of resected Dukes C colorectal carcinoma. Lancet 343: 1177–1183

Schlimok G, Funke I, Holzmann B, Göttlinger HG, Schmidt G, Häuser H, Swierkot S, Warnecke HH, Schneider B, Koprowski H, Riethmüller G (1987) Micrometastatic cancer cells in bone marrow: in vitro detection with anti-cytokeratin and in vivo labeling with anti-17-1A monoclonal antibodies. Proc Natl Acad Sci USA 84: 8672–8676

Schlimok G, Funke I, Pantel K, Strobel F, Lindemann F, Witte J, Riethmüller G (1991) Micrometastatic tumor cells in bone marrow of patients with gastric cancer: methodological aspects of detection and prognostic significance. Eur J Cancer 27: 1461–1465

Schlimok G, Lindemann F, Holzmann K, Witte J, Renner D, Riethmüller G (1992) Prognostic significance of disseminated tumor cells detected in bone marrow of patients with breast and colorectal cancer: a multivariate analysis. Proc Am Soc Clin Oncol 11: 102

Schwarz MA, Owaribe K, Kartenbeck J, Franke WW (1990) Desmosomes and hemidesmosomes: Constitutive molecular components. Annu Rev Cell Biol 6: 461–491

Slamon JD, Godolphin W, Jones LA, Holt JA, Wong SG, Keith DE, Levin WJ, Stuart SG, Udove J, Ullrich A, Press MF (1989) Studies of the HER-2/neu proto-oncogene in human breast cancer and ovarian cancer. Science 244: 707–712

Springer TA (1990) Adhesion receptors of the immune system. Nature 346: 425–434

Taylor-Papadimitrou J (1994) Oncogene signalling epithelial polarity and metastatic phenotype. Proceedings of the Lancet Conference, Brugge, April 21–22, 1994. Lancet, London, p 7

Tiwari RK, Borgen PI, Wong GY, Cordon CC, Osborne MP (1992) HER-2/neu amplification and overexpression in primary human breast cancer is associated with early metastasis. Anticancer Res 12: 419–425

Tomita Y, Nishiyama T, Watanabe H, Fujiwara M, Sato S (1990) Expression of intercellular adhesion molecule-1 (ICAM-1) on renal-cell cancer: possible significance in host immune responses. Int J Cancer 46: 1001–1006

Varley JM, Swallow JE, Brammer WJ, Whittaker JL, Walker RA (1987) Alterations to either c-erbB-2(neu) or c-myc proto-oncogenes in breast carcinomas correlate with poor short-term prognosis. Oncogene 1: 423–430

Venter DJ, Tuzi NL, Kumar S, Gullick WJ (1987) Overexpression of the c-erb-B-2 oncoprotein in human breast carcinomas: immunhistochemical assessment correlates with gene amplification. Lancet 11: 69–72

Model Systems of Carcinoma Cell Dispersion

B. Boyer, A.M. Vallés, and J.P. Thiery

1 Introduction

The progressive loss of expression of a fully differentiated epithelial phenotype is a hallmark of carcinoma progression in both humans and animals. In the mouse skin system, for example, the loss of epithelial differentiation is accompanied by the appearance of malignant properties. In this well-studied system, chemical carcinogenesis generates multiple benign tumors, a proportion of which develop to form squamous cell carcinomas and, in extreme cases, spindle cell carcinomas that are highly invasive and metastatic tumors (Stoler et al. 1993). The transformation from squamous into spindle cells is characterized by the loss of epithelial features and the acquisition of a fibroblastic appearance. Analogous changes from an epithelial morphology to a fibroblast-like phenotype, correlating with the progression to malignancy, have also been observed in epithelial tumors derived from other sites, including colon and breast. Whether or not these phenotypic alterations can be regarded as authentic epithelium-to-mesenchyme transitions (EMTs), similar to those observed during embryogenesis, remains to be clarified. Nevertheless, since the maintenance of epithelial cell differentiation acts as a

CNRS UMR 144, Institut Curie, Section de Recherche, 26 rue d' Ulm, 75231 Paris Cedex 05, France suppressor of malignancy (HARRIS 1990), it is essential to study the genetic and epigenetic events controlling the cellular program driving the transformation of epithelial cells into fibroblast-like cells. In that context, numerous in vitro models have been designed to clarify some aspects of tumor progression. Using this approach, several genes have been identified as potential tumor activators or suppressors: for example, E-cadherin, an epithelial cell adhesion molecule, has been demonstrated to act in vitro as a tumor suppressor gene (FRIXEN et al. 1991), while the mutant H-*ras* gene is responsible for the increased invasiveness of epithelial cell lines (MAREEL et al. 1991). Similarly, the mechanisms which govern the reversible, nongenetic changes from an epithelial phenotype to a more fibroblastic one have been analyzed by using normal or transformed cells in culture. Identifying by in vitro methods the signals and target molecules responsible for carcinoma and, more generally, for epithelial cell scattering is a first and necessary step before in vivo studies can be devised to validate the findings made in vitro (Fig. 1).

2 Are Epithelium-to-Mesenchyme Transitions Inductive Events?

Considering that cancer cells are subjected to chromosomal instability which generates cells with different genotypes, the question arises as to whether EMT-

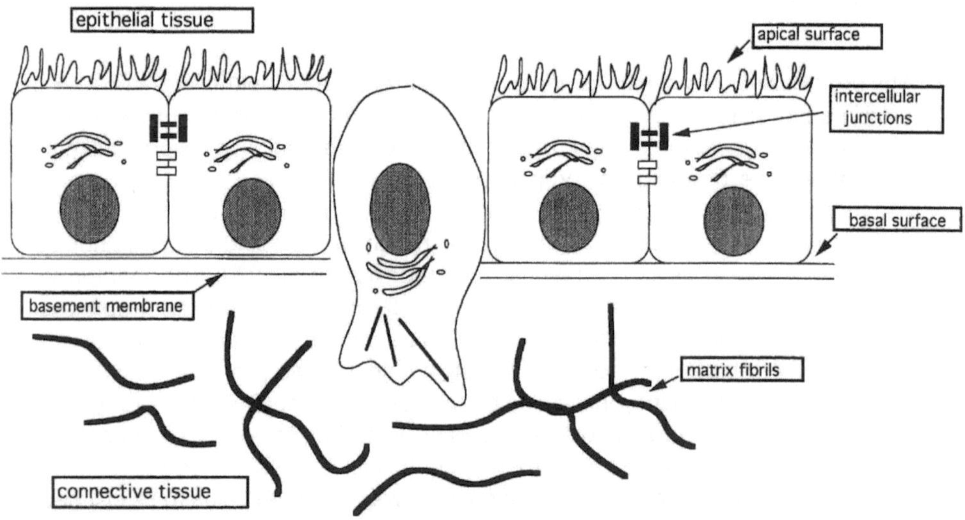

Fig. 1. Epithelium-to-mesenchyme transition. In response to extrinsic or intrinsic signals, epithelial cells turn off genes encoding junction-associated proteins, and start to express vimentin intermediate filaments. Cells now display the characteristic mesenchymal front end-back end polarity that allows them to elongate and move out through gaps formed in the basement membrane

like events arising in carcinoma cells are "spontaneous" processes, which are achieved in certain cells having acquired a certain genotype, or whether they are induced by external factors to which cells are responsive. If the first hypothesis is true, the genetic modifications leading to the acquisition of a motile, fibroblast-like phenotype are expected to be irreversible. If the second one is correct, EMT processes are expected to be reversible, cells ceasing to express a fibroblastic phenotype as soon as the inducing molecule disappears. The validity of one of these hypotheses is still a matter of debate, since observations arguing for each of them have been reported. On the one hand, as stressed before, the loss of genes involved in cell-cell interactions or the mutational activation of several oncogenes is accompanied by the acquisition of an invasive and metastatic phenotype. On the other hand, metastases derived from carcinomas often exhibit epithelial characteristics similar to those of the parental tumor, suggesting that the process of tumor cell dispersion might be reversible. However, the two hypotheses can be reconciled in a unifying scheme, in which one could postulate that the genetic alterations resulting in cell dispersion involve primarily genes coding for proteins implicated in the cellular events leading to EMT.

The in vitro studies devoted to the analysis of EMT of normal and transformed cells have led to the description of different modes of induction. Interestingly, there is a network of potential interplays between the distinct types of inducers that could explain how EMT is regulated. It should be noted, however, that in vitro-induced EMTs are generally reversible. Therefore, EMT can be regarded, in many aspects, as a model system for carcinoma cell dispersion, but it does not produce cells irreversibly "dedifferentiated" into fibroblast-like cells.

3 The Different Modes of Induction

3.1 Growth Factors

There is increasing evidence that growth factors, discovered on the basis of their growth-promoting actions, have multiple effects and elicit a wide range of biological responses. Interestingly, their activities on cell differentiation appear to involve requirements opposed to those implicated in mitogenesis, since it is well established that cell proliferation and differentiation are two antagonistic processes: cells usually cease dividing when they differentiate and, conversely, differentiation events are inhibited when cells enter the cell cycle. The EMT-inducing effect of growth factors can be related to a differentiating activity, since, in general, growth factors cannot elicit simultaneously mitogenesis and EMT responses (Shiota et al. 1992; Vallés et al. 1990b; Geimer and Bade 1991; Miettinen et al. 1994). Similarly, the platelet-derived growth factor (PDGF)-stimulated motogenic response of fibroblasts is maximal in growth-arrested cultures (Grotendorst 1984; Hughes and McCulloch 1991). However, there are exceptions to this rule.

Epitaxin, a fibroblast-derived motility factor for epithelial cells, simultaneously increases cell proliferation and motility (SHIMONAKA and YAMAGUCHI 1994).

Many growth factors have been demonstrated to promote in vitro EMTs. Epidermal growth factor (EGF) promotes the in vitro migration of normal keratinocytes (BARRANDON and GREEN 1987), intestinal epithelial cells (BLAY and BROWN 1985), liver epithelial cells (BADE and FEINDLER 1988), and pancreatic carcinoma cells (KLEMKE et al. 1994) and induces the generation of fibroblast-like cells from normal mammary epithelial cells in culture (STAMPFER and BARTLEY 1988). Transforming growth factor (TGF)-α which is a ligand of EGF receptors, converts bladder carcinoma cells into fibroblasts (GAVRILOVIC et al. 1990). We also found that fibroblast growth factor (FGF)-1 elicits a scattering response of cultured bladder carcinoma cells (VALLÉS et al. 1990a) (Fig. 2). The prototype of growth factors endowed with a scattering activity toward epithelial cells is hepatocyte growth factor (HGF), also named scatter factor (SF). This factor, which is secreted predominantly by mesenchymal cells, is capable of dispersing a variety of normal and tumor epithelial cells (ROSEN et al. 1994). Recently, challenging discoveries have modified the current ideas on the biological role of SF. When mouse fibroblasts are cotransfected with cDNAs encoding SF and its cognate receptor (c-*met*), they are converted into epithelial-like cells (TSARFATY et al. 1994). Accordingly, anti-SF antibodies inhibit the differentiation of metanephric mesenchymal cells into epithelial precursors of metanephric rudiments grown in culture (WOOLF et al. 1995). If the positive effects of SF on the regulation of two opposite processes (i.e. epithelium-mesenchyme and mesenchyme-epithelium transitions) are to be reconciled, it implies that SF would rather act as an "interconverting factor" between epithelial and mesenchymal programs, suggesting that the two apparently antagonistic transdifferentiation events could share common denominators. Elucidating the precise contribution of growth/scatter factors during morphogenetic events implying transdifferentiation processes between the epithelial and mesenchymal programs is an ongoing area of research.

3.2 Extracellular Matrix

As for growth factors, the effects of matrix macromolecules on cell motility is well documented. The requirement of cell adhesion to extracellular matrix (ECM) in the generation of cell movement is absolute, since cells have to exert tractional forces on their substrate in order to migrate. However, it is clearly established that cell-matrix interactions made during cell locomotion must be labile, in order to allow cell translocation. Requirements of cell-matrix interactions have been demonstrated in many types of migratory cells: for example, neural crest cells utilize matrix fibronectin and tenascin for their oriented emigration from the neural tube (ERICKSON 1990), and the migration of retinal ganglion axons on astroglial cells is mediated in part by laminin (NEUGEBAUER et al. 1988). However, the requirement of matrix molecules in promoting and sustaining cell movements is not necessarily associated with a role for these matrix components in the induction of EMT.

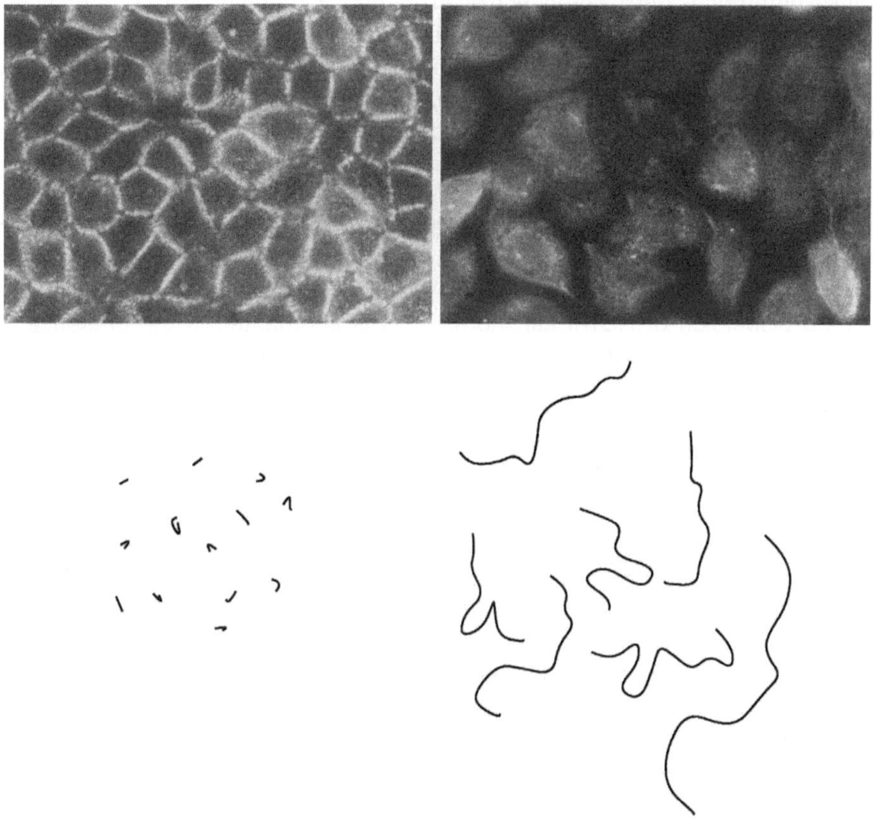

A B C D

Fig. 2A–D. Fibroblast growth factor (FGF-1)-induced of epithelium-to-mesenchyme transition (EMT) NBT-II cells is accompanied by cell dissociation and acquisition of cell motility. The NBT-II rat bladder carcinoma cells were processed for immunofluorescence microscopy with a desmoplakin monoclonal antibody (**A,B**). NBT-II cells were cultured under standard conditions (**A**) or in the presence of 30 ng/ml FGF-1 (**B**) for 18 h. **A** cells are linked by numerous desmosomes, as revealed by the bright desmoplakin cortical staining. **B** After incubation with FGF-1, desmoplakin immunoreactivity disappears from the cell surface and is now found in the cytoplasm, suggesting that desmosomes are internalized. **C, D** Video microcinematography recordings of NBT-II cells under standard conditions (**C**) or after addition of 30 ng/ml FGF-1 (**D**). **C** The tracks are very short, showing that NBT-II epithelial cells are not motile, while the length of the tracks in **D** is dramatically increased, demonstrating that FGF-1 induces cell movement

Nevertheless, the role played by ECM components in EMT processes has been demonstrated in few cases.

As in the case of growth factors, it is well established that the ECM can exert differentiating activities (HAU 1993). Remarkably, transdifferentiation from an epithelial program of differentiation to a mesenchymal one, which defines EMT events, is switched on by culturing a variety of differentiated epithelial cells in collagen gels (HAY 1993). This EMT event can be inhibited by retinoic acid, which promotes basement membrane formation, suggesting that the basal lamina, which determines the maintenance of an epithelial phenotype, can counteract the EMT-promoting effects of collagen. Basement membranes could therefore serve as negative regulators of EMT, explaining why epithelial tissues are normally protected from scattering. Collagen could therefore exert its scattering effects only when the basal lamina is destroyed, as happens during tumor progression.

We demonstrated that collagen fibrils, and not only tridimensional collagen gels, can induce EMT of carcinoma cells in culture (TUCKER et al. 1990). Furthermore, the scattering activity of growth factors can be modulated by specific matrix molecules: SF-mediated scattering of MDCK cells is enhanced in the presence of fibronectin (CLARK 1994). We also demonstrated that FGF-1-promoted dispersion of NBT-II carcinoma cells is augmented in the presence of collagen (TUCKER et al. 1991). The cooperation between matrix and growth factors might reflect a positive interference between matrix-mediated cellular signals leading to EMT and those induced by growth factors, or it might simply result from the fact that matrix elements display positive effects on cell dispersion by acting mechanically as permissive substrates for cell locomotion.

In other cases, matrix molecules are capable of repressing the cytokine-promoted EMT, as illustrated by the fact that fibronectin inhibits the EGF-induced scattering of liver epithelial cells (BADE and NITZGEN 1985). Whether ECM components exert their negative effect by interfering with cytokine-mediated signals or by simply acting as nonpermissive substrates for cell translocation remains unclear. Interestingly, the proliferative response of several types of normal and tumor cells to various growth factors has been shown to be inhibited by collagen fibrils (NISHIYAMA et al. 1991), suggesting that the composition of the ECM might also influence the development of tumors.

4 Signaling Cascades and Downstream Targets in Cytokine- and Matrix-Mediated Epithelium-to-Mesenchyme Transitions

4.1 Cell Adhesion Molecules

Adhesion molecules are obvious targets of growth factors and ECM-mediated scattering signals. Epithelial cells are interconnected by intercellular junctions in which cell adhesion molecules (CAMs) assume recognition and adhesion roles (Fig. 3). During EMT, intercellular adhesion is weakened or abrogated, allowing cell dispersion to occur. In the NBT-II carcinoma cell line, FGF-1 induces the breakdown and internalization of desmosomes, correlating with cell dispersion. Among the different components of desmosomes, desmoglein, a cadherin-like cell-adhesion molecule, is rapidly down-regulated (BOYER et al. 1989). Apart from desmosomal cadherin-like components, much attention has been paid to E-cadherin modulation during EMT processes, since it was long recognized that this

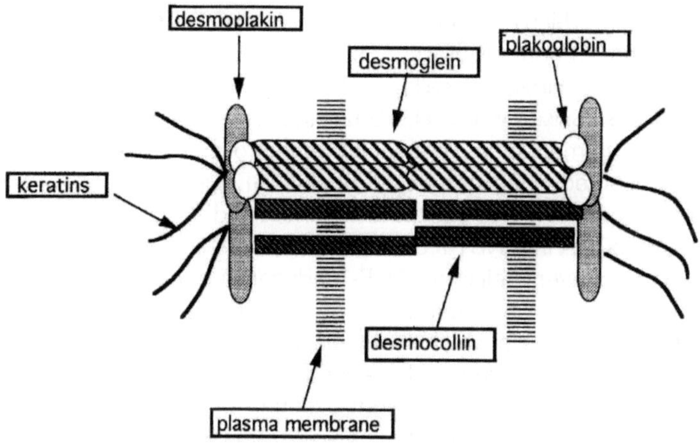

A

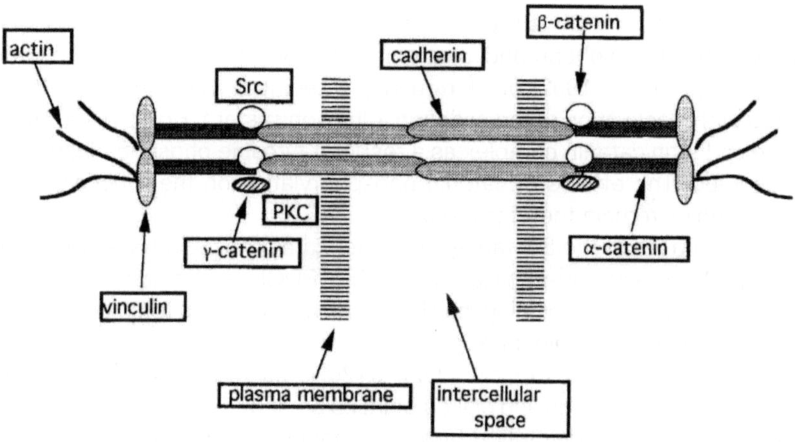

B

Fig. 3A,B. The structure of epithelial junctions. **A** Desmosome; **B** adherent junction. Only the major proteins interacting in the junctions are represented. The location of some signaling molecules represented in this figure is tentative

calcium-dependent cell adhesion molecule plays a fundamental role in the establishment and maintenance of epithelial structures (MÈGE et al. 1988; MCNEILL et al. 1990). In agreement with the current view on adhesion systems, E-cadherin is lost from colocarcinoma cells undergoing phorbol-ester induced cell scattering (FABRE and DE HERREROS 1993). Accordingly, a partial or complete absence of E-cadherin expression has been observed in several types of carcinomas of high invasive potential associated with a dedifferentiated phenotype (MAYER et al. 1993; BRINGUIER et al. 1993; OKA et al. 1993; GAMALLO et al. 1993; NAVARRO et al. 1991). However, cell scattering, correlating with an increase in invasive and metastatic potentials, is not necessarily associated with a down-modulation of E-cadherin, although the adhesive properties of the carcinoma cells are clearly altered. Cadherin function depends on its interactions with cytoplasmic molecules, the catenins, which mediate the binding of the COOH-terminal portion of cadherins with actin filaments (GUMBINER 1993). In few reported cases, alteration in cell-cell adhesion in carcinoma cells results from the loss of α- or β-catenin, due to gene mutations (SHIMOYAMA et al. 1992; KAWANISHI et al. 1995). In some puzzling cases, no down-modulation of any component of the cadherin adhesive complex could be observed. For example, during the FGF-1-induced EMT of NBT-II carcinoma cells, E-cadherin expression is not modified. Furthermore, transfection of E-cadherin cDNA into NBT-II cells, leading to stable overexpression of the molecule, does not prevent cells from FGF-1-induced dissociation and scattering, indicating that EMT does not result from a down-regulation of E-cadherin (BOYER et al. 1991). In the cases in which the levels of expression of cadherins and cadherin-associated molecules are not decreased, posttranscriptional modifications of the cadherin adhesion system might account for the reduced cell-cell adhesion observed during in vitro EMT or in vivo scattering of carcinoma cells. SHIBAMOTO and colleagues found β-catenin and plakoglobin/γ-catenin tyrosine phosphorylated during EGF and SF-stimulation of in vitro carcinoma cell scattering (SHIBAMOTO et al. 1994). Concurrently, $pp60^{v\text{-}src}$ was shown to cause tyrosine phosphorylation of E-cadherin and β-catenin (BEHRENS et al. 1993), of β-catenin alone (MATSUYOSHI et al. 1992) or of α- and β-catenins (HAMAGUCHI et al. 1993), associated with inactivation of the cadherin adhesion system. However, a disruption of the cadherin-catenin complex as a result of tyrosine phosphorylation was never observed. The effects of catenin phosphorylation on the function of the adhesive complex remain therefore to be clarified.

Tyrosine kinases of the Src family phosphorylate other substrates related to β-catenin: p120, previously implicated in EGF-, PDGF- and colony-stimulating factor-1α (CSF-1)-induced signaling, has been shown to contain four copies of the Arm repeat found in β-catenin, plakoglobin, and armadillo, the Drosophila homolog of plakoglobin (REYNOLDS et al. 1992). p120 is associated with E-cadherin complexes and, like β-catenin and plakoglobin, its Src-induced tyrosine phosphory lation does not cause its disruption from the cadherin-based adhesion complex (REYNOLDS et al. 1994).

In the context of cancer progression, it is noteworthy that β-catenin associates directly with the tumor suppressor protein adenomatous polyposis coli (APC), which is frequently found mutated in colocarcinomas (RUBINFELD et al. 1993; SU et al. 1993). Association with APC occurs via the Arm domain of β-catenin, suggesting that Arm is an important motif for mediating protein-protein interactions in adhesion complexes. Therefore, it is critical to determine whether tyrosine phosphorylation of the catenin family, mediated by scatter factors and oncogenic Src kinase, affects the association between APC and catenins and how this could lead to alterations in cell-cell adhesion.

The cadherin complex can also undergo positive regulations, leading to increased adhesiveness. In that respect, recent reports have provided evidence that the signaling molecule Wnt-1 up-regulates the expression of E-cadherin and plakoglobin (BRADLEY et al. 1993), stabilizes the association between β-catenin and E-cadherin (HINCK et al. 1994), and induces an epithelioid transformation of PC12 pheochromocytoma cells (SHACKELFORD et al. 1993). Understanding how different modes of regulation of E-cadherin complexes are orchestrated during EMT events is a central goal in the field.

4.2 Integrins

Matrix macromolecules capable of triggering in vitro EMTs are also essential for sustaining the migratory behavior of mesenchymal cells generated by EMT processes. Therefore, integrins that are the cellular receptors for ECM components are likely candidates for participating in EMT. Cell surface integrins are composed of an α and a β subunit linked by noncovalent bonds. The 13 α and eight β subunits can combine to form the 21 known integrins with different binding specificities. Integrins mediate cell-matrix as well as cell-cell interactions (GUMBINER 1993). In fibroblasts, integrins are primarily localized at the basal cell surface, where they often cluster in complexes called focal adhesions or focal contacts (BURRIDGE et al. 1988) (Fig. 4). Integrins were first considered as mechanical links between matrix and cells. It was then recognized that integrins participate in signaling processes, eventually leading to cell remodeling (HYNES 1992). In some cases, integrins have to be activated in order to exert their adhesive function. The molecular steps that couple cell activation to changes in the ability of an integrin to bind its extracellular ligand are far from being elucidated.

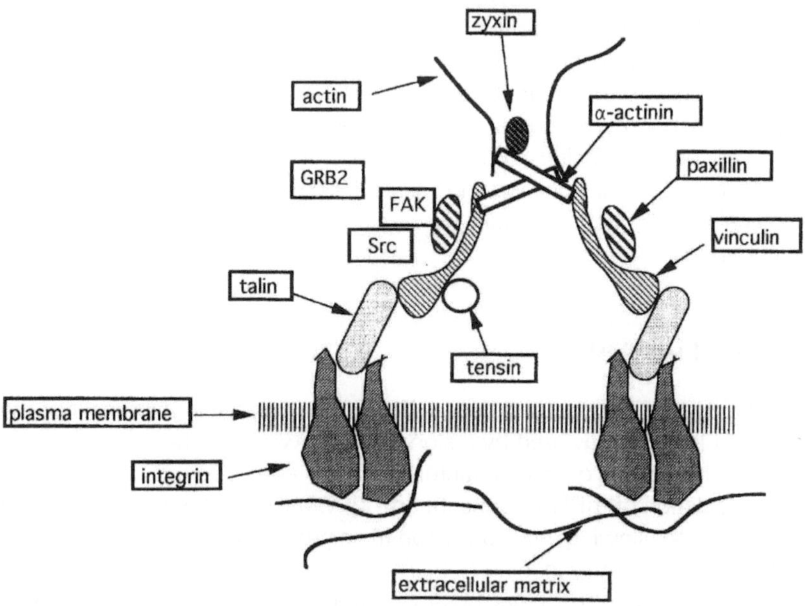

Fig. 4. Focal contact structure. The integrins bind to matrix macromolecules and at the same time interact with cytoskeletal components, which connect the transmembrane integrins with the actin filaments. The illustrated location of signaling molecules is tentative

However, changes in integrin adhesiveness are likely to arise from modifications in the affinity of the integrins for their ligands. The effector mechanisms that mediate this change in integrin affinity are not understood; they might include phosphorylation of the cytoplasmic tail of the α and β subunits (Diamond and Springer 1994).

Among matrix macromolecules, collagen is a major component capable of inducing in vitro EMT (Hay 1993). As demonstrated in our laboratory, while collagen and FGF-1 can induce separately the EMT of NBT-II carcinoma cells, the combination of both inducers greatly enhances the scattering response, suggesting that the two inducing machineries share common denominators (Tucker et al. 1991). Integrins, and particularly the collagen-specific integrins, are obvious candidates for playing this role. In that context, it is interesting to note that the EGF-induced EMT of a normal mammary epithelial cell line can be specifically inhibited by blocking antibodies directed against the integrin α2β1, which is a major collagen-binding integrin (Matthay et al. 1993). α2β1 might therefore represent one key element in the regulation of matrix and growth factor-induced EMT. How the signal caused by integrin ligation to collagen can produce the same effects as those generated by growth factor binding to its cellulat receptors, and how the two types of inductions can cooperate are unsolved questions. However, signaling molecules involved in transducing the message emanating from activated growth factor receptors have recently been shown to participate in integrin-mediated signaling pathways and could therefore represent converging points in the growth factor and ECM-induced EMTs (see below).

In addition to its role in EMT processes, α2β1 has been implicated in collagen-induced morphogenesis of normal mammary epithelial cells (Berdichevsky et al. 1992) and is involved in the contraction of collagen gels induced by fibroblasts in culture (Schiro et al. 1991). It is also implicated in metastasis of rhabdomyosarcoma cells (Chan et al. 1991) and is a marker of melanoma cancer progression (Klein et al. 1991). In contrast, its expression is often altered in carcinomas (Juliano and Varner 1993). Therefore, the role of α2β1 in pathological EMT processes leading to carcinoma cell invasion and metastasis remains uncertain.

4.3 Signaling Cascades

With respect to growth factor-induced EMT, two related questions can be formulated: (1) Is EMT regulated by a signaling pathway, initiated at the plasma membrane by the ligation of growth factor receptors and achieved in the nucleus by specific regulation of mRNA transcription? (2) What is the molecular basis for the cellular decision allowing cells to escape from growth factor-induced mitogen esis and to enter a pathway leading to EMT?

As a first approach, we have observed that confluent cultures of NBT-II carcinoma cells are stimulated to enter mitosis but not EMT, and, conversely, that sparse cultures undergo EMT but are growth-inhibited upon stimulation with

FGF-1 (VALLÉS et al. 1990b). The culture conditions in which cells are placed can therefore give information capable of directing cells toward the scattering or the mitogenic response. Since epithelial differentiation depends on the degree of cell compaction and is maximized in high density cultures, it is reasonable to speculate that this culture condition, reflecting a high degree of epithelial differentiation, can prevent cells from undergoing EMT. By using sparse and confluent cultures as tools for discriminating the two effects of FGF-1, we have determined that the scattering response is controlled by a signaling pathway different from that regulating mitogenesis (BOYER and THIERY 1993). The FGF-1-induced EMT is inhibited by a rise in intracellular cAMP concentrations, which does not pertub FGF-1- mediated entry into cell division. The inhibitory role of cAMP has also been demonstrated in EGF-induced cell scattering of rat liver epithelial cells (MANSKE et al. 1990).

The signaling pathways generated by different scattering factors might vary according to the type of inducing molecule. In that respect, the EMT occurring during embryonic heart formation is controlled by the activity of a member of the TGFβ family (POTTS et al. 1991). This process is positively regulated by tyrosine kinases but , in contrast to the FGF-1-induced EMT of NBT-II cells, it is not affected by an increase in cAMP levels (RUNYAN et al. 1990). Tyrosine kinase activities are generally implicated in cytoskeletal modifications leading to cell motility. This might result from the involvement of $pp125^{FAK}$, a focal contact-associated tyrosine kinase, in signal transduction during cell interactions with ECM. FAK is considered as a point of convergence between integrin and growth factor signaling pathways (ZACHARY and ROZENGURT 1992): it is activated by integrin ligation to matrix macromolecules and by a number of cytokines (ZACHARY et al. 1992; HORDIJK et al. 1994; ZACHARY and ROZENGURT 1992; SEUFFERLEIN and ROZENGURT 1994). Interestingly, integrin-mediated activation of FAK induces its association with GRB2 adaptor protein and with c-src protein kinase, resulting in the activation of mitogen-activated protein kinase (SCHLAEPFER et al. 1994). This might be the point at which integrin- and mitogen-mediated pathways converge. The association of FAK with the Src family of tyrosine kinases can induce the targeting of Src molecules to focal contacts, where they are preferentially localized (COBB et al. 1994).

In addition, HGF/SF stimulates the Ras-guanine nucleotide exchanger (GRAZIANI et al. 1993), which is required for the scattering activity of SF (HARTMANN et al. 1994). In agreement with these results, Ras activity is continuously required throughout the process of endothelial cell motility (FOX et al. 1994). Taken together, these data suggest that Ras activity, stimulated by either growth factor binding to the cell surface or integrin engagement, might be a central regulator of processes involving cell motility. Interestingly, the other small GTP-binding proteins rho and rac have been implicated in growth factor-induced actin reorganization (RIDLEY and HALL 1992; RIDLEY et al. 1992; NOBES and HALL 1995). These proteins could also participate in cytokine-mediated motogenic responses which also involve the reorganization of actin cytoskeleton.

Phospholipase C, another signaling molecule that binds to a number of activated growth factor receptors, is necessary for EGF-induced motility of NR6 fibroblastic cells (CHEN et al.1994), but not for FGF-mediated motility of L6 myoblasts or CHO fibroblasts (CLYMAN et al. 1994).

A signaling cascade thus appears to be required for EMT to proceed. The specificity of it, as opposed to signals leading to mitogenesis, remains obscure. The same type of cellular decision applies in the biological systems in which mitogenesis and cell differentiation can be alternatively induced by the same growth factors. During the stimulation of PC12 cells by insulin or EGF, MAP kinase is activated in the two pathways leading to either cell division or neuronal differentiation. However, the intensity and the duration of MAP kinase activation, associated with the entry of the enzyme into the nucleus, are high in insulin- or EGF-mediated cell differentiation and weak during mitogenesis. The duration of MAP kinase activation is directly correlated with the level of activation of insulin or EGF receptors and can be modulated by modifying the number of receptors expressed at the cell surface (DIKIC et al. 1994; TRAVERSE et al. 1994). The choice between EMT and mitogenesis could be governed by mechanisms similar to those for proliferation and differentiation.

5 Are Epithelium-to-Mesenchyme Transitions Regulated by Master Genes?

In order to proceed, EMT clearly requires transcriptional activity. Gene expression is modified to a large extent. For example, several secreted proteins are specifically associated with the EGF-induced fibroblastic state of differentiation (BADE and FEINDLER 1988), and the secretion of collagenases is rapidly induced during TGF-α-mediated EMT (GAVRILOVIC et al. 1990). The synthesis of cytoskeletal proteins is also affected: cytokeratins gradually disappear and vimentin starts to be expressed (BOYER et al. 1989). Given the coordination and large extent of changes in gene expression, it is tempting to assume that a group of genes under the control of a putative master gene exist for the mesenchymal phenotype and/or another for the epithelial phenotype. EMT would be therefore associated with the switch-on of the mesenchymal program and/or the switch-off of the epithelial program.

The nature of master genes, if they exist, remains elusive. E-cadherin, which is capable by itself to induce some aspects of epithelial differentiation (MÈGE et al. 1988; MCNEILL et al. 1990), does not bring about the full range of transformations of mesenchyme to epithelium and therefore cannot be considered as the unique master gene controlling the epithelial program. A recent report has shown that the function of Pax-2, a transcription factor belonging to the paired-box (Pax)-containing family of morphogenetic regulatory genes, is required for the mesenchyme-to-epithelium conversion arising in the developing kidney (ROTHENPIELER and DRESSLER 1993). Pax-2 might be important for the condensation of kidney mesenchyme and, like other Pax genes, for the subsequent delineation of boundaries between induced and noninduced cells. However, Pax-2 is not the only gene controlling mesenchyme-to-epithelium conversion, since N-myc is also an important factor during the early stages of epithelium formation (STANTON et al. 1992). The same kind of genetic cascade, controlling a complex series of events might also operate in EMT. In that respect, Slug, a zinc finger transcription factor homologous to the Drosophila snail gene, is expressed in epiblast cells shortly before they emerge and migrate as mesenchyme during the process of gastrulation. Slug is also expressed by precursors of neural crest cells shortly before the onset of cell emigration from the neural tube. Moreover, treatment of embryos with Slug antisense oligonucleotides results in the impairment or lack of EMT of neural crest cells and of mesodermal cells (NIETO et al. 1994). Thus, Slug appears as a key factor in the regulation of EMT. In addition, the activation of the product of the cellular immediate-early gene c-fos, but not of c-myc, is capable of triggering the EMT of normal mammary epithelial cells (REICHMANN et al. 1992). Therefore, it is tempting to postulate that EMT is not under the control of a single master gene, but that a series of genetic modifications, temporarily coordinated, bring about the morphogenetic process of EMT. It would be interesting to elucidate whether dispersion of carcinoma cells is controlled by transcription factors similar to those believed to operate during embryonic EMTs.

Acknowledgments. We are indebted to D. Morineau for his excellent photographic work. The work in the authors' laboratory is supported by the Centre National de la Recherche Scientifique (CNRS), the Association pour la Recherche contre le Cancer (ARC 64-65), the Ligue Française contre le Cancer (National Committee and Committee of Paris), the National Cancer Institute of the NIH (2R01 CA 49417-04), and the Human Frontier Science Program Organization.

References

Bade EG, Feindler S (1988) Liver epithelial cell migration induced by epidermal growth factor or transforming growth factor alpha is associated with changes in the gene expression of secreted proteins. In vitro Cell Dev Biol 24: 149–154

Bade EG, Nitzgen B (1985) Extracellular matrix (ECM) modulates the EGF-induced migration of liver epithelial cells in serum-free, hormone-supplemented medium. In Vitro Cell Dev Biol 21: 245–248

Barrandon Y, Green H (1987) Cell migration is essential for sustained growth of keratinocyte colonies: the roles of transforming growth factor-alpha and epidermal growth factor. Cell 50: 1131–1137

Behrens J, Vakaet L, Friis R, Winterhager E, Van Rοy F, Mareel MM, Birchmeier W (1993) Loss of epithelial differentiation and gain of invasiveness correlate with tyrosine phosphorylation of the E-cadherin/β-catenin complex in cells transformed with a temperature-sensitive v-SRC gene. Cell Biol 120: 757–766.

Berdichevsky F, Gilbert C, Shearer M, Taylor-Papadimitriou J (1992) Collagen-induced rapid morphogenesis of human mammary epithelial cells: the role of the α2β1 integrin. J Cell Sci 102: 437–446.

Blay J, Brøwn KD (1985) Epidermal growth factor promotes the chemotactic migration of cultured rat intestinal epithelial cells. J Cell Physiol 124: 107–112

Boyer B, Thiery JP (1993) Cyclic AMP distinguishes between two functions of acidic FGF in a rat bladder carcinoma cell line. J Cell Biol 120: 767–776

Boyer B, Tucker GC, Vallés AM, Franke WW, Thiery JP (1989) Rearrangements of desmosomal and cytoskeletal proteins during the transition from epithelial to fibroblastoid organization in cultured rat bladder carcinoma cells. J Cell Biol 109: 1495–1509

Boyer B, Dufour S, Thiery JP (1991) E-cadherin expression during the acidic FGF-induced dispersion of a rat bladder carcinoma cell line. Exp Cell Res 201: 347–357
Bradley RS, Cowin P, Brown AMC (1993) Expression of Wnt-1 in PC12 cells results in modulation of plakoglobin and E-cadherin and increased cellular adhesion. J Cell Biol 123: 1857–1865
Bringuier PP, Umbas R, Schaafsma HE, Karthaus HFM, Debryune FMJ, Schalken JA (1993) Decreased E-cadherin immunoreactivity correlates with poor survival in patients with bladder tumors. Cancer Res 53: 3241–3245
Burridge K, Fath K, kelly T, Nuckolls G, Turner C (1988) Focal adhesions: transmembrane junctions between the extracellular matrix and the cytoskeleton. Annu Rev Cell Biol 4: 487–525
Chan BMC, Matsuura N, Takada Y, Zetter BR, Hemler ME (1991) In vitro and in vivo consequences of VLA-2 expression on rhabdomyosarcoma cells. Science 251: 1600–1602
Chen P, Xie H, Sekar MC, Gupta K, Wells A (1994) Epidermal growth factor receptor-mediated cell motility: phospholipase C activity is required, but mitogen-activated protein kinase activity is not sufficient for induced cell movement. J Cell Biol 127: 847–857
Clark P (1994) Modulation of scatter/hepatocyte growth factor activity by cell-substratum adhesion. J Cell Sci 107: 1265–1275
Clyman RI, Peters KG, Chen YQ, Escobedo J, Williams LT, Ives HE, Wilson E (1994) Phospholipase Cγ activation, phosphotidylinositol hydrolysis, and calcium mobilization are not required for FGF receptor-mediated chemotaxis. Cell Adhesion Commun 1: 333–342
Cobb BS, Schaller MD, Leu T-H, Parsons JT (1994) Stable association of pp60^{src} and pp60^{fyn} with the focal adhesion-associated protein tyrosine kinase, pp125^{FAK}. Mol Cell Biol 14: 147–155
Diamond MS, Springer TA (1994) The dynamic regulation of integrin adhesiveness. Curr Opin Cell Biol 4: 506–517
Dikic I, Schlessinger J, Lax I (1994) PC12 cells overexpressing the insulin receptor undergo insulin-dependent neuronal differentiation. Curr Opin Cell Biol 4: 702–708
Erickson CA (1990) Cell migration in the embryo and adult organism. Curr Opin Cell Biol 2: 67–74
Fabre M, de Herreros AG (1993) Phorbol ester-induced scattering of HT-29 human intestinal cancer cells is associated with down-modulation of E-cadherin. J Cell Sci 106: 513–522
Fox PL, Sa G, Dobrowolski SF, Stacey DW (1994) The regulation of endothelial cell motility by p21ras. Oncogene 9: 3519–3526
Frixen UH, Behrens J, Sachs M, Eberle G, Voss B, Warda A, Lochner D, Birchmeier W (1991) "E-cadherin-mediated cell-cell adhesion prevents invasiveness of human carcinoma cells". J Cell Biol 113: 173–185
Gavrilovic J, Moens G, Thiery JP, Jouanneau J (1990) Expression of transfected transforming growth factor alpha induces a motile fibroblast-like phenotype with extracellulat matrix-degrading potential in a rat bladder carcinoma cell line. Cell Regul 1: 1003–1014
Geimer P, Bade EG (1991) The epidermal growth factor-induced migration of rat liver epithelial cells associated with a transient inhibition of DNA synthesis. J Cell Sci 100: 349–355
Graziani A, Gramaglia D, dalla Zonca P, Comoglio PM (1993) Hepatocyte growth factor/scatter factor stimulates the Ras-guanine nucleotide exchanger. J Biol Chem 268: 9165–9168
Grotendorst GR (1984) Alteration of the chemotactic response of NIH/3T3 cells to PDGF by growth factors, transformation, and tumor promotors. Cell 36: 279
Gumbiner BM (1993) Proteins associated with the cytoplasmic surface of adhesion molecules. Neuron 11: 551–564
Hamaguchi M, Matsuyoshi N, Ohnishi Y, Gotoh B, Takeichi M, Nagai Y (1993) P60$^{v\text{-}src}$ causes tyrosine phosphorylation and inactivation of the N-cadherin-catenin cell adhesion system. EMBO J 12: 307–314
Harris H (1990) The role of differentiation in the suppession of malignancy. J Cell Sci 97: 5–10
Hartmann G, Weidner KM, Schwarz H, Birchmeier W (1994) The motility signal of scatter factor/hepatocyte growth factor mediated through the receptor tyrosine kinase Met requires intracellular action of Ras. J Biol Chem 269: 21936–21939
Hay ED (1993) Extracellular matrix alters epithelial differentiation. Curr Opin Cell Biol 5: 1029–1035
Hinck L, Nelson WJ, Papkoff J (1994) Wnt-1 modulates cell-cell adhesion in mammalian cells by stabilizing β-catenin binding to the cell adhesion protein cadherin. J Cell Biol 124: 729–741
M Cano, A (1993) Correlation of E-cadherin expression with differentiation grade and histological type in breast carcinoma. Am J Pathol 142: 987–993
Hordijk PL, Verlaan I, van Corven EJ, Moolenaar WH (1994) Protein tyrosine phosphorylation induced by lysophosphatidic acid in Rat-1 fibroblasts. J Biol Chem 269: 645–651
Hugues FJ, McCulloch AG (1991) Quantification of chemotactic response of quiescent and proliferating fibroblasts in Boyden chambers by computer-assisted image analysis. J Histochem Cytochem 39: 243–246

Hynes RO (1992) Integrins: versatility, modulation, and signalling in cell adhesion. Cell 69: 11–25

Juliano RL, Varner JA (1993) Adhesion molecules in cancer: the role of integrins. Curr Opin Cell Biol 5: 812–818

Kawanishi J, Kato J, Sasaki K, Fujii S, Watanabe N, Niitsu Y (1995) Loss of E-cadherin-dependent cell-cell adhesion due to mutation of the β-catenin gene in a human cancer cell line, HCS-39. Mol Cell Biol 15: 1175–1181

Klein CE, Steinmayer T, Kaufmann D, Weber L, Bröcker E-B (1991) Identification of a melanoma progression antigen as integrin VLA-2. J Invest Dermatol 96: 281–284

Klemke RL, Yebra M, Bayna EM, Cheresh DA (1994) Receptor tyrosine kinase signalling required for integrin αVβ5 cell motility but not adhesion on vitronectin. J Cell Biol 127: 859–866

Manske M, Feindler S, Bade EG (1990) The epidermal growth factor-induced cell migration and expression of the 47000 M_r secreted glycoprotein EIP-1 of rat liver epithelial cells are down-regulated by cyclin. AMP Eur J Cell Biol 52: 201–206

Mareel MM, Behrens J, Birchmeier W, De Bruyne GK, Vleminckx, Hoogewijs A, Fiers WC, Van Roy FM (1991) Down-regulation of E-cadherin expression in Madin Darby canine kidney (MDCK) cells inside tumors of nude mice. Int J Cancer 47: 922–928

Matsuyoshi N, Hamaguchi M, Taniguchi S, Nagafuchi A, Tsukita S, Takeichi M (1992) Cadherin-mediated cell-cell adhesion is perturbed by v-src tyrosine phosphorylation in metastatic fibroblasts. J Cell Biol 118: 703–714

Matthay MM, Thiery JP, Lafont F, Stampfer MF, Boyer B (1993) Transient effectof epidermal growth factor on the motility of an immortalized mammary epithelial cell line. J Cell Sci 106: 869–878

Mayer B, Johnson JP, Leitl F, Jauch KW, Heiss MM, Schildberg FW, Birchmeier W, Funke I (1993) E-cadherin expression in primary and metastatic gastric cancer: down-regulation correlates with cellular dedifferentiation and glandular disintegration. Cancer Res 53: 1690–1695

McNeill H, Ozawa M, Kemler R, Nelson WJ (1990) Novel function of the cell adhesion molecule uvomorulin as an inducer of cell surface polarity. Cell 62: 309–316

Mège RM, Matsuzaki F, Gallin W, Goldberg JI, Cunningham B, Edelman GM (1988) Construction of epithelioid sheets by transfection of mouse sarcoma cells with cDNAs for chicken cell adhesion molecules. Proc Natl Acad Sci USA 85: 7274–7278

Miettinen PJ, Ebner R, Lopez AR, Derynck R (1994) TGF-β induced transdifferentiation of mammary epithelial cells to mesenchymal cells: involvement of type I receptors. J Cell Biol 127: 2021–2036

Navarro P, Gomez M, Pizarro A, Gamallo C, Quintanilla M, Cano A (1991) A role for E-cadherin cell-cell adhesion molecule during tumor progression of mouse epidermal carcinogenesis. J Cell Biol 115: 517–533

Neugebauer KM, Tomaselli KJ, Lilien J, Reichardt LF (1988) N-cadherin, NCAM, and integrins promote retinal neurite outgrowth on astrocytes in vitro. J Cell Biol 107: 1177–1187

Nieto MA, Sargent MG, Wilkinson DG, Cooke J (1994) Control of cell behavior during vertebrate development by Slug, a zinc finger gene. Science 264: 835–839

Nishiyama T, Akutsu N, Horii I, Nakayama Y, Ozawa T, Hayashi T (1991) Response to growth factors of human dermal fibroblasts in quiescent state owing to cell-matrix contact inhibition. Matrix 11: 71–73

Nobes CD, Hall H (1995) Rho, rac, and Cdc42 GTPases regulate the assembly of multimolecular focal complexes associated with actin stress fibers, lamellipodia, and filopodia. Cell 81: 3–62

Oka H, Shiozaki H, Inoue M, Kobajashi K, Tahara H, Kobayashi T, Takatsua Y, Matsuyoshi N, Hirano S, Takeichi M, Mori T (1993) Expression of E-cadherin cell adhesion molecules in human breast cancer tissues and its relationship to metastasis. Cancer Res 53: 1696–1701

Potts JD, Dagle JM, Walder JA, Weeks DL, Runyan RB (1991) Epithelial-mesenchymal transformation of embryonic cardiac endothelial cells is inhibited by a modified antisense oligodeoxynucleotide to transforming growth factor β3. Proc Natl Acad Sci USA 88: 1516–1520

Reichmann E, Schwarz H, Deiner EM, Leitner I, Eilers M, Berger J, Busslinger M, Beug H (1992) Activation of an inducible c-FosER fusion protein causes loss of epithelial polarity and triggers epithelial-fibroblastoid cell conversion. Cell 71: 1103–1116

Reynolds AB, Herbert L, Cleveland JL, Berg ST, Gaut JR(1992) p120, a novel substrate of protein tyrosine kinase receptors and of $p60^{v\text{-}src}$, is related to cadherin-binding factors β- catenin, plakoglobin and armadillo. Oncogene 7: 2439–2445

Reynolds AB, Daniel J, McCrea PD, Wheelock MJ, Wu J, Zhang Z(1994) Identification of a new catenin: the tyrosine kinase substrate $p120^{cas}$ associates with E-cadherin complexes. Mol Cell Biol 14: 8333–8342

Ridley AJ, Paterson HF, Johnston CL, Diekmann D, Hall A (1992) The small GTP-binding protein rac regulates growth factor-induced membrane ruffling. Cell 70: 401–410

Rosen EM, Nigam SK, Goldberg ID (1994) Scatter factor and the c-met receptor: a paradigm for mesenchymal/epithelial interaction. J Cell Biol 127: 1783–1787
Rothenpieler UW, Dressler GR (1993) Pax-2 is required for mesenchyme-to-epithelium conversion during kidney development. Development. 119: 711–720
Rubinfeld B, Souza B, Albert I, Müller O, Chamberlain SH, Masiarz FR, Munemitsu S, Polakis P (1993) Association of the APC gene product with β-catenin. Science 262: 1731–1734
Runyan RB, Potts JD, Sharma RV, Loeber CP, Chiang JJ, Bhalla RC (1990) Signal transduction of a tissue interaction during embryonic heart development. Cell Regul 1: 301–313
Schiro JA, ChanBMC, Roswit WT, Kassner PD, Pentland AP, Hemler ME, Eisen AZ, Kupper TS (1991) Integrin α2β1 (VLA-2) mediates reorganization and contraction of collagen matrices by human cells. Cell 67: 403–410
Schlaepfer DD, Hanks SK, Hunter T, van der Geer P (1994) Integrin-mediated signal transduction linked to Ras pathway by GRB2 binding to focal adhesion kinase. Nature 372: 786–791
Seufferlein T, Rozengurt E (1994) Lysophosphatidic acid stimulates tyrosine phosphorylation of focal adhesion kinase, paxillin, and p130. J Biol Chem 269: 9345–9351
Shackelford GM, Willert K, Wang J, Varmus HE (1993) The Wnt-1 proto-oncogene induces changes in morphology, gene expression, and growth factor responsiveness in PC12 cells. Neuron 11: 865–875
Shibamoto S, Hahakawa M, Takeuchi K, Hori T, Oku N, Miyazawa K, Kitamura N, Takeichi M, Ito, F (1994) Tyrosine phosphorylation of β-catenin and plakoglobin enhanced by hepatocyte growth factor and epidermal growth factor in human carcinoma cells. Cell Adhesion Commun 1: 295–305
Shimonaka M, Yamaguchi Y (1994) Purification and biological characterization of epitaxin, a fibroblast-derived motility factor for epithelial cells. J Biol Chem 269: 14284–14289
Shimoyama Y, Nagafuchi A, Fujita S, Gotoh M, Takeichi M, Tsukita S, Hirohashi S (1992) Cadherin dysfunction in a human cancer cell line: possible involvement of loss of α-catenin expression in reduced cell-cell adhesiveness. Cancer Res 52: 5770–5774
Shiota G, Rhoads DB, Wang TC, Nakamura T, Schmidt EV (1992) Hepatocyte growth factor inhibits growth of hepatocellular carcinoma cells. Proc Natl Acad Sci USA 89: 373–377
Stampfer MR, Bartley JC (1988) Human mammary epithelial cells in culture: differentiation and transformation. In: Lippman M, Dickson R, Breast Cancer: cellular and molecular biology. Kluwer Academic, Boston, pp 1–24
Stanton BR, Perkins AS, Tessarollo L, Sassoon DA, Parada LF (1992) Loss of N-myc function results, in embryonic lethality and failure of the epithelial component of the embryo to develop. Genes Dev 6: 2235–2247
Stoler AB, Stenback F, Balmain A (1993) The conversion of mo skin squamous cell carcinomas to spindle cell carcinomas is a recessive event. Cell Biol 122: 1103–1117
Su L-K, Vogelstein B, Kinzler KW (1993) Association of the APC tumor suppressor protein with catenins. Science 262: 1734–1737
Traverse S, Seedorf K, Paterson H, Marshall CJ, Cohen P, Ullrich A (1994) EGF triggers neuronal differentiation of PC12 cells that overexpress the EGF receptor. Curr Opin Cell Biol 4: 694–701
Tsarfaty I, Rong S, Resau JH, Rulong S, Pinto da Silva P, Vande Woude GF (1994) The met proto-oncogene mesenchymal to epithelial cell conversion. Science 263: 98– 101
Tucker GC, Boyer B, Gavrilovic J, Emonard H, Thiery JP (1990) Collagen-mediated dispersion of NBT–II rat bladder carcinoma cells. Cancer Res 50: 129–137
Tucker GC, Boyer B, Vallés AM, Thiery JP (1991) Combined effects of extracellular matrix and growth factors on NBT-II rat bladder carcinoma cell dispersion. Cell Sci 100: 371–380
Vallés AM, Boyer B, Badet J, Tucker GC, Barritault D, Thiery JP (1990a) Acidic fibroblast growth factor is a modulator of epithelial plasticity in a rat bladder carcinoma cell line. Proc Natl Acad Sci USA 87: 1124–1128
Vallés AM, Tucker GC, Thiery JP, Boyer B (1990b) Alternative patterns of mitogensis and cell scattering induced by acidic FGF as a function of cell density in a rat bladder carcinoma cell line. Cell Regul 1: 975–988
Woolf AS, Kolatsi-Joannou M, Hardman P, Andermacher E, Moorby C, Fine LG, Jat PS, Noble MD, Gherardi E (1995) Roles of hepatocyte growth factor/scatter factor and the met receptor in the early development of the metanephros. J Cell Biol 128: 171–184
Zachary I, Rozengurt E (1992) Focal adhesion kinase (p125FAK): a point of convergence in the action of neuropeptides, integrins, and oncogenes. Cell 71: 891–894
Zachary I, Sinnet-Smith J, Rozengurt E (1992) Bombesin, vasopressin, and endothelin stimulation of tyrosine phosphorylation in Swiss 3T3 cells. Chemotherapy 267: 19031–19034

Involvement of α6 and αv Integrins in Metastasis

B.A. Imhof[1], L. Piali[1], R.H. Gisler[1], and D. Dunon[2]

1 Introduction

The occurrence of metastasis requires detachment of metastatic tumor cells of the primary tumor. It is known that loss of several adhesion molecules by cells of the primary tumor is involved in this event, for example, E-cadherin in carcinomas (Frixen et al. 1991) and α4 integrin in melanomas (Qian et al. 1994). Migrating metastatic cells enter the blood vasculature or the lymphatics through which they are transported to peripheral organs. In the periphery the cells must adhere in some way to the vascular endothelium before they can extravasate and form a secondary tumor. This chapter describes the adhesion molecules α6 and αv integrins and shows at what point they are involved in the adhesion and migration of metastatic tumor cells.

Adhesion of circulating cells to vascular endothelium followed by extravasation is best studied with leukocytes of the immune system. Recently it has been suggested that similar mechanisms can be found in tumor cells undergoing metastasis. Adhesion to endothelium is not merely a one-molecular event: it is a complex multimolecular mechanism (Imhof and Dunon 1995). Three laboratories recently presented a model describing how hemopoietic cells from the immune system can interact with the endothelium before they finally extravasate (Butcher 1991; Schweighoffer and Shaw 1992; Springer 1994). This model shows in three

[1]Institute for Immunology, 4002 Basel, Switzerland
[2]Université Pierre et Marie Curie, CNRS URA 1135, 75005 Paris, France

steps how circulating cells, including metastasizing cells, interact with the endothelium. There is a first, low-affinity interaction with the endothelium which leads to rolling of loose adherent cells along the vascular surface mediated by adhesion molecules of the selectin or integrin families (MACKAY 1995). It then requires a second step, which is called triggering, the activation of integrins present on the circulating cell. This mechanism consists of chemokines (BAGGIOLINI and DAHINDEN 1994; DURUM and OPPENHEIM 1993; TANAKA et al. 1993) and/or by platelet/endothelial cell adhesion molecule 1 (PECAM-1/CD31 (PIALI et al. 1993; TANAKA et al. 1992). We know that almost 20 different chemokines can be expressed by endothelial or other cells; upon contact with the rolling cell on the endothelium chemokines activate the integrins. Only after this triggering mechanism does a circulating cell come to a halt on the vascular endothelium before the cell extravasates.

2 Integrins

Integrins are classical adhesion molecules that can interact with ligands, which are mostly components of the extracellular matrix (HYNES 1992). However, some integrins can also mediate cell-cell interactions. For instance, α4β1 integrin is a receptor for fibronectin, but it also binds to the transmembrane vascular cell adhesion molecule 1 (VCAM-1); α4β7 integrin binds to the mucosal endothelial addressin cell adhesion molecule 1 (MAdCAM-1); αEβ7 binds to epithelial E-cadherin, and αLβ2 and αMβ2 integrins bind to the family of intercellular cell adhesion molecules. It is important to remember that the ligand specificity of β1 and β2 integrins depends mostly on the associated α-chain, and that the ligand-binding site is not always active. Upon intracellular signals integrins undergo a rapid transition from a nonadhesive, low-affinity state to a transient, high-affinity state. It is likely that the increased adhesiveness is due to a conformational change of the integrin molecules caused by activation of the cells, for instance, by chemokines. On the other hand, integrins can also transduce the information of ligand binding from the outside into the cell. For example, α6β1 integrin occupancy leads to the tyrosine phosphorylation of 52 and 90 kDa proteins; α3β1 integrin signaling leads to a 55 kDa phosphoprotein whereas signal transduction by α5β1 integrin has no effect on the phosphorylation of these proteins (JEWELL et al. 1995). Engagement of α6 integrin by the ligand laminin has been shown to regulate chemotaxis receptors (BLOOD and ZETTER 1993). The signaling may be transmitted by cytoskeletal elements which are associated with the integrin β chain. Interestingly, signal transduction by integrins is involved in the regulation of the cell adhesion process. Two such mechanisms take part in tumor metastasis, one mediated by the α6 integrin and the other by the αvβ3 integrin.

3 α6 Integrin

We recently discovered that α6 integrins expressed by vascular endothelium and by cells of the immune system are involved in the colonization of the mouse embryonic thymus by pro-T lymphocytes (RUIZ et al. 1995). A closer investigation of this mechanism revealed that it clearly resembled what is seen during metastasis. We used a B16 melanoma cell subline which is highly metastatic for the lung and found that these cells effectively adhered to frozen sections of a mouse lung (RUIZ et al. 1993). When this adhesion assay was performed in the presence of an antibody (EA-1) which recognizes the α6 chain of the α6β1 and α6β4 integrins, the melanoma cells no longer adhered to the lung tissue. We were then interested in determining whether the antibody would react in a similar way in vivo. For this reason we injected B16 melanoma cells into the tail veins of C57Bl/6 mice; after 12 days we observed metastatic melanoma nodules in the lung. When the same experiment was carried out in the presence of the EA-1 anti-α6 integrin antibody, no nodules appeared in the lung, i.e., the tumor cells were not able to colonize the lung. The next question to address was how and where the α6 integrin exerts its role as an adhesion molecule. It was found that α6 integrin is expressed by almost all blood vessels on the luminal and basal cell membrane, hence accessible for circulating cells, and is used by the blood vessel itself to adhere to the surrounding extracellular matrix. We found that blocking of the α6 integrin on the luminal side of the blood vessel can prevent the colonization of the lung by metastatic melanoma cells (RUIZ et al. 1993). This suggests that the α6 integrin may have a ligand on the circulating metastatic cell. To study this we made soluble recombinant α6β1 integrin and removed the transmembrane and cytoplasmic portions. The α-chain was linked to an immunoglobulin Cκ domain which was used as a tag for affinity purification by anti-Cκ antibodies. The soluble α6β1 integrin was coated as a substrate onto the surface of culture dishes. Preliminary experiments suggested that in an adhesion assay the cells can adhere to this recombinant molecule. From this we concluded that the α6β1 integrin may have a putative ligand which mediates cell-cell contact between the circulating metastatic cells and the vascular endothelium. Cellular ligands for the α6 integrin expressed by the human egg have recently been found on sperm. They are involved in the binding and fusion of sperms to the egg and have partial homologies to disintegrins (ALMEIDA et al. 1995). It remains to be seen whether molecules with disintegrin domains also play a role in α6 integrin–endothelial interactions.

When adhesion assays using the α6 integrin were performed, it was apparent that in addition to the vascular endothelium many tumor cells also expressed the α6 integrin on their cell surface. When α6 integrin expressed by B16 melanoma cells was blocked by antibodies, the melanoma cells no longer formed experimental metastases in vivo (RUIZ et al. 1993). Together with the experiments described above this suggested that α6 integrins on both the vascular endothelium and the tumor cells were actively involved in the metastatic invasion process. The presence of α6 integrin on cancer cells prompted us to investigate several primary human tumors. In collaboration with K. FRIEDRICHS (Hamburg), F. FRANKE (Giessen), and H.J. TERPE (Giessen), invasive breast tumors of 119 patients were analyzed, and the disease course was followed over 70 months: 105 patients survived this period while 14 died from the effects of metastasis.

It was clear from the beginning that in normal breast tissue the α6 integrin was strongly expressed by epithelial cells from the mammary gland, and that the expression was enriched at contact regions of epithelial cells and the connective tissue; thus α6 integrin expression was polarized (TERPE et al. 1994). The most striking finding with the tumors was that many lost the expression of α6 integrin (24%; FRIEDRICHS et al. 1995). As shown by many laboratories, benign tumors from different tissue origin can downregulate α6 integrin expression. In contrast, metastatic head, neck, and skin carcinomas highly express the α6 integrin (TENNENBAUM et al. 1992; WOLF and CAREY 1992; WOLF et al. 1990). After inspection of all the 119 tumors in our study a large number was found that still expressed the α6 integrin at different levels. However, the expression of α6 integrin was no longer polarized and directed to the extracellular matrix as in normal tissue, but it was expressed on all the tumor cells in a nonordered, nonpolarized way. Since the tumors were observed at long term, it was possible to compare the survival rate of the patients with the expression of α6 integrin. All the patients with non-α6 integrin-expressing tumors survived. In the group of 63 patients who showed tumors with high α6 integrin expression 10 patients died from metastasis. This indicates that α6 integrin expression is related to enhanced metastatic spread of tumor cells which ultimately leads to death.

The mechanism by which the α6 integrin provoke the tumor cell to interact with the vascular endothelium seemed to be complex. The α6 integrins are classical adhesion molecules. However, we found that upon ligand occupancy of the α6 integrin on the endothelium, tumor cells adhered better by a factor of two to three. This increased adhesion could be obtained by crosslinking part of the endothelial α6 integrin with a low concentration of antibodies. As this effect occurred only at 37°C and was energy dependent, we suggest that α6 integrin can activate an additional adhesion molecule by intracellular signaling.

4 αvβ3 Integrin

As mentioned above, benign tumors can downregulate α6 integrin expression, and metastatic tumors highly express α6 integrins (FRIEDRICHS et al. 1995; TENNENBAUM et al. 1992; WOLF and CAREY 1992; WOLF et al. 1990). Similar modulation of expression by benign and malign tumors can also be found with the integrin αvβ3. It shows a low expression in benign tumors but high expression in malignant metastases (ALBELDA and BUCK 1990; FILARDO et al. 1995; NIP et al. 1992). We recently developed an antibody that blocked the adhesion of tumor cells to endothelium and recognized the adhesion molecule PECAM-1/CD31. This is expressed by most immune cells and by all vascular endothelial cells. PECAM-1 is a transmembrane glycoprotein of the immunoglobulin gene superfamily (IgSF), with six Ig-like homology units mediating leukocyte-endothelial interactions. The adhesive interactions mediated by PECAM-1 are complex and include homophilic (PECAM-1/PECAM-1) or heterophilic (PECAM-1/X) contacts. It was assumed that the heterophilic ligands for PECAM-1 would bind to the Ig domain 2. Recently we found that domain 2 can bind to the αvβ3 integrin which, as like PECAM-1, is expressed by leukocytes, endothelial cells, tumor cells, and cells from various other tissues (ALBELDA and BUCK 1990; GIANCOTTI and MAINIERO 1994; PIALI et al. 1995). PECAM-1 has been shown to be a ligand for αvβ3 integrin by the construction of recombinant, soluble PECAM-1 fusion proteins (PIALI et al. 1995). These consisted of the extracytoplasmic Ig domains 1–6 (CD31-6D), 1–3 (CD31-3D), and 1–6 lacking domain 2 (CD31-6DΔ2). Similar to the α6 integrin constructs, all three PECAM-1/CD31 molecules were linked to an immunoglobulin Cκ domain. Upon coating these molecules to culture dishes, cells adhered to the CD31-6D and the CD31-3D forms, but no adhesion was obtained to the CD31-6DΔ2. Further experiments suggested that cell binding to CD31-6D is due to homo- and heterophilic interactions, whereas binding to CD31-3D is exclusively heterophilic. Because the CD31-3D form required a heterophilic ligand and did not interact homophilically with intact PECAM-1, we investigated the cellular interaction with the CD31-3D form to identify the heterophilic ligand. Two monoclonal antibodies against the αv integrin chain and one against β3 blocked cell adhesion to CD31-3D coated surfaces. Many antibodies against adhesion molecules, such as the αL, α6, α4, and β1 integrins, had no effect. Furthermore, after covalently coupling onto sepharose beads the CD31-3D molecule was able to precipitate αvβ3 integrin from a lysate of ^{125}I-iodinated surface proteins of endothelial cells and lymphocytes.

In conclusion, we demonstrated that αvβ3 integrin is a new ligand for PECAM-1 (PIALI et al. 1995). This integrin seems to interact directly with the Ig domain 2 of cellular PECAM-1. However PECAM-1 domain 2 is not active in all PECAM-1 molecules, and it may depend on activation of PECAM-1 itself. Such regulation may depend on signaling mediated by the cytoplasmic residue of the molecule since it has been shown that heterophilic, calcium-dependent adhesion is hampered when PECAM-1 is truncated at its cytoplasmic domain (DELISSER et al. 1994; KIRSCHBAUM et al. 1994). Several PECAM-1 molecules are expressed by vascular endothelium; they are generally identical in their extracellular domains but vary in their cytoplasmic tails due to alternative splice products of the mRNA. Thus expression of a membrane adhesion molecule by a cell does not necessarily imply that this molecule is active and can bind to its potential ligand. Identifying a PECAM-1 which can bind the αvβ3 integrin will require assaying of the function, for example, by testing cell adhesion to coated αvβ3 integrin molecules.

We have recently found that RGD peptides can block the interaction of the αvβ3 integrin with the PECAM-1 molecule. Several laboratories have shown that RGD peptides also block metastasis, in particular the homing of metastatic tumor cells to the lung in an experimental system (GIANCOTTI and MAINIERO 1994; HUMPHRIES et al. 1986). Until now it was thought that RDG peptides interfere with platelets' interaction with the tumor cell or with the adhesion of tumor cell integrins with the extracellular matrix. We are now testing the hypothesis that these peptides can block the first contact of the tumor cell with the vascular endothelium; this may explain why RGD peptides are extremely potent inhibitors of metastasis.

Further evidence for the involvement of αvβ3 integrin in extravasation comes from a recent study which showed that antibodies to integrin-associated protein (IAP) block transendothelial migration (Cooper et al. 1995). Integrin-associated protein is a 5 transmembrane protein which pairs with and regulates αvβ3 integrin (Lindberg et al. 1993).

5 Cell Migration Along the Endothelial Cell Precedes Extravasation

Before a metastatic cell can leave the vascular system it must adhere to the endothelium and then migrate along the endothelial cells until it reaches the endothelial cell-cell junctions. Extravasation starts by breaching the endothelial junctions. Cell migration along the endothelial cell wall is mediated by several endothelial molecules simultaneously serving as ligands for adhesion molecules expressed by the migrating cell. The question was how a migrating cell can integrate the interactions with several different adhesion molecules simultaneously. We investigated cell migration mediated by endothelial VCAM-1 and its ligand α4β1 integrin, known to be involved in tumor metastasis (Rice and Bevilacqua 1989), and the pair of adhesion molecules PECAM-1 and αvβ3 integrin. The speed of cell locomotion was studied by coating culture dishes with recombinant VCAM-1 or PECAM-1 (CD31-3D). Analysis was performed by measuring the tracks of migrating cells recorded with a video-microscope equipped with a time-lapse machine. The more CD31-3D molecules that were coated, the better the cells migrated (concentrations 0.1–2.4 μ*M*). In contrast, a more complex result was obtained with cells migrating along VCAM-1. Upon coating low concentrations of VCAM-1 (0.1 μ*M*) the cells migrated slowly. Coating with higher amounts of V-CAM (0.3 μ*M*) caused the cells to migrate with a higher optimal speed, but they became immobilized when the coated VCAM-1 concentrations reached saturation at 2.4 μ*M*. Briefly, VCAM-1, the ligand for the α4β1 integrin, provoke optimal cell migration when only a small number of the cellular integrin molecules are occupied; interaction of all α4β1 integrins by VCAM-1 leads to immobilization. Preliminary experiments suggest that the two molecular adhesion pairs of α4β1 integrin/VCAM-1 and αvβ3 integrin/PECAM-1 influence each other, for example, ligand occupancy of the αvβ3 integrin modulates the speed of locomotion of cells migrating along VCAM-1 coated surfaces. This illustrates that integrins can transduce signals into cells (see above), and that they can influence the function of other adhesion molecules (in this case the α4β1 integrin).

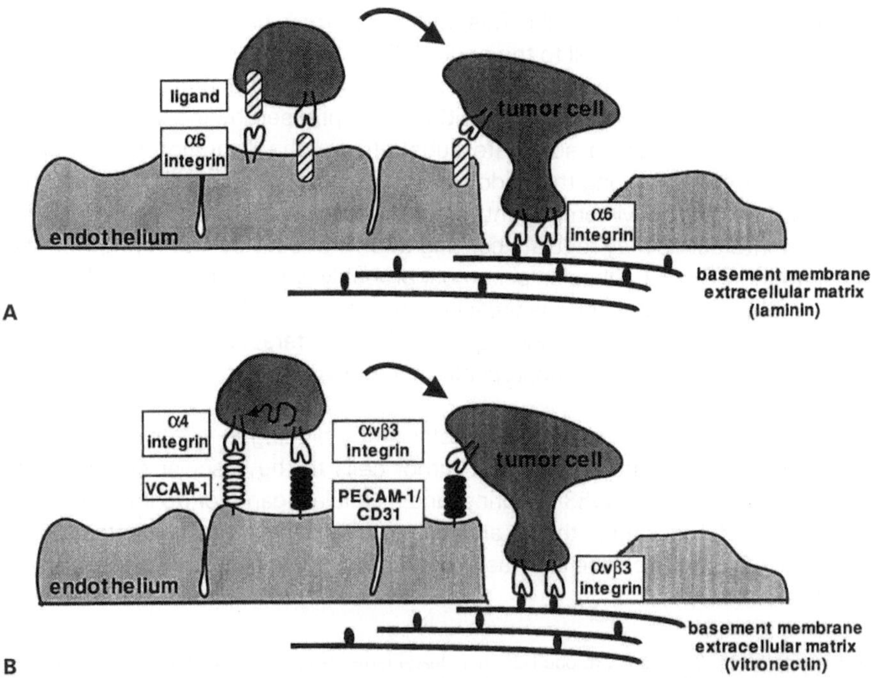

Fig. 1A, B. The adhesion and migration steps of a metastatic tumor cell before it enters a target tissue by extravasation. **A** The involvement of α6 integrins, **B** The involvement of αvβ3 integrin in this process

6 Conclusion

To form a growing secondary tumor, a metastasis, the tumor cell which has detached from the primary tumor must leave the vascular system by transendothelial migration. Transendothelial migration can be initiated only after trapping the circulating cell from the blood stream by tight adhesion to the vascular endothelium. This process is mediated mainly by integrins and their ligands. We recently found that α6β1 integrins are mediators for metastatic tumor cell adhesion to the vascular endothelium of the lung (Ruiz et al. 1993). Furthermore, we discovered that the ligands for α6 integrins which mediate this cell-cell interaction are different from the ECM component laminin. However, the interaction of tumor cells with laminin by α6 integrins is necessary for the adhesion of the cell to the subendothelial ECM during extravasation. After leaving the blood vessel α6 integrins may again mediate cell-cell contacts, as shown with metastatic carcinomas which adhere to hepatocytes in the liver (Kemperman et al. 1993). Thus, α6 integrins may trap a circulating tumor cell on the vascular endothelium, mediate contact of the tumor cell with the endothelial basement membrane, and establish a final cell-cell contact in the target tissue (Fig. 1a).

Furthermore, our results show that PECAM-1 present on the surface of vascular endothelial cells can interact with αvβ3 integrins expressed by circulating cells, including metastatic tumor cells. This interaction may simply contribute to the tight adhesion of a tumor cell to the endothelium. Previous research has demonstrated that the αvβ3 integrin plays an active role in cell migration along ECM components (LEAVESLEY et al. 1992). PECAM-1 expressed by the vascular endothelium might represent a surrogate substrate for the migration of an αvβ3-expressing tumor cell along the endothelium towards the intercellular junctions, where immobilized ECM components are generally absent. In addition, the αvβ3/PECAM-1 interaction may regulate cell migration indirectly by interfering with the α4β1 integrin/VCAM-1 adhesion pair, most probably with a signaling mechanism. Once a metastatic tumor cell has breached the endothelial layer, the cellular αvβ3 integrin has access to the subendothelial ECM. This interaction may then promote invasive migration into the appropriate microenvironment of a target tissue (Fig. 1b).

To conclude, tumor cells can metastasize by binding to α6 or αv integrins. Both molecules mediate adhesion of tumor cells to the vascular endothelium; PECAM-1 is a ligand for αvβ3 integrins, and a cellular ligand for α6 integrin is still to be identified. Blocking of the α6 and αv integrins can prevent metastasis, and, finally, the α6 integrin can be used as a marker for aggressive tumors especially in breast carcinoma.

Acknowledgments. We thank Philippe Hammel, Jean-Pierre Dangy, and Barbara Ecabert for expert technical assistance. The Basel Institute was founded and is supported by F. Hoffmann-La Roche AG, CH-4005 Basel, Switzerland. Part of this work was supported by the Association pour la Recherche contre le Cancer.

References

Albelda SM, Buck CA (1990) Integrins and other cell adhesion molecules. FASEB J 4: 2868–2880

Almeida EAC, Huovila APJ, Sutherland AE, Stephens LE, Calarco PG, Shaw LM, Mercurio AM, Sonnenberg A, Primakoff P, Myles DG, White JM (1995) Mouse egg integrin α6β1 functions as a sperm receptor. Cell 81: 1095–1104

Baggiolini M, Dahinden CA (1994) CC chemokines in allergic inflammation. Immunol Today 15: 127–133

Blood CH, Zetter BR (1993) Laminin regulates a tumor cell chemotaxis receptor through the laminin-binding integrin subunit α6. Cancer Res 53: 2661–2666

Butcher EC (1991) Leukocyte-endothelial cell recognition: three (or more) steps to specificity and diversity. Cell 67: 1033–1036

Cooper D, Lindberg FP, Gamble JR, Brown EJ, Vadas MA (1995) Transendothelial migration of neutrophils involves integrin-associated protein (CD47). Proc Natl Acad Sci USA 92: 3978–3982

DeLisser HM, Chilkotowsky J, Yan HC, Daise ML, Buck CA, Albelda SM (1994) Deletions in the cytoplasmic domain of platelet-endothelial cell adhesion molecule-1 (PECAM-1, CD31) result in changes in ligand binding properties. J Cell Biol 124: 195–203

Durum SK, Oppenheim JJ (1993) Proinflammatory cytokines and immunity. In: Paul WE (ed) Fundamental immunology, 3 edn. Raven, New York, pp 801–835

Filardo EJ, Brooks PC, Deming SL, Damsky C, Cheresh DA (1995) Requirement of the NPXY motif in the integrin β3 subunit cytoplasmic tail for melanoma cell migration in vitro and in vivo. J Cell Biol 130: 441–450

Friedrichs K, Ruiz P, Franke F, Gille I, Terpe HJ, Imhof BA (1995) High expression level of α6 integrin in human breast carcinoma is correlated with reduced survival. Cancer Res 55: 901–906

Frixen UH, Behrens J, Sachs M, Eberle G, Voss B, Warda A, Lochner D, Birchmeier W (1991) E-cadherin-mediated cell-cell adhesion prevents invasiveness of human carcinoma cells. J Cell Biol 113: 173–185

Giancotti FG, Mainiero F (1994) Integrin-mediated adhesion and signaling in tumorigenesis. Biochim Biophys Acta 1198: 47–64

Humphries MJ, Olden K, Yamada KM (1986) A synthetic peptide from fibronectin inhibits experimental metastasis of murine melanoma cells. Science 233: 467–470

Hynes RO (1992) Integrins: versatiliy, modulation, and signaling in cell adhesion. Cell 69: 11–25

Imhof BA, Dunon D (1995) Leukocyte migration and adhesion. Adv Immunol 58: 345–416

Jewell K, Kapron-Bras C, Jeevaratnam P, Dedhar S (1995) Stimulation of tyrosine phosphorylation of distinct proteins in response to antibody-mediated ligation and clustering of $\alpha3$ and $\alpha6$ integrins. J Cell Sci 108: 1165–1174

Kemperman H, Wijnands Y, de-Rijk D, Roos E (1993) The integrin alpha 6 beta 4 on TA3/Ha mammary carcinoma cells is involved in adhesion to hepatocytes. Cancer Res 53: 3611–3617

Kirschbaum NE, Gumina RJ, Newman PJ (1994) Organization of the gene for human platelet/endothelial cell adhesion molecule-1 shows alternatively spliced isoforms and a functionally complex cytoplasmic domain. Blood 84: 4028–4037

Leavesley DI, Ferguson GD, Wayner EA, Cheresh DA (1992) Requirement of the integrin $\beta3$ subunit for carcinoma cell spreading or migration on vitronectin and fibronectin. J Cell Biol 117: 1101–1107

Lindberg FP, Gresham HD, Schwarz E, Brown EJ (1993) Molecular cloning of integrin-associated protein: an immunoglobulin family member with multiple membrane-spanning domains implicated in $\alpha v\beta3$-dependent ligand binding. J Cell Biol 123: 485–496

Mackay C (1995) A new spin on lympocyte homing. Curr Biol 5: 733–736

Nip J, Shibata H, Loskotoff DJ, Cheresh DA, Brodt P (1992) Human melanoma cells derived from lymphatic metastasis use integrin $\alpha v\beta3$ to adhere to lymph node vitronectin. J Clin Invest 90: 1406–1413

Piali L, Albelda SM, Baldwin HS, Hammel P, Gisler RH, Imhof BA (1993) Murine platelet endothelial cell adhesion molecule (PECAM-1)/CD31 modulates $\beta2$ integrins on lymphokine-activated killer cells. Eur J Immunol 23: 2464–2471

Piali L, Hammel P, Uherek C, Bachmann F, Gisler RH, Dunon D (1995) CD31/PECAM-1 is a new ligand for $\alpha v\beta3$ integrin involved in adhesion of leukocytes to endothelium. J Cell Biol 130: 451–460

Qian F, Vaux DL, Weissman IL (1994) Expression of the integrin $\alpha4\beta1$ on melanoma cells can inhibit the invasive stage of metastasis formation. Cell 77: 335–347

Rice GE, Bevilacqua MP (1989) An inducible endothelial cell surface glycoprotein mediates melanoma adhesion. Science 246: 1303–1306

Ruiz P, Dunon D, Sonnenberg A, Imhof BA (1993) Suppression of mouse melanoma metastasis by EA-1, a monoclonal antibody specific for alpha 6 integrins. Cell Adh Comm 1: 67–81

Ruiz P, Wiles MV, Imhof BA (1995) $\alpha6$ integrins participate in pro-T cell homing to the thymus. Eur J Immunol 25: 2034–2041

Schweighoffer T, Shaw S (1992) Adhesion cascades: diversity through combinatorial strategies. Curr Opin Cell Biol 4: 824–829

Springer TA (1994) Traffic signals for lymphocyte recirculation and leukocyte emigration: the multistep paradigm. Cell 76: 301–314

Tanaka Y, Albelda S, Horgan KJ, van Seventer GA, Shimizu Y, Newman W, Hallam J, Newman PJ, Buck CA, Shaw S (1992) CD31 expressed on distinctive T cell subsets is a preferential amplifier of $\beta1$ integrin-mediated adhesion. J Exp Med 176: 245–253

Tanaka Y, Adams DH, Hübscher S, Hirano H, Siebenlist S, Shaw S (1993) T-cell adhesion induced by proteoglycan-immobilized cytokine MIP-1β. Nature 361: 79–82

Tennenbaum T, Yuspa SH, Grover A, Castronovo V, Sobel ME, Yamada Y, De LL (1992) Extracellular matrix receptors and mouse skin carcinogenesis: altered expression linked to appearance of early markers of tumor progression. Cancer Res 52: 2966–2976

Terpe HJ, Stark H, Ruiz P, Imhof BA (1994) Alpha6 integrin distribution in human embryonic and adult tissues. Histochemistry 101: 41–49

Wolf GT, Carey TE (1992) Tumor antigen phenotype, biologic staging, and prognosis in head and neck squamous carcinoma. Monogr Natl Cancer Inst

Wolf GT, Carey TE, Schmaltz SP, McClathey KD, Poore J, Glaser L, Hayashida SJS, Hsu S (1990) Altered antigen expression predicts outcome in squamous cell carcinoma of the head and neck. J Natl Cancer Inst 82: 1566–1572

The 67-kDa Laminin Receptor and Tumor Progression

N. MONTUORI and M.E. SOBEL

1 Introduction

Penetration of host basement membranes by maligant cells is the critical event in tumor invasion and metastasis. Basement membranes are specialized extracellular matrices that underlie epithelia and endothelia and surround muscle, nerve,and fat. The major glycoprotein of the basement membrane is laminin, a large, multidomain protein (BECK et al.1990). Cancer cells need to cross host basement membranes several times during the metastatic cascade: when detaching from the primary tumor, during intra- and extravasation, and at the metastatic site during the penetration of organ parenchyma, muscles, and nerves. It has been hypothesized that the attachment of tumor cells via cell-surface receptors to components of the basement membrane, and in particular to laminin, is the prerequisite for basement membrane penetration (LIOTTA 1986). Normal and cancer cells express a variety of cell-surface proteins that are capable of binding to laminin (MERCURIO1990). Laminin receptors include several integrins (ABELDA and BUCK 1990), as well as a variety of nonintegrin proteins (MERCURIO1990), including a 67-kDa high-affinity laminin receptor.

In 1983 three independent laboratories reported the isolation and the purification of a 67-kDa protein that specifically bound to laminin. The 67-kDa laminin-

Molecular Pathology Section, Laboratory of Pathology, Room 2A33, National Cancer Institute, National Institutes of Health, Bethesda, MD 20892, USA

binding protein was purified by laminin affinity chromatography from murine fibrosarcoma cell extracts (Malinoff and Wicha 1983), from detergent membrane extracts of mouse skeletal muscle (Lesot et al. 1983), and from mouse melanoma cells (Rao et al. 1983). Cell-surface iodination studies showed that the 67-kDa laminin-binding protein was a cell-surface protein (Malinoff and Wicha 1983), therefore it was designated the 67-kDa laminin receptor (67LR). By 1986, the 67LR had been isolated and identified from a large number of cell sources (For review see Castronovo 1993). In addition, there was evidence that expression of the 67LR was increased in malignant cells and that its overexpression directly correlated with metastatic aggressiveness of cancer cells (Liotta 1986).

The 67LR may also bind to type IV collagen (Senior et al. 1989), type V collagen (Minafra et al. 1992), and elastin (Mecham et al. 1989). Alternatively, there may exist a family of similar, but not identical 67-kDa receptors that bind to different extracellular matrix ligands (Sobel 1993).

2 Molecular Cloning and Biosynthesis

Attempts to isolate and to determine the coding sequence of the 67LR resulted in cDNA clones that encoded a putative precursor, whose synthetic relationship with the 67LR is still the object of several controversies. The original cDNA clones (Wewer et al.1986) were isolated from a human umbilical vein endothelial cell expression cDNA library using a monoclonal antibody raised against a 67-kDa protein that had been purified by laminin affinity chromotography from a detergent extract of human breast cancer metastatic tissue cell membranes. The antibody recognized a 67-kDa protein on immunoblots of cancer cell extracts and specifically inhibited the binding of laminin to the cancer cell surface (Liotta et al. 1985). The identity of the cDNA clones as encoding a 67LR sequence was demonstrated by the complete homology between a cDNA-deduced 12 amino acid sequence and microsequenced cyanogen bromide-generated peptide fragments of purified human placental 67LR (Wewer et al. 1986).

Full length mouse cDNA clones encoding the 67LR were isolated by hybridizing cDNA libraries with restriction fragments from the original partial clones (Rao et al. 1989). Primer extension experiments demonstrated that such full-length clones contained the complete 5' end of the corresponding mRNA. Interestingly, human 67LR cDNA clones were isolated by differential expression in colon carcinoma cells (Yow et al. 1988; Kondoh et al. 1992), as well as by using an antibody that had been raised against a highly expressed protein on the surface of human small-cell lung carcinoma (Van den Ouweland et al. 1989).

Surprisingly, the full-length cDNA clones coded for a polypeptide that was only 295 amino acids long, with a calculated molecular mass of 32 kDa (Rao et al. 1989; Yow et al. 1988; Makrides et al. 1988). The rabbit reticulocyte lysate cell-free translation product of selectively hybridized laminin receptor mRNA or of in vitro synthesized RNA encoded by the clones exhibited, on sodium dodecylsulfate (SDS)-polyacrylamide gels, an apparent molecular mass ranging from 32 to 43 kDa (RAO et al. 1989; CASTRONOVO et al. 1991a; MAKRIDES et al. 1988; MCCAFFREY et al. 1990; RABACCHI et al. 1990; AUTH and BRAWERMAN 1992). In addition, cell extracts demonstrated an immunologically reactive protein with the same size range which was smaller than expected (RAO et al. 1989; CASTRONOVO et al.1991a,b; MCCAFFREY et al. 1990).

In our laboratory, the nascent polypeptide had an apparent mobility on SDS-polyacrylamide gels of 37 kDa (RAO et al.1989; CASTRONOVO et al. 1991a,b). Pulse-chase experiments suggested that the 37-kDa molecule was incorporated into the mature 67LR (CASTRONOVO et al. 1991a). Therefore, the existence of a precursor–product relationship between the 37-kDa molecule and the 67LR was postulated and the former was designated 37-kDa laminin receptor precursor (37LRP). On human cancer cell immunoblots, cDNA-deduced antisynthetic peptide antibodies recognized either the 67LR, the 37LRP, or both molecules (RAO et al. 1989; CASTRONOVO et al. 1991a).

Posttranslational modifications responsible for the transformation of the 37LRP into the mature 67LR are still unknown. The 37LRP sequence does not contain asparagine-linked oligosaccharide attachment sites (RAO et al.1989; GROSSO et al. 1991), microsomal membranes failed to increase the size of in vitro translated 37LRP (GROSSO et al.1991). and purified 67LR does not contain sufficient carbohydrate to be stained by periodic acid schiff reagent (CASTRONOVO et al. 1991a). Transfection of the 37LRP cDNA into eukaryotic cells, which are constitutively able to produce the 67LR, resulted in an increased expression of 37LRP but not of the 67LR. Therefore it was proposed that the 67LR might be the product of the association of the cytoplasmic 37LRP with another gene product that confers lectin-binding properties on the membrane-bound 67LR and is a rate-limiting factor for its synthesis (CASTRONOVO et al. 1991a). Recently, it has been found that laminin is able to induce synthesis of both the 37LRP and the 67LR in tumor cells and that this effect requires protein synthesis (ROMANOV et al. 1995). Further insights into the existence of a synthetic relationship between the 37LRP and the 67LR have been obtained by epitope-tagging experiments showing that epitope-tagged 37LRP is incorporated into a tagged, membrane-bound 67LR (MONTUORI et al., submitted for publication).

The cytoplasmic 37LRP polypeptide may also have functions other than laminin binding. Some studies have suggested that p40, the mouse protein homologous to the 37LRP, interacts with ribosomes and may play a role in polysome formation (AUTH and BRAWERMAN 1992). It is worthwhile noting that 37LRP homologs in mammalian species (e.g., bovine, rat, mouse, human and hamster) all have an intact laminin-binding domain (see below), and share 99% identity. Proteins with partial (less than 60%) homology to the mammalian 37LRP/p40 have been identified in nonmammalian species such as yeast (DAVIS et al. 1992), hydra (KEPPEL and SHALLER 1991), Drosophila (KIM and BAKER 1991), and Arabidopsis (GARCIA-HERNANDEZ et al. 1994). The latter proteins do not possess a laminin-binding domain (AXELOS et al.1993) and show some homology (less than

30% for the Arabidopsis homolog) to acidic ribosomal proteins. It is an intriguing feature that the homology of all the nonmammalian 37LRP/p40 homologs so far characterized with the mammalian polypeptides is limited to the amino-terminal part of the molecule. The carboxy-terminal part, which has been described as containing the laminin-binding domain in the mammalian proteins, shows no sequence homology (AXELOS et al. 1993). Due to their lack of an intact laminin-binding domain, nonmammalian 37LRP/p40 homologs possess other cell functions distinct from laminin binding, e.g., in the translation process (GARCIA-HERNANDEZ et al. 1994) or in interaction with the cytoskeleton (KEPPEL and SHALLER 1991). These functions might be conserved to at least some extent in the mammalian protein (AUTH and BRAWERMAN 1992). Importantly, we propose that in mammalian cells, a new function appears, namely, laminin binding due to the presence of a laminin-binding domain in the carboxy-terminal part of the molecule (CASTRONOVO et al. 1991b).

3 Membrane-Associatedand Laminin-Binding Domains in the 37LRP

In addition to its coding capacity being shorter than expected, other unexpected features of the 37LRP include the absence of a traditional signal peptide and the lack of a classic 20 amino acid-long transmembrane domain. Computer analysis of the 37LRP sequence defined a nonclassical membrane domain that was 16 amino acids long in a nonhelical region of the receptor (CASTRONOVO et al. 1991b). Immunofluorescence studies using antisynthetic peptide antibodies to distinct regions of the 37LRP identified the computer-predicted membrane domain (residues 86–101) as an actual membrane-associated domain, and showed that the carboxy two-thirds of the 37LRP were in the extracellular space, while domains in the amino-terminal third of the 37LRP were either buried in the membrane or were localized in the cytoplasm (CASTRONOVO et al. 1991b).

Peptide G, a 37LRP-derived 20 amino acid-long synthetic peptide containing residues 161–189, specifically bound to laminin with nanomolar affinity (K_d=50 n*M*) and eluted the 67LR from a laminin-affinity column (CASTRONOVO et al.1991b). Immunofluorescence studies with antipeptide G antibodies confirmed that peptide G was expressed on the cancer cell surface (CASTRONOVO et al. 1991b). In addition, a monoclonal antibody to a 37LRP-derived sequence overlapping the peptide G region competed with laminin for binding to living human metastatic breast cancer cells in vitro (RAHMAN et al. 1989). These data confirmed the hypothesis that the 37LRP contained a membrane-associated domain and a cell-surface laminin-binding region, characteristics that one would expect of the 67LR. Furthermore, a trpE-37LRP fusion protein expressed in bacteria resulted in increased laminin binding (SIYANOVA 1992), and eukaryotic transfection of antisense 37LRP RNA inhibited laminin-mediated adhesion of aggressive human colon carcinoma cells, while sense RNA had the opposite effect (MAFUNE and RAVIKUMAR 1992).

4 Mediation of Cancer Cell Attachment by the Laminin-Binding Domain

Attachment of cancer cells to endothelium is an important step in the metastatic cascade. Endothelial cells express, on their apical surface, the 67LR. Therefore, cancer cells that are often coated with secreted laminin can adhere to endothelium via a direct interaction between the laminin on their surface and the 67LR on the endothelial cells. In fact, 37LRP-derived peptide G was able to inhibit the adherence of laminin-coated melanoma cells to endothelial cells, in vitro (CASTRONOVO et al. 1991c). But, more importantly, peptide G enhanced tumor cell adherence to exposed subendothelial matrix in vivo, with an increase in lung metastases when peptide G-treated tumor cells were injected into nude mice (TARABOLETTI et al.1993). These studies suggested that direct adhesion of tumor cells to the subendothelial matrix is a major pathway for hematogenous metastases, and that tumor cell–matrix interactions may be more important than tumor cell–endothelial cell attachment in the metastatic cascade.

Peptide G has also been reported to inhibit the binding of laminin to heparin and heparan sulfate, and heparin was reported to inhibit the direct binding of peptide G to laminin (GUO et al. 1992), leading to the suggestion that interaction between peptide G and laminin might be heparin-dependent. However, recent studies have demonstrated that peptide G can bind to heparanase-treated laminin (D. BELOTTI, G. TARABOLETTI, V. CASTRONOVO, and M.E. SOBEL, in preparation).

5 Interaction with a Specific Region of Laminin

Laminin is a large cross-shaped molecule with three short arms and one long arm (ENGEL et al. 1981). Initially, experiments using laminin fragments suggested that the interaction between cells and laminin was mediated by a domain of laminin located around the intersection of the four arms (BARSKY et al. 1984). Sequences such as YIGSR (GRAF et al. 1987) or LGTIPG (MECHAM et al. 1989), located on the short arm of laminin close to the intersection, have been proposed as the binding site for the 67LR. By rotary shadowing method and electron microscopy, the site of the specific interaction between purified 67LR and laminin was localized on the long arm of laminin, close to the intersection (CIOCE et al. 1993). In addition, iodinated 67LR specifically bound to a chymotrypsin-resistant fragment of laminin that contained a small piece of the long arm, but not to a pepsin-resistant laminin fragment that did not contain that segment of the long arm (CIOCE et al. 1993).

Recently, immunogold electron microscopy studies showed that the 67LR and the laminin integrin receptor alpha 6 beta 1 were colocalized in cytoplasmic granules that are exported to the cell surface upon exposure to laminin ((ROMANOV et al. 1994).

6 Expression in Normal and Cancer Tissues

Initial studies showing increased expression of the 67LR in cancer tissues compared to normal adjacent tissues were performed by immunoperoxidase staining with a monoclonal antibody raised against human metastatic breast cancer 67-kDa receptor (HORAN-HAND et al. 1985).

Affinity purified polyclonal antibodies raised against synthetic peptides have also been used in immunohistochemical studies, demonstrating increased expression of the 67LR in a wide variety of human adenocarcinomas, including those of colon, breast, stomach, and liver (DAIDONE et al. 1991; D'ERRICO et al. 1991; GRIGIONI et al. 1991; CIOCE et al. 1991). Immunocytochemical studies performed on fine-needle aspiration biopsies of suspect breast lesions with anti-37LRP synthetic peptide antibodies made it possible to predict the malignant nature of most breast carcinomas (CASTRONOVO et al. 1990).

Two monoclonal antibodies directed against the 67LR of small-cell lung carcinoma cells have also been used to study breast carcinoma progression (MARTIGNONE et al. 1992,1993). Infiltrating carcinomas showed high immunoreactivity, while in situ carcinomas and benign lesions were poorly reactive (MARTIGNONE et al. 1992). A statistically significant correlation was found between poor prognosis and high expression of the 67LR protein (MARTIGNONE et al. 1993). Recently, it has been shown that the expression of the 67LR in breast tumors can be predictive of their metastatic potential. The presence of cells expressing the 67LR in bone marrow micrometastases has been shown to be correlated with its expression in the primary tumor (MENARD et al. 1994). A multiparametric study conducted on node-negative breast cancers showed in a multivariate analysis for relapse-free survival that the joint variable of neovascularization and 67LR expression was the strongest prognostic factor (GASPARINI et al. 1995). Furthermore, the simultaneous expression of c-erbB-2 oncoprotein and 67LR showed a predicting potential analogous to that of the lymph node status in breast cancer (COLNAGHI 1994).

In colorectal adenocarcinomas, western blot, northern blot, and in situ hybridization experiments also showed an increased expression of the 67LR compared with normal tissue (CIOCE et al. 1991; CASTRONOVO et al. 1992; CAMPO et al. 1992). In particular, on northern blots 37LRP mRNA levels were increased 3- to 23-fold over adjacent normal mucosa (CASTRONOVO et al. 1991d). In situ hybridization studies demonstrated that colon adenocarcinomas had increased expression of the 37LRP mRNA, while adenomas had normal levels of the mRNA (CAMPO et al. 1992). Thus, it was suggested that in colon cancers the increased expression of the 67LR is not an early event in disease progression (CAMPO et al. 1992). Interestingly, studies on 67LR expression in fetal and adult rat intestine have shown that overexpression of the 37LRP mRNA is a feature of the fetal intestine and of the undifferentiated, mitotically active crypt cells (RAO et al. 1994).

The pattern of expression of the 67LR in cervical neoplasm was clearly different from that found in colon cancers. In adenomatous lesions, increased mRNA levels were a very early event in disease progression, while mRNA expression in squamous lesions did not differ from normal cells (DEMETER et al. 1992). Furthermore, 67LR expression appeared to be correlated with cell proliferation rather than invasiveness (DEMETER et al. 1992).

In ovarian cancers, the expression of the 67LR was also increased and directly correlated to the invasive phenotype of tumor cells (VAN DEN BRULE et al.1994a). Invasive trophoblast showed, by northern and western blot analysis, an increased 67LR expression from 7 weeks to 12 weeks of gestational age, when invasiveness is maximal; afterwards the levels were decreased (VAN DEN BRULE et al.1994b). Interestingly, the same studies showed that the 67LR and another laminin-binding protein, recently named galectin 3,are inversely modulated as the invasive phenotype of cancer cells progresses, with upregulation of the former, and down-regulation of the latter (VAN DEN BRULE et al.1994a,b).

Recently, a new function has been hypothesized for the 67LR. When lung cancers infiltrated by gamma/delta T lymphocytes were compared with lung cancers infiltrated by alpha/beta T lymphocytes, the former were characterized by overexpression of the 67LR, which led to the suggestion that the 67LR might account for gamma/delta T-cell recognition (FERRARINI et al. 1994).

7 Conclusions

In a large variety of adenocarcinomas, increased expression of the 67LR is a molecular marker of metastatic potential and aggressiveness. The selective overexpression of this nonintegrin receptor suggests that the cancer cell modulates its use of the various extracellular matrix receptors during tumor progression. Future studies to define the regulatory regions of the receptor genes will further elucidate the molecular mechanism involved during the metastatic process.

References

Abelda S, Buck CA (1990) Integrins and other adhesion molecule. FASEB J 4: 2868–2880

Auth D, Brawerman G (1992) A 33 kDa polypeptide with homology to the laminin receptor: component of the translation machinery. Proc Natl Acad Sci USA 89: 4368–4372

Axelos M, Bardet C, Lescure B (1993) An Arabidopsis cDNA encoding a 33-kilodalton laminin receptor homolog. Plant Physiol 103: 299–300

Barsky SH, Rao CN, Williams JE, Liotta LA (1984) Laminin molecular domains which alter metastasis in a murine model. J Clin Invest 74: 843–848

Beck K,Hunter I, Engel J (1990) Structure and function of laminin: anatomy of a multidomain glycoprotein. FASEB J 4: 148–160

Campo E, Monteagudo C, Castronovo V, Claysmith AP, Fernandez PL, Sobel ME (1992) Detection of laminin receptor mRNA in human cancer cell lines and colorectal tissues by in situ hybridization. Am J Pathol 141: 1073–1083

Castronovo V (1993) Laminin receptors and laminin binding proteins during tumor invasion and metastasis. Invasion Metastasis 13: 1–30

Castronovo V, Colin C, Claysmith AP, Chen PH, Lifrange E, Lambotte R, Krutzsch HC, Liotta LA, Sobel ME (1990) Immunodetection of the metastasis-associated laminin receptor in human breast cancer cells obtained by fine needle aspiration biopsy. Am J Pathol 137: 1373–1381

Castronovo V, Claysmith AP, Barker KT, Cioce V,Krutzsch HC, Sobel ME (1991a) Biosynthesis of the 67 kD high affinity laminin receptor. Biochem Biophys Res Commun 177: 177–183

Castronovo V, Taraboletti G, Sobel ME (1991b) Functional domains of the 67kDa laminin receptor precursor. J Biol Chem 226: 20440–20446

Castronovo V, Taraboletti G, Sobel ME (1991c) Laminin receptor complementary DNA-deduced synthetic peptide inhibits cancer cell attachment to endothelium. Cancer Res 51: 5672–5678

Castronovo V,Campo E , Van den Brule FA, Claysmith AP, Cioce V, Liu FT, Fernandez P, Sobel ME (1992) Inverse modulation of steady-state messenger RNA levels of two nonintegrin laminin-binding proteins in human colon carcinoma. J Natl Cancer Inst 84: 1161–1169

Cioce V, Castronovo V, Shmookler BM, Garbisa S, Grigioni WF, Liotta LA, Sobel ME (1991) Increased expression of the laminin receptor in colon cancer. J Natl Cancer Inst 83: 29–36

Cioce V, Marguiles IMK, Sobel ME, Castronovo V (1993) Interaction between the 67 kilodalton metastasis-associated laminin receptor and laminin. Kidney Int 43: 30–37

Colnaghi MI (1994) The simultaneous expression of c-erbB-2 oncoprotein and laminin receptor on primary breast tumors has a predicting potential analogous to that of the limph node status. Adv Exp Med Biol 353: 149–154

Daidone MG, Silvestrini R,D'Errico A, Di FG, Benini E, Mancini AM, Garbisa S, Liotta LA, Grigioni WF (1991) Laminin receptors, collagenase IV and prognosis in node negative breast cancers. Int J Cancer 48: 529–532

Davis SC, Tzagoloff A, Ellis SR (1992) Characterization of a yeast mitochondrial ribosomal protein structurally related to the mammalian 68 kDa high affinity laminin receptor. J Biol Chem 267: 5508–5514

Demeter LM, Stoler MH, Sobel ME, Broker TR, Chow LT (1992) Expression of high affinity laminin receptor mRNA correlates with cell proliferation rather than invasion in human papillomavirus–associated cervical neoplasm. Cancer Res 52: 1561–1567

D'Errico A, Garbisa S, Liotta LA, Castronovo V, Stetler-Stevenson WG, Grigioni WF (1991) Augmentation of type IV collagenase, laminin receptor and Ki67 proliferation antigen associated with human colon, gastric and breast carcinoma progression. Mol Pathol 4: 239–246

Engel J, Odermatt E, Engel A, Madri JA, Furthmayr H, Rohde H, Timpl R (1981) Shapes, domain organization and flexibility of laminin and fibronection, two multifunctional proteins of the extracellular matrix. J Mol Biol 150: 97–120

Ferrarini M, Pupa SM, Zocchi MR, Rugarli C, Menard S (1994) Distinct pattern of HSP72 and monomeric laminin receptor expression in human lung cancers infiltrated by gamma/delta T lymphocytes. Int J Cancer 57: 486–490

Gasparini G, Barbareschi M, Boracchi P, Bevilacqua P, Verderio P, Dalla Palma P, Menard S (1995) 67 kDa laminin receptor expression adds prognostic information to intra-tumoral microvessel density in node-negative breast cancer. Int J Cancer 60: 604–610

Garcia-Hernandez M, Davies E, Staswick PE (1994) Arabidopsis p40 homologue: a novel acidic protein associated with the 40 S subunit of ribosomes. J Biol Chem 269: 20744–20749

Graf J, Ogle RC, Robey FA, Sasaki M, Martin GR, Yamada Y, Kleinman HK (1987) A pentapeptide from the laminin beta1 chain mediates cell adhesion and binds to the 67,000 laminin receptor. Biochemistry 26: 6896–6900

Grigioni WF,Garbisa S, D'Errico A, Baccarini P, Stetler-Stevenson WG, Liotta LA, Mancini AM (1991) Evaluation of hepatocellular carcinoma aggressiveness by a panel of extracellular matrix antigens. Am J Pathol 138: 647–654

Grosso LE, Park PW, Mecham RP (1991) Characterization of a putative clone for the 67 kDa elastin-laminin receptor suggests that it encodes a cytoplasmic protein rather than a cell surface receptor. Biochemistry 50: 3346–3350

Guo N, Krutzsch HC, Vogel T, Roberts DD (1992) Interaction of a laminin binding peptide from a 33 kDa protein related to the 67 kDa laminin receptor with laminin and melanoma cells are heparin dependent. J Biol Chem 267: 17743–17747

Horan-Hand P,Thor A, Schlom J, Rao CN, Liotta LA (1985) expression of laminin receptor in normal and carcinomatous tissues as defined by a monoclonal antibody. Cancer Res 45: 2713–2719

Keppel E, Schaller HC(1991) A 33 kDa protein with homology to the "laminin binding protein" is associated with cytoskeleton in hydra and in mammalian cells. J Cell Sci 100: 789–797

Kim YJ, Baker BS (1991) EMBL Databank Accession Number M77133

Kondoh N, Schweinfest CW, Henderson KW, Papas TS (1992) Differential expression of S19 ribosomal protein, laminin binding protein and human lymphocyte antigen class I messenger RNAs associated with colon carcinoma progression and differentiation. Cancer Res 52: 791–796

Lesot LA, Kuhl U, Von der Mark K (1983) Isolation of a laminin binding protein from muscle cell membrane. EMBO J 2: 861–865

Liotta H (1986) Tumor invasion and metastasis—role of the extracellular matrix. Rhoads Memorial Award lecture. Cancer Res 46: 1–7

Liotta LA, Horan-Hand P, Rao CN, Bryant G, Barsky SH, Schlom J (1985) Monoclonal antibodies to the human laminin receptor recognize structurally distinct sites. Exp Cell Res 156: 117–126

Mafune K, Ravikumar TS (1992) Antisense RNA of 32kDa laminin binding protein inhibits attachment and invasion of a human colon carcinoma cell line. J Surg Res 52: 340–346

Makrides S, Chitpatima ST, Bandyopadhyay R, Brawerman G (1988) Nucleotide sequence for a major messenger RNA for a 40 kilodalton polypeptide that is under translational control in mouse tumor cells. Nucleic Acids Res 16: 2349

Malinoff HL, Wicha MS (1983) Isolation of a cell surface receptor protein for laminin from murine fibrosarcoma cells. J Cell Biol 96: 1475–1479

Martignone S, Pellegrini R, Villa E, Tandon NN, Mastoiani A, Tagliabue E, Menard S, Colnaghi MI (1992) Characterization of two monoclonal antibodies directed against the 67 kDa high affinity laminin receptor and application for the study of breast carcinoma progression. Clin Exp Metastasis 10: 379–386

Martignone S, Menard S, Bufalino R, Cascinelli N, pellegrini R, Tagliabue E, Andreola S, Rilke F, Colnagh MI (1993) Prognostic significance of the 67 kilodalton laminin receptor expression in human breast carcinomas. J Natl Cancer Inst 85: 398–402

Mc Caffrey P, Neve RL, Drager UC (1990) A dorso-ventral asymmetry in the embryonic retina defined by protein conformation. Proc Natl Acad Sci USA 87: 8570–8574

Mecham RP, Hinek A, Griffin GL, Senior RM, Liotta LA (1989) The elastin receptor shows structural and functional similarities to the 67 kDa tumor laminin receptor. J Biol Chem 264: 16652–16657

Menard S, Squicciarini P, Luini A, Sacchini V, Rovini D,Tagliabue E, Veronesi P, Salvadori B, Veronesi U, Colnaghi MI (1994) Immunodetection of bone marrow micrometastases in breast carcinoma patients and its correlation with primary tumor prognostic features. Br J Cancer 69: 1126–1129

Mercurio AM (1990) Laminin: multiple forms, multiple receptors. Curr Opin Cell Biol 2: 845–849

Minafra S, Luparello C, Pucci-Minafra I, Sobel ME, Garbisa S (1992) Adhesion of 8701-BC breast cancer cells to type V collagen and 67 kDa receptor. J Cell Sci 102: 323–328

Rabacchi SA, Neve RL, Drager UC (1990) A positional marker for the dorsal embryonic retina is homologous to the high affinity laminin receptor. Development 109: 521–531

Rahman A, Panneerselvam M, Guirguis R, Castronovo V, Sobel ME, Abraham K, Daddona PE, Liotta LA (1989) Anti-laminin receptor antibody targeting of liposomes with encapsulated doxorudicin to human breast cancer cells in vitro. J Natl Cancer Inst 81: 1794–1800

Rao CN, Barsky SH, Terranova VP, Liotta LA (1983) Isolation of a tumor cell laminin receptor, Biochem Biophys Res Commun 111: 804–808

Rao CN, Castronovo V, Schmitt MC, Wewer UM, Claysmith AP, Liotta LA, Sobel ME (1989) Evidence for a precursor of the high affinity metastasis-associated murine laminin receptor. Biochemistry 28: 7476–7486

Rao M, Manishen WJ, Maheswari Y, Sykes DE, Sijanova EY, Tyner AL, Weiser MM (1994) Laminin receptor expression in rat intestine and liver during development and differentiation. Gastroenterology 107: 764–772

Romanov VI, Sobel ME, Pinto da Silva P, Menard S, Castronovo V (1994) Cell localization and redistribution of the 67kDa laminin receptor and alpha 6 beta 1 integrin subunits in response to laminin stimulation: an immunogold electron microscopy study. Cell Adhes Commun 2: 201–209

Romanov VI, Wrathall LS, Simmons TD, Pinto da Silva P,Sobel ME (1995) Protein synthesis is required for laminin-induced expression of the 67 kDa laminin receptor and its 37 kDa precursor. Biochem Biophys Res Commun 208: 637–643

Senior RM, Hinek A, Griffin GL, Pipoly DJ, Crouch EC, Mecham RP (1989) Neutrophils show chemotaxis to type IV collagen and its 7S domain and contain a 67-kD type IV collagen-binding protein with lectin properties. Am J Resp Cell Mol Biol 1: 479–489

Siyanova EY (1992) Expression of LBP32/67 human gene in E. coli and analysis of its binding with laminin. Bull Exp Biol Med 113: 101–104

Sobel ME (1993) Differential expression of the 67 kDa laminin receptor in cancer. Semin Cancer Biol 4: 311–317

Taraboletti G, Belotti D, Giavazzi R, Sobel ME, Castronovo V (1993) Enhancement of metastatic potential of murine and melanoma cells by laminin receptor peptide G: attachment to subendothelial matrix as a pathway for hematogenous metastasis. J Natl Cancer Inst 85: 235–240

Van den Brule FA, Price J, Sobel ME, Lambotte R, Castronovo V (1994a) Inverse expression of two laminin binding proteins, 67LR and galectin-3, correlates with the invasive phenotype of throphoblastic tissue. Biochem Biophys Res Commun 201: 388–393

Van den Brule FA, Berchuck A, Bast RC, Tong Liu F, Gillet C, Sobel ME, Castronovo V (1994b) Differential expression of the 67-kD laminin receptor and 31 kD human laminin-binding protein in human ovarian carcinomas. Eur J Cancer 30A: 1096–1099

Van den Ouweland AMW, Van Duijnhoven HLP, Deichman KA, van Groningen JJM, de Leij L, Van de Ven WJM (1989) Characteristics of a multicopy gene family predominantly consisting of processed pseudogenes. Nucleic Acids Res 17: 3829–3843

Wewer UM, Liotta LA, Jaye M, Ricca GA, Drohan WN, Claysmith AP, Rao CN, Wirth P, Coligan JE, Albrechtsen R, Mudryi M, Sobel ME (1986) Altered levels of laminin receptor mRNA in various human carcinoma cells that have different abilites to bind laminin. Proc Natl Acad Sci USA 83: 7137–7141

Yow H, Wong JM, Chen HS, Lee C, Steel GDJ, Chen LB (1988) Increased mRNA expression of a laminin binding protein in human colon carcinoma: complete sequence of a f ull length cDNA encoding protein. Proc Natl Acad Sci USA 85: 6394–6398

Joint Features of Metastasis Formation and Lymphocyte Maturation and Activation

M. ZÖLLER[1,2]

1 Introduction

Adhesion molecules are an essential prerequisite for the development of multicellular organisms (EDELMAN 1983, 1986; YAP et al. 1994). They are now known to guide morphogenesis and organogenesis and are involved in the maintenance of organ structures (ANDERSON 1990; HYNES 1992a; KNUDSON and KNUDSON 1993; NATHKE et al. 1993). Adhesion molecules are of the utmost importance for most functional activities of the immune system (SPRINGER 1990; DUSTIN and SPRINGER 1991; SHIMIZU et al. 1992; VESTWEBER 1992; RATNER 1992; CLARK et al. 1992; PARDI et al. 1992; SIMMONS et al. 1994) and are thought to be involved in tumor progression (HEMLER 1990; DEDHAR 1990; JOHNSON 1991; MCCARTHY et al. 1991; RUSCIANO and BURGER 1992; LESTER and MCCARTHY 1992; EVANS 1992; BEHRENS 1993; ZETTER 1993;

[1]Department of Tumor Progression and Immune Defense, German Cancer Research Center, 69120 Heidelberg, Germany
[2]Department of Applied Genetics, University of Karlsruhe, Karlsruhe, Germany

GLINSKY 1993; MIYASAKA and TOYAMA-SORIMACHI 1993; GIANCOTTI and MAINIERO 1994). The number of adhesion molecules discovered is still growing and at present they are grouped in five families: integrins, selectins, cadherins, the immunoglobulin (Ig) superfamily, and hematopoietic cell adhesion molecules (HCAM), the latter group including CD44 (WILLIAMS and BARCLAY 1988; RUTISHAUSER et al. 1988; YONG and KHWAJA 1990; ALBELDA and BUCK 1990; HEMLER 1990; SPRINGER 1990; GEIGER and AYALAN 1992; POSTIGO and SANCHEZ-MADRID 1993; TUCKWELL et al. 1993; CARLOS and HARLAN 1994; BIRCHMEIER and BEHRENS 1994). Especially integrins, the Ig superfamily, and HCAM consist of a multitude of members, a number which is increased by combinatorial variations of heterodimers and by changes in glycosylation and in the protein structure, frequently due to alternative splicing of pre-mRNA (HOFFMAN and EDELMAN 1983; WILLIAMS and BARCLAY 1988; SPRINGER 1990; HAYNES et al. 1991a; HYNES 1992b; SONNENBERG 1993; LESLEY et al. 1993a; SCREATON et al. 1993). This complexity of structures correlates with a multitude of functions. Cell adhesion molecules mediate either cell–cell or cell–matrix interactions or both. Above and beyond this, they are frequently involved in signal transduction, resulting in altered patterns of gene expression (EDELMAN 1986; WILLIAMS and BARCLAY 1988; SPRINGER 1990; JOHNSON 1991; RUOSLATHI 1991; LASKY 1992; GORIDIS and BRUNET 1992; HYNES 1992b; BOSMAN 1993; HUMPHRIES et al.1993; JULIANO and HASKILL 1993; GIANCOTTI and MAINIERO 1994; ROSALES and JULIANO 1995). In view of the complexity of structure–function relationships of the whole array of adhesion molecules, this report will only summarize some of the known features of CD44 and point out possible links between the functions of CD44 isoforms in lymphocytes and metastasizing tumor cells.

CD44 comprises a family of glycoproteins with variable N- and O-linked glycosylation sites (HUGHES et al. 1981; JALKANEN et al. 1986; OMARY et al. 1988; GOLDSTEIN et al. 1989; KANSAS et al. 1989; ZHOU et al. 1989; QUACKENBUSH et al.1990; LOKESHWAR and BOURGUIGNON 1991; CAMP et al. 1991a). The so-called standard or hematopoietic form (CD44s) spans a region of seven extracellular exons, a transmembrane exon, and a cytoplasmic exon (IDZERDA et al. 1989; WOLFFE et al.1990; BOSWORTH et al. 1991), which can be short (exon 9) or long (exon 10) (GOLDSTEIN and BUTCHER 1990). Between exon 5 and exon 6, up to ten additional so-called variant exons can be inserted (GÜNTHERT et al. 1991; JACKSON et al. 1992; SCREATON et al. 1992; TÖLG et al. 1993; SCREATON et al.1993). Although multiple combinations of variant exons have been described, there are some which are differentially expressed or at least predominantly found on defined tissues such as the epithelial form (COOPER et al. 1992), the keratinocyte type (BROWN et al. 1991; HAGGERTY et al. 1992), or the reticulocyte type (ISOLA et al. 1991). Translation of the variant exons has been suggested to follow the 3'- to 5'-end rule. However, there are exceptions inasmuch as individual cells can express a multitude of combinations of splice variants, in which the individual combinations do not necessarily contain sequential exons. Depending on the state of activation, individual cells can repeatedly change the splicing of CD44 pre-mRNA. The mechanisms which regulate alternative splicing of CD44 are as yet unknown, although there are preliminary reports on regulation of the CD44 promoter by ras and of altered splicing after hyaluronidase treatment (HOFMANN et al. 1993; TANABE et al. 1993b).

CD44 is an adhesion molecule with binding domains for hyaluronate (HA) (MIYAKE et al. 1990b; MIYAKE and KINCADE 1990; ARUFFO et al. 1990; CULTY et al. 1990; UNDERHILL 1992; YANG et al. 1994). Two HA-binding sites have been described (PEACH et al. 1993). HA binding is thought to be influenced by the cytoplasmic tail (LIAO et al. 1993; NEAME and ISACKE 1993), but the membrane-proximal domain appears not to be involved (HE et al. 1992). Not all $CD44^+$ cells bind to HA, but HA binding can be induced by cross-linking (LESLEY et al. 1993b), which is thought to result either in conformational changes or in a redistribution of CD44 in the cell membrane (LESLEY et al.1992). Furthermore, O-glycosylation sites are important for CD44–HA interaction (LOKESHWAR and BOURGUIGNON1991). CD44 also binds to fibronectin (LOKESHWAR et al. 1994; JALKANEN and JALKANEN 1992), laminin and type IV collagen (ISHII et al. 1993), and gycosaminoglycans (TOYAMA-SORIMACHI and MIYASAKA 1994). The molecule is known to be involved in the assembly of the extracellular matrix (KNUDSON et al. 1993; NEAME and BARRY 1993; LESLEY et al.1993). For some functions, binding to the cytoskeleton via ankyrin is essential (LOKESHWAR et al. 1994; LOKESWAR and BOURGUIGNON 1992; BOURGUIGNON et al. 1992). CD44 variant isoforms (CD44v), in particular, are linked via the esrin-radixin-moesin (ERM) family to the actin-based cytoskeleton (TSUKITA et al. 1994). There are three phosphorylation sites at the intracytoplasmatic tail, and binding to the cytoskeleton is not observed in the phosphorylated state (NEAME and ISACKE 1992).

CD44 displays a large array of functional activities, most of them related to the hematopoetic system; these will be described in more detail in the following sections. CD44 is essential in hematopoiesis (LEWINSOHN et al. 1990), and it is known to be involved in lymphocyte homing (SHIMIZU and SHAW 1991). During lymphocyte activation, CD44 functions as a costimulatory molecule (HAYNES et al. 1991b). Finally, it has been reported to augment lytic activity of cytotoxic T cells (SETH et al. 1991), of double-negative thymocytes (WANG 1993), and of natural killer (NK) cells (YANG and BINNS 1993; TAN et al. 1993).

More recently, attention has been paid to the expression of CD44 during ontogeny (WHEATLEY et al. 1993; SRETAVAN et al. 1994), in particular in view of its function as one of the more abundant hyaluronan receptors. CD44 is expressed early during ontogeny and the pattern of expression partially follows the hyaluronan profiles (WHEATLEY et al. 1993); it has been suggested that hyaluronan is involved in the formation of the early mesoderm, the differentiation of the craniofacial mesenchyme, and the morphogenesis of the axial skeleton (FENDERSON et al. 1993). Interestingly, it was also reported to degrade hyaluronan (HUA et al. 1993; UNDERHILL et al. 1993; CULTY et al. 1992; Sampson et al.1992), a function which could be of importance in the formation of ducts, cavities, and canniculi as, for example, required in the formation of the respiratory tract, the homeostasis of cartilage tissue (HUA et al.1993; EDWARDS et al. 1994), and the formation of dermal condensations (UNDERHILL 1993). Independent of the concomitant presence of hyaluronan, expression of CD44 has also been noted on instructive epithelia (WHEATLEY et al. 1993).

CD44v isoforms are less abundantly expressed. Although most epithelia and the hematopoietic organs are CD44v⁺ during ontogeny (Wirth et al. 1993; Weber et al. 1996; Terpe et al. 1994), in adults expression of CD44v is mainly restricted to the skin, the epithelium of the gut, and a variety of glands (Wirth et al. 1993; Kennel et al. 1993; Hirano et al. 1994; Fox et al. 1994). In all instances, expression of CD44v was linked to a high rate of cell division (Wirth et al.1993; Terpe et al. 1993). Interestingly, even within the CD44v⁺ tissues, different cell layers express distinct CD44v isoforms. This implies a strictly regulated mechanism of splicing and is indicative of divergent functions (Günthert 1993; Hardingham and Fosang 1992; Haynes et al. 1991a).

CD44v isoforms were first described on a metastasizing rat tumor line (Matzku et al.1983). This pancreatic adenocarcinoma predominantly expressed the variant exons v4–v7 (Günthert et al. 1991). Since metastasis formation of locally growing tumors could be initiated by the transfer of CD44v4-v7 or CD44v6-v7 cDNA (Günthert et al. 1991; Rudy et al. 1993) and metastatic spread could be inhibited (Reber et al.1990; Seiter et al. 1993) by an antibody, 1.1ASML, recognizing an epitope on exon v6 (Matzku et al. 1989; Günthert et al. 1991), it was hypothesized that CD44v and in particular exon v6 may be of special importance for tumor progression (Tarin and Matsumara 1993a,b; Thomas 1993; Koopman et al. 1993a; Herrlich et al. 1993). Screening of human tumors did not unequivocally support this assumption (Fox et al. 1994). However, as described in detail below, for some human malignancies expression of CD44v could be proven to be linked to tumor progression.

Taking the physiological and pathological patterns of expression of CD44, it can be assumed that distinct isoforms of the molecule may play a role in cell–cell and cell–matrix adhesion as well as in cell motility leading in concert to organ-specific homing and to signal transduction initiating cell proliferation, differentiation, and/or cytokine, protease, and enzyme activation. We have started to evaluate whether this hypothesis can hold true and whether defined isoforms can be linked to distinct functions. As a physiological model system, we chose lymphocyte maturation and activation. Finally, published and preliminary evidence was viewed from the aspect of compatible requirements in lymphocyte maturation/activation and tumor progression.

2 CD44 in Hematopoiesis

2.1 Expression of CD44 on Hematopoietic Precursor Cells During Ontogeny and in Adults

The main hematopoietic organ during development is the liver. Later on hematopoiesis is observed in the spleen and, starting around birth, bone marrow becomes (and remains for life) the major origin of hematopoietic cells, which –

with the exception of T cells – leave the bone marrow as naive, but mature cells. T cells behave exceptionally inasmuch as they require a further step of maturation within the thymus.

Bone marrow cells of humans and mice are known to express CD44 (CLARK et al. 1992; REUSS-BORST et al. 1992). More detailed analyses revealed that the latest pluripotent stem cell likely will be CD44⁺ (ANTICA et al. 1994). The earliest myeloid precursors are also CD44 high; when committed to the monocyte lineage the cells remain CD44 high, while commitment to granulopoiesis is accompanied by transient downregulation. The erythroid lineage remains CD44 high throughout development (KANSAS et al. 1990; SUGIMOTO et al. 1994). Expression on B and T cells also varies with the stage of maturation. At early stages they are CD44⁺, including prothymocytes in the liver (HORST et al. 1990a; RYAN et al. 1990; BELL and ZAMOYSKA 1991; HAIG et al. 1992; LUND-JOHANSEN and TERSTAPPEN 1993); pre-B cells are largely negative and during intrathymic maturation T cells also downregulate CD44 (HORST et al. 1990a; COLLADO et al. 1991; WILKINSON et al. 1993), while mature T and B cells again express CD44. Monocytes and mast cells are also CD44 ⁺ (VALENT et al. 1990; PRIETO et al. 1994), and maturation of pre-NK to NK cells requires interaction of CD44 with HA (DELFINO et al. 1994). Since maturation of hematopoietic cells is essentially dependent on stroma cell interactions (LEWINSOHN et al. 1990), it is interesting to note that CD44 is also expressed on bone marrow stromal cells (CLARK et al. 1992; REUSS-BORST et al. 1992) as well as thymic stromal cells (LEPESANT et al. 1990).

Our own preliminary data on bone marrow stromal cells and stem cells/progenitor cells of the mouse suggest that both express CD44s and CD44v, whereby the patterns of CD44 splice variants in stromal cells and hematopoietic stem cells/progenitor cells appear to be related, exons v5 and v9 mRNA being the most prominent (V. ZAWADZKI, unpublished observation). These variant exons are also found on hematopoietic precursor cells of the rat, which in addition hybridize with exon v6-specific probes (R. ARCH, unpublished observation). In the rat, we also investigated surface expression of CD44s and CD44v6 on hematopoietic organs throughout fetal development by immunohistology starting at day 8 of gestation. Expression CD44s was already noted at this early stage in primitive vessels and blood cells. Expression of variant isoforms was not observed before establishment of hematopoiesis in the liver. Later on, the thymic epithelium as well as the immigrating prothymocytes stained brightly with anti-CD44s and anti-CD44v6. In the newborn rat, hematopoietic cells of the liver, thymus, and spleen still expressed CD44s and CD44v6, although only a few cells stained with 1.1ASML and only with low intensity. Bone marrow (both stromal elements and the majority of stem cells/progenitor cells) was CD44s⁺ for life. Expression of variant isoforms was only detected after birth and was restricted to a small population of large cells (Table 1).

Table 1. Expression of CD44 standard (s) and variant 6 (v6) isoforms in hematopoetic cells during ontogeny of the rat

Organ	Anti-CD44	CD44 expression								
		Day of gestation							Age	
		14	15	16	17	18	19	21	NB	AD
Liver	s	+	++	++	++	++	+	+	±(25.1)	(ntl)
	v6	±	+	++	++	+	+	±	– (2.9)	(nt)
Thymus	s		+	++	++	++	+	+	++(98.3)	++ (97.8)
	v6		+	++	++	+	+	+	+(55.6)	– (2.1)
BM	s				–	±	±	±	+(35.2)	++ (77.8)
	v6				–	–	–	–	-(2.7)	±(18.3)
Spleen	s				–	±	±	+	+(47.3)	++(64.3)
	v6				–	–	–	±	±(14.5)	– (5.2)

The figures in parentheses indicate the percentage of positive cells found by fluorescence-activated cell sorter (FACS). BM, bone marrow; NB, newborn; AD, adult; nt, not tested; ++, strong; +, distinct; ±, weak; – absent

2.2 Evidence for Functional Requirement of CD44 Standard and Variant Isoforms in Hematopoiesis

CD44 is known to play important roles in the differentiation and proliferation of hematopoietic progenitor cells at different maturation stages in the bone marrow microenvironment (KOBAYASHI et al. 1994). As early as 1990 MIYAKE et al.described that in long-term bone marrow cultures of the mouse, no cobblestone areas appeared in the presence of anti-CD44s and nonadherent progenitors did not develop. CD44 could be shown to be required for myelopoiesis as well as for lymphopoiesis (MIYAKE et al. 1990a; KINCADE 1991, 1992; KINCADE et al. 1993). Since antibodies which did not interfere with HA binding of CD44 were inhibitory as well, it can be assumed that prohibition of binding to HA of the stroma layer was at least not the dominating function of CD44 in hematopoiesis (MIYAKE et al. 1990). Similarly, it was not an Fc-receptor-mediated effect, since cobblestone formation was also inhibited by anti-CD44 F(ab')$_2$ fragments (S. KHALDOYANIDI et al., submitted). Inhibition of hematopoiesis by anti-CD44s was mainly restricted to stem cells and early progenitors i.e., anti-CD44s displayed no (MIYAKE et al. 1990) or only minor effects (S. KHALDOYANIDI et al., submitted) on colony formation of committed progenitors in soft agar cultures. Since anti-CD44s interfered predominantly with the maturation/expansion of stem cells and /or early progenitor cells, we can assume that expression of CD44s was required either for interactions between stem cells/progenitor cells and stromal elements or that by ligand binding growth-promoting signals were transferred into the hematopoietic stem cell/precursor cell. These two assumptions are not mutually exclusive.

In fact, at least part of the anti-CD44s-mediated blockade relies on inhibition of cell division. When freshly harvested bone marrow cells were incubated with an anti-CD44s antibody (IM-7) and then treated with [^{3}H]-thymidine, upon transfer into lethally irradiated mice a significant decrease in the number of stem cells which had undergone suicide was noted. There is, however, evidence that anti-CD44 also blocks an interaction between stem cells and stromal elements, which appears to be essential in hematopoiesis (see below).

Due to the lack of CD44v-specific antibodies in the mouse, there is little information available about whether and at what stage of maturation expression of CD44s or CD44v may be required in hematopoiesis. Competition studies with glutathione-*S*-transferase (GST)–CD44v fusion proteins did not reveal strong effects in long-term bone marrow cultures (LTBMC) of the Dexter type. Thus granulopoiesis appears to be largely independent of CD44v expression. Considering B cell maturation in LTBMC, it appears that fusion proteins containing the exon v9 influence B cell progenitor maturation. In line with this was the observation that after sublethal irradiation and intravenous application of GST–CD44v9 fusion protein, a higher number of B cell progenitors were recovered in bone marrow and spleen after 7–14 days (M.ZOLLER unpublished results).

As already mentioned, in the rat there is at least one CD44v-specific antibody available, which recognizes an epitope on exon v6, and a small population of rat (and human) bone marrow cells stains brightly with anti-CD44v6. Establishment of rat LTBMC in the presence of anti-CD44v6 revealed that in particular maturation of the adherent stem cell population required expression of CD44v. As in mouse LTBMC, anti-CD44s prohibited the development of nonadherent progenitors at least during the first 5–7 weeks of culture. Anti-CD44v6 displayed only a minor effect during the first 2–3 weeks. Thereafter, however, the cultures contained exclusively stromal cells, and hematopoiesis did not recover on omission of anti-CD44v6, while it did recover in cultures transiently containing anti-CD44s (Fig. 1).

Maturation of hematopoietic stem cells essentially requires interaction with the stromal environment (LEWINSOHN et al. 1990). Hence, the question arose of whether expression of CD44s and CD44v is required for stroma formation and/or whether CD44s and CD44v are involved in stem cell/progenitor cell–stroma interactions. To answer the first question, mouse and rat LTBMC were irradiated and the surviving stromal cells were cultured in the presence of anti-CD44s (mouse and rat) or anti-CD44v6 (rat only). Neither anti-CD44s nor anti-CD44v6 displayed any macroscopically visible influence on stroma cell survival. Cultures even contained lipocytes, one of the constituents critically involved in stem cell maturation (S. KHALDOYANIDI et al.; submitted). In addition, during establishment of LTBMC, no influence of mouse or rat anti-CD44s on stroma formation was noted. However, in the presence of anti-CD44v6, stroma formation of rat bone marrow was significantly delayed. The same observation was made for stroma formation of human bone marrow cells. Taking into account that the stromal cells did not express CD44v6, it is tempting to speculate that as a result of CD44v6 (stem cells/progenitor cells)–ligand (stroma cells) interaction signals are transduced which facilitate stroma formation. Since, however stem cell/progenitor maturation was also completely inhibited in the presence of anti-CD44v6, a binary mode of signal transduction has to be assumed.

While granulocytes, monocytes, and B cells can mature in the microenvironment of bone marrow, maturation of T cells requires passage through the thymus.

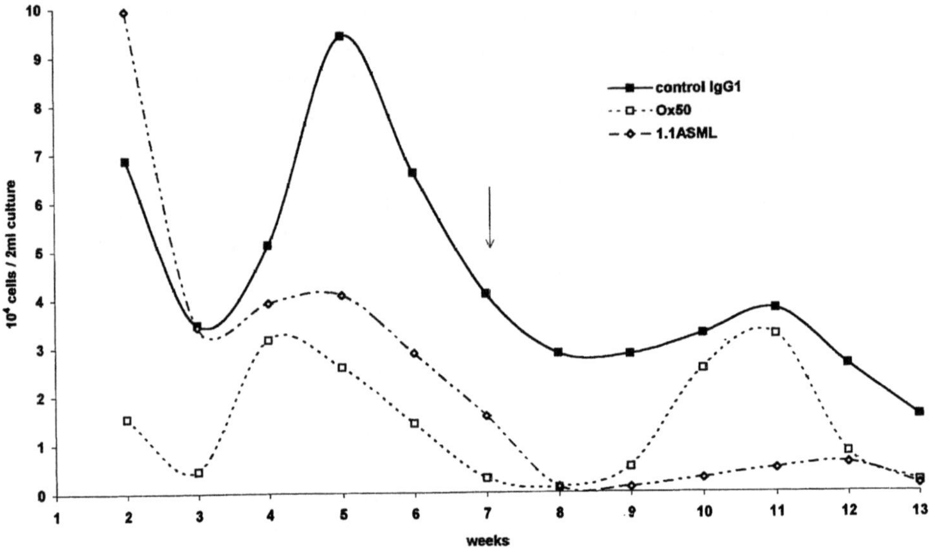

Fig. 1. Influence of anti-CD44 standard(*s*) and anti-CD44 variant 6 (*v6*) isoforms on stem cell proliferation and maturation. Rat bone marrow cells were kept for 13 weeks in culture and the number of nonadherent cells in the culture supernatant was evaluated at weekly intervals. In cultures containing anti-CD44s (*Ox50*), significantly lower numbers of nonadherent cells were detected. A similar, although less pronounced effect was seen with anti-CD44v6 (*1.1ASML*). Cultures kept with anti-CD44s or anti CD44v6, but not those with an isotype-matched control antibody went into crisis after 6–8 weeks of culture, cells dying with features of apoptosis. When anti-CD44s and anti-CD44v6 were removed from the culture (*arrow*), cultures originally containing anti-CD44s recovered, while in those originally containing anti-CD44v6 nonadherent cells did not reappear. The stroma layer appeared unaltered in all instances. *Ig*, immunoglobulin

A possible influence of CD44v in T cell maturation was evaluated in transgenic mice, which express rat CD44v4-v7 under the control of the Thy1 promoter on thymocytes and peripheral T cells (MOLL et al.1996). When syngeneic, lethally irradiated mice were reconstituted with bone marrow cells from control mice or CD44v4-v7 transgenic mice, no significant difference in the recovery of bone marrow and spleen cells was noted. However, repopulation of the thymus was significantly accelerated. The same was true when transgenic and non-transgenic mice were sublethally irradiated and expansion of surviving stem cells was investigated. In both situations the reconstitution advantage of bone marrow cells from transgenic mice was abolished in the presence of anti-CD44v6, but was not affected by anti-CD44s (M.ZÖLLER, unpublished results). Since the transgene is not expressed before the $CD4^+/CD8^+$ stage of T cell maturation, it is likely that expression of CD44v6 provides a growth advantage during the intrathymic selection processes.

Taken together and notwithstanding that CD44 might be replaced by distinct adhesion molecules, CD44 apparently plays an essential role in stem cell proliferation, expansion, and maturation. According to published evidence and in line with our findings, CD44s provides (upon ligand interaction) proliferation-initiating signals for early progenitors of all three hematopoietic lineages. CD44v, on the other hand, may be primarily involved in transducing signals (between stromal cells and stem cells and in late maturation steps of B cells and T cells) that initiate differentiation.

3 CD44 and Immune Response

3.1 Expression of CD44 on Mature Lymphocytes and During Lymphocyte Activation

The vast majority of peripheral lymphocytes in human, mouse, and rat express CD44s. CD44 is expressed on T cells, B cells (especially the $CD5^+$ subset), and antigen presenting cells (GABRILOVICH et al. 1994; ABE et al. 1994). Furthermore, CD44s is upregulated during lymphocyte activation (HAMILTON et al. 1991; LESLEY and HYMAN 1992). This is a rather late event (PRINCE et al. 1992). It is accompanied either by post-translational modifications or by different association with other cellular components (HATHCOCK et al. 1993). Differences in N-linked glycosylations have also been described for activated macrophages. In contrast to resting macrophages, which may or may not be phosphorylated, in activated macrophages CD44 is always phosphorylated and not linked to the cytoskeleton (CAMP et al. 1991a). Furthermore, upregulation accompanies transient activation of the HA receptor function (LESLEY et al. 1994), which requires protein synthesis but no association with the cytoskeleton (MURAKAMI et al. 1994), although binding to HA apparently leads to linking of CD44 to the actinomycin contractile system via ankyrin (BOURGUIGNON et al. 1993). Finally, upregulation of CD44s during the activation process is permanent, i.e., memory cells are defined by bright staining with anti-CD44s (BUDD et al. 1987; CAMP et al. 1991b). It is thought that the high expression of CD44s is important for the particular adherence and traffic requirements of the memory cells (MURAKAMI et al. 1990).

Expression of CD44v, on the other hand, is rare in the adult animal and in the absence of an activating stimulus. However, expression of CD44v is noted when an immune response is mounted (ARCH et al. 1992; KOOPMAN et al. 1993a). Activation-dependent changes are seen with dendritic cells (SALLUSTO and LANZAVECCHIA 1994), which start to express at least CD44v9; alveolar macrophages express multiple CD44 isoforms, and peripheral blood monocytes start to express CD44v upon activation (CULTY et al. 1994a). T cells were reported to express CD44v6 and CD44v9 after stimulation with tumor necrosis factor (TNF)-α and interferon (IFN)-γ (MACKAY et al. 1994). For Epstein-Barr virus-transformed B cell lines, expression of CD44 receptors 1 and 2 has been described, and both forms carry inserts of different lengths at the common CD44 splice site. Interestingly, no changes in CD44 isoforms have been described on myeloid cell lines (DOUGHERTY et al. 1991).

Analysis of de novo synthesis of CD44v in the rat during an immune response revealed that different subpopulations of lymphocytes appear to express distinct variant isoforms and the pattern of expression changes during the activation period. According to preliminary findings, B cells and antigen-presenting cells exhibit similar features of CD44v mRNA, i.e., early during the activation period signals were seen with exon v1-, v3-v7-, and v9-specific probes, and later with exon v4-v7- and v9-specific probes. T cells transiently contained mRNA of exons v4–v7, and later exclusively of exons v6 and v7 (R. ARCH, unpublished observation). Expression of CD44v was restricted to the activation stage, i.e., there was no evidence for expression of CD44v on memory T or B cells. Considering the time course of CD44v surface expression as judged by fluorescence staining with anti-CD44v6, expression was first noted in antigen-presenting cells, thereafter in T and B cells, reached its peak around day 4–5 after antigenic stimulation, and declined thereafter (Fig. 2). This was similar after in vivo and in vitro stimulation (ARCH et al. 1992; R. ARCH, unpublished observation).

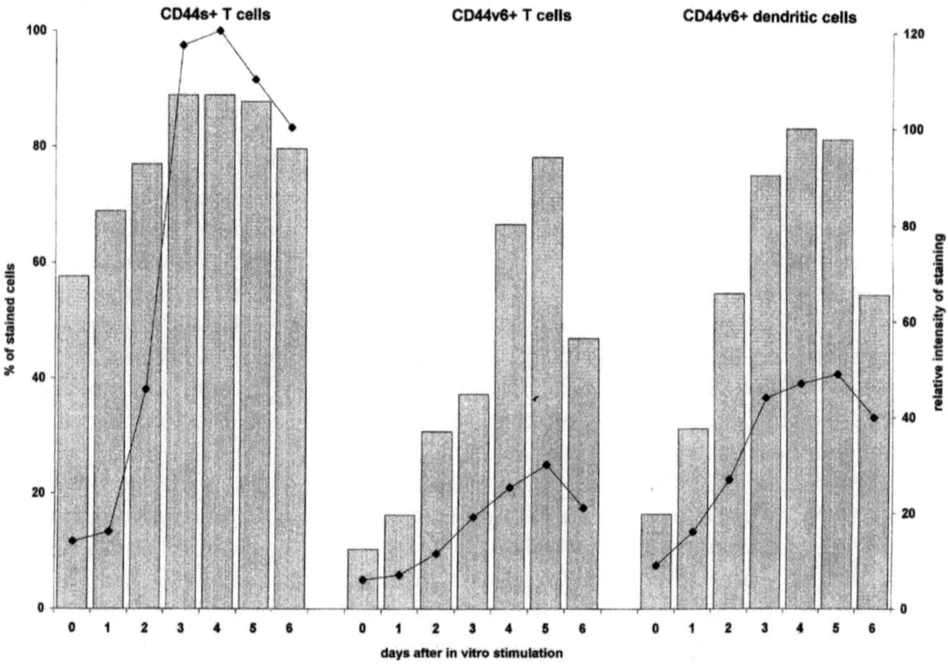

Fig. 2. Upregulation of CD44 standard(*s*) and variant 6(*v6*) isoforms during lymphocyte activation in vitro. Rat lymphocytes were stimulated in vitro with trinitrophenyl-bovine serum albumin and expression of CD44s and CD44v6 (percentage of stained cells and relative intensity) was determined at daily intervals. While upregulation of CD44s on T cells was noted as early as day 2 of stimulation and remained at an elevated level, a significant increase in the number of T cells expressing CD44v6 was not seen before 3–4 days of culture. The relative increase in intensity of expression followed the same time schedule, i.e. it was transient. Expression of CD44v6 was more rapidly upregulated on antigen-presenting cells, although it was, transient as on T lymphocytes

3.2 Function of CD44 in Lymphocyte Activation

Upregulation of CD44s or de novo expression of CD44v is an active process and has pathophysiological consequences. The latter can be deduced from the observation that altered patterns of CD44 expression are noted in many autoimmune diseases such as Grave's disease (HEUFELDER et al. 1993), rheumatoid arthritis (JOHNSON et al. 1993; TAKAHASHI et al. 1992, HAYNES et al. 1991b), in MLR-lpr mice (WANG et al. 1992), autoimmune uveitis (FOETS et al. 1992), experimental allergic encephalomyelitis (ZEINE and OWENS 1992), multiple sclerosis (GIGRAH et al. 1991), and chronic graft versus host disease, which indicates the importance of CD44 not only in T cell, but also in B cell responsiveness (MURAKAMI et al. 1991). In multiple sclerosis, CD44 was not only upregulated, but expression of variant isotypes was noted on astrocytes and infiltrating T lymphocytes, which exhibited augmented adhesiveness (HÄGEL et al. 1993). Considering the function of CD44 in immune response, there is ample evidence that, like most adhesion molecules, CD44 functions as a costimulatory molecule in T cell activation (SHIMIZU et al. 1989; HUET et al. 1989; DENNING et al. 1990; ROTHMAN et al. 1991; CONRAD et al. 1992; HALE and HAYNES 1992; PIERRES et al. 1992; KRAKAUER 1994). The underlying mechanism has not yet been clarified. Interestingly, some antibodies are stimulatory together with anti-CD2, but not with anti-CD3. The latter have been found to stimulate palmitoylation of CD44 (GUO et al. 1994a). It was shown that CD44 is involved in enhanced binding of dendritic cells to T cells (STJOHN et al. 1990). CD44 promotes homotypic adhesion via lymphocyte function-associated antigen (LFA)-1 (KOOPMAN et al. 1990) and triggers the chondroitin sulfate form of the invariant chain to function as a constimulus (NAUJOKAS et al. 1993). There is evidence that by HA binding interleukin (IL)-2 production, and release of trypsin-like esterase by cytotoxic T lymphocytes (CTL) may be triggered (GALANDRINI et al. 1994a). These processes are clearly phospho-tyrosine-kinase (PTK) dependent (GALANDRINI et al. 1993; FURANO et al. 1994). Cross-linking via anti-CD44 leads to activation of cytolytic T cells and is a trigger for NK cells, the pathway of activation being much like activation via the T cell receptor (SETH et al. 1991; SCONOCCHIA et al. 1994; GALANDRINI et al. 1994a). It is also known that for T cell activation via cross-linking of CD44, association with the cytoskeleton is especially important (GEPPERT and LIPSKY 1991).

Taken together, there is no question about the functional importance of CD44 as a costimulatory molecule in T and probably also B cell activation. However, the mechanism of function remains to be elucidated. One of the first questions to be answered might depend on the differentiation between CD44s- and CD44v-mediated effects. We have started to unravel this question, focusing in particular on T cell responses in the rat, because rat T cells upon antigenic or mitogenic stimulation are known to express only two of the ten variant exons, namely CD44v6-v7 and thus the anti-CD44v6 monoclonal antibody could be helpful in a first trial to associate CD44 functions to defined isoforms. In the rat, functional activity of CD44 during the activation process is linked to the expression of CD44v and is apparently independent of the upregulation of CD44s. This assumption is based on the following observations: (a) T cell-dependent and T cell-independent immune responses in vivo (Arch et al. 1992) as well as in vitro (R. Arch, unpublished observation) are significantly inhibited in the presence of anti-CD44v6, but are not altered by anti-CD44s; (b) proliferation and cytotoxicity assays set up under limiting dilution conditions revealed that the number of responding cells was significantly reduced by anti-CD44v6; when anti-CD44v6 was added during the pulsing period or during the cytotoxicity assay only, no inhibition of proliferative activity or cytotoxic potential was noted; (c) when purified T cells are cultured on anti-T cell receptor-coated plates, a strong costimulatory effect of anti-CD44v6 is observed; in the rat, no costimulatory function with CD2 was observed, and proliferation in response to anti-T cell receptor (TCR) or to anti-CD2 was not influenced by anti-CD44s (M. Zöller, unpublished observation). These data are interpreted to mean that CD44v is required for the activation process itself, but, in contrast to CD44 activity in the mouse, not for effector functions. Since upon TCR occupancy, ligand binding or cross-linking of CD44v6 at the cell surface initiates signals leading to lymphocyte proliferation and maturation, there are principally two modes of CD44v6 function in lymphocyte activation. Either signals are transduced into the antigen-presenting cell, which becomes activated. This could result in increased cytokine production, as described for IL-1β, TNF-α, TNF-β, insulin-like growth factor-1, macrophage colony-stimulating factor (Webb et al. 1990; Chong et al. 1992; Gruber et al. 1992; Noble et al. 1993), and IL-2 (Guo et al. 1993; Chong et al. 1994) by activated CD44 in the mouse, or in an augmentation of presentation, as seen after binding of CD44 to the chondroitin sulfate form of the invariant chain (Naujokas et al. 1993). The observation that, after preincubation of antigen-presenting cells with anti-CD44v6, lymphocyte activation appears severely impaired supports this assumption. Alternatively, but not mutually exclusive, cross-linking or ligand binding of CD44v6 causes signals which initiate proliferation or activation of genes associated with immune responses to be transferred within the lymphocyte. The latter assumption is supported by the observation that, upon cross-linking of CD3, anti-CD44v6 supports T cell proliferation. It thus appears that in the rat predominantly CD44v (v6 or v6-v7) is involved in the process of T cell activation and that CD44v6/CD44v6-v7 fulfils divergent functions, i.e., modulates the activity of antigen-presenting cells and lymphocytes. Experiments are in progress to clarify the underlying molecular events.

4 CD44 and Lymphocyte Migration

4.1 Homing and Migration of Progenitor Cells and Naive Lymphocytes

It has long been known that CD44 facilitates homing of lymphocytes into lymph nodes by binding to high endothelial venules (Jalkanen et al. 1988; Gallatin et al.

1989; BERG et al. 1989; PICKER et al. 1989; MIYAKE et al. 1980b). This function is restricted to CD44s and is not mediated by the epithelial isoform of CD44, for example (SCHEEREN et al. 1991). CD44 plays a major role in lymphocyte–cell and lymphocyte-extracellular matrix interactions (SHIMIZU and SHAW 1991). This is now known for a large array of lymphocyte–cell and lymphocyte–extracellular matrix interactions. It is involved in binding of bone marrow cells to stromal elements, where binding of myeloid cells in particular seems to function via HA binding (MORIMOTO et al. 1994). On the other hand, seeding of stem cells on stroma layers requires CD44, but can be inhibited by antibodies which do not block HA binding (S. KHALDOYANIDI et al., submitted). CD44 is involved in binding of colony-forming cells to fibronectin (VERFAILLIE et al. 1994), in plasmocytoma cell–stroma interactions (DEGRASSI et al. 1993), in binding of lymphocytes to human umbilical vein endothelial cells (MUNZIG et al. 1994), and in lymphocyte–endothelial cell interactions in general (TOYAMA-SORIMACHI et al. 1993). It should be mentioned that lymphocyte binding is inducible (OPPENHEIMER-MARKS et al. 1990) and in most instances, especially in the context of HA binding, is observed only after induction (LESLEY et al. 1990). Recently, evidence has been presented that migration of prothymocytes into the thymus is also guided by CD44, but not via HA binding (WU et al. 1993). CD44 also plays a role in the reappearance of T cells in the periphery after depletion protocols (GUO et al. 1994b).

In line with published evidence, experiments in the rat revealed that stem cell seeding, migration of prothymocytes, and homing of nonactivated lymphocytes could be partially inhibited by anti-CD44s, but not by either anti-CD44v6 or GST–CD44v fusion proteins covering the variant exons v4–v10 (M. ZÖLLER., unpublished observations), which has been demonstrated for bone marrow cell homing (Fig. 3). Although further experiments are required, all data available so far indicate that homing and migration of hematopoietic progenitor cells as well as of mature lymphocytes into lymphoid organs is independent of the expression of CD44v, but is influenced by CD44s.

4.2 Homing of Activated Lymphocytes

In addition to its function in lymphocyte homing into lymphoid organs, CD44 is also involved in homing into nonlymphoid organs (FROGNER and O'NEILL 1992), which is especially important in infectious, allergic, and autoimmune reactions. CD44 has been thought to be involved in particular in the extravasation of lymphocytes, but not in the migration process (CAMP et al. 1993). Thus, it has been described that T cell–keratinocyte binding is strengthened by anti-CD44 (BRUYNZEEL et al. 1993); infiltration of B cells in the lacrimal gland is CD44 dependent (O'SULLIVAN et al. 1994), and as already mentioned, T cell–astrocyte interactions are also CD44 mediated (HÄGEL et al. 1993). Furthermore, CD44 induces cell aggregation (STJOHN et al. 1990), which depends on its interaction with the cytoskeleton (BELISOS et al. 1990). In addition, upon lymphocyte–endothelial cell interaction, syncaping of CD44 has been noted, which might play a critical role

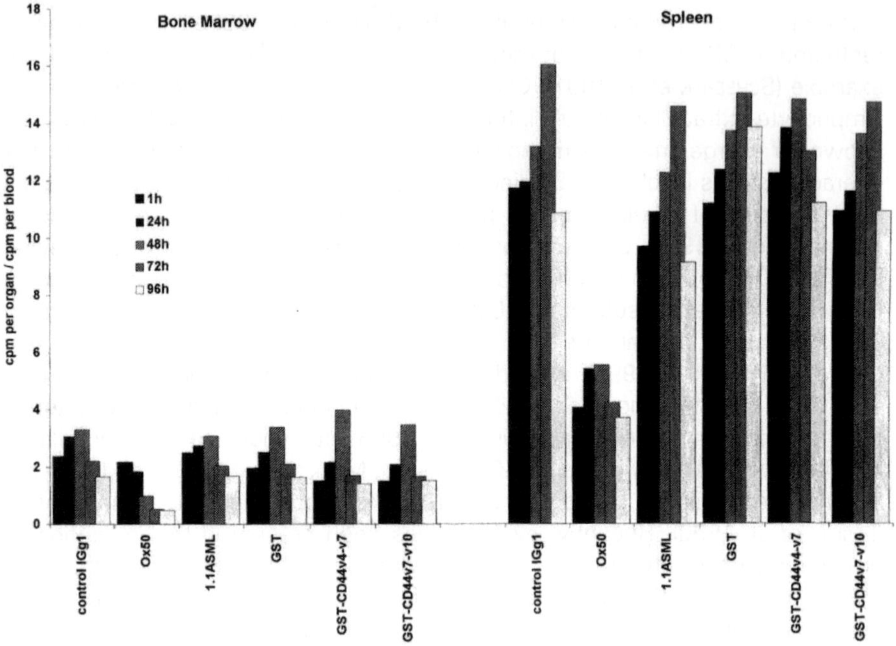

Fig. 3. Homing of rat bone marrow cells. Rat bone marrow cells were depleted of mature T and B cells, labeled with ^{51}Cr, and injected (5×10^6) intravenously into sublethally irradiated rats together with an isotype-matched control antibody, anti-CD44 standard(*s*) isoform (*Ox50)*, anti-CD44 variant 6(*v6*) isoform (*1.1ASML*), glutathione-*s*-transferase (GST) fusion protein, GST–CD44v4-v7 fusion protein, or GST–CD44v7-v10 fusion protein. Rats were killed after1–96 h, lymphoid organs were excised, and remaining radioactivity was determined in a γ-counter. The ratio of the counts in the bone marrow and the spleen to the counts in the peripheral blood are shown. Only in the presence of anti-CD44s was homing of bone marrow cells in the bone marrow and the spleen clearly inhibited

during recirculation and homing of activated lymphocytes in injured organs (Rosenman et al. 1993). It has been described that by immobilizing macrophage inflammatory protein-1β CD44 induces chemotaxis and adhesion of T cells to vascular cell adhesion molecule (VCAM)-1 in inflammatory processes. Finally, after tissue injury, binding of platelets to endothelial cells also appears to be mediated by CD44 (Koshiishi et al. 1994).

While homing of progenitors and of naive lymphocytes in hematopoietic organs appears to be influenced exclusively by CD44s there is evidence that CD44v is involved in nonhematopoietic, tissue-specific homing of activated lymphocytes. The human skin abundantly expresses the so-called keratinocyte form of CD44, which contains the variant exons v3-v10 (Brown et al. 1991). Although expression of exon v10 was not noted in either the bone marrow or during lymphocyte activation lymphocytes infiltrating the skin, irrespective of whether they are malignantly transformed or in the course of infectious or allergic reactions, strongly expressed exon v10. Concomitantly, expression of CD44v10 was also noted on capillary walls in the surrounding tissue (S.N. Wagner et al., submitted). It is thus tempting to speculate that expression of CD44v10 may

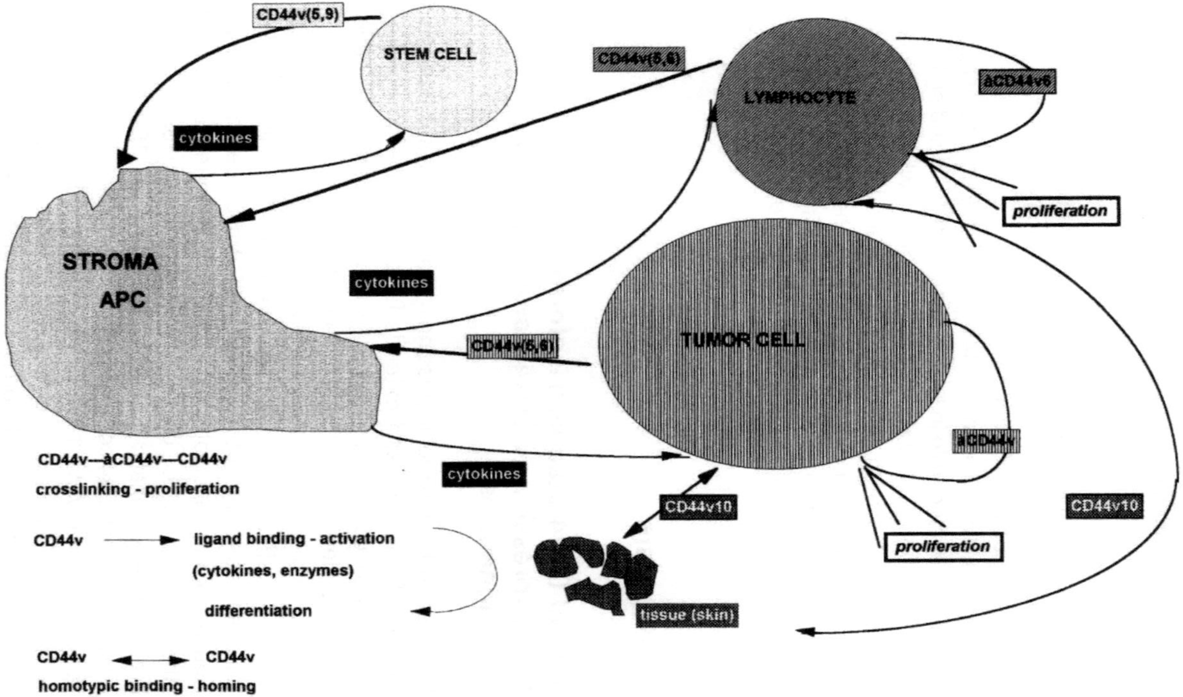

Fig. 4. Array of possible joint functional activities of CD44 variant (*v*) isoforms in lymphocyte maturation, lymphocyte activation and tumor progression. According to experimental evidence, CD44v predominantly functions in cell–cell interactions rather than being involved in cell–matrix interactions. These interactions appear to be binary in all instances evaluated so far, i.e. signal transduction involves the CD44-expressing cell as well as the ligand-bearing cell. The nature of the ligand has not yet been elucidated. *APC*, antigen-presenting cells; àCD44, anti-CD44

essentially be required for lymphocyte homing into the skin. Blocking studies with anti-CD44v10 (available for human and murine CD44v) and transfection experiments may prove this assumption. Furthermore, it will be interesting to evaluate whether the requirement of CD44v10 expression for homing into the skin represents a unique situation or whether infiltration of nonlymphoid organs in general may depend on expression of defined CD44v.

Considering the functional activities of CD44v as currently defined, it appears that CD44v plays no or only a minor role in lymphocyte motility, including homing into lymphoid organs. However, CD44v appears to be required when cells change their functional program, i.e., in stem cell differentiation and lymphocyte activation. Furthermore, preliminary evidence supports the hypothesis that CD44v may allow lymphocytes to infiltrate distinctly organized tissues. The question arises as to whether these functions might play a role in tumor progression.

5 CD44 and Metastasis

5.1 Upregulation of CD44 Standard Isoform and De Novo Expression of CD44 Variant Isoform on Human Malignancies

The connection between adhesion molecules and metastasis is a subject in itself (ZETTER 1993; HONN and TANG 1992) and extends far beyond the scope of this article. As far as CD44 is concerned, in line with its history, where knowledge of the standard isoform preceded knowledge of variant isoforms, a variety of reports suggest upregulation of expression on tumor metastasis (BIRCHMEIER et al. 1991; MATSUMARA and TARIN 1992; EAST and HART 1993; FOX et al. 1993), in particular in hematopoietic malignancies (GROSSI et al. 1992), e.g., it has been noted that in lymphoma and leukemia the level of CD44 expression correlates with the dissemination rather than the degree of dedifferentiation (ROOS 1991). This has been found in B cell acute lymphatic leukemia (CSANAKY et al. 1993), multiple myeloma, where expression of CD44 correlates with homotypic adhesion (BARKER et al. 1992), Burkitt lymphoma (WALTER et al. 1991), non-Hodgkin's lymphoma (NHL) (HORST et al. 1990b,c; QUACKENBUSH et al. 1990; JALKANEN et al. 1991), and T lymphoma, where expression of CD44 correlated with an increase in tumorigenicity (GJERSET et al. 1992). For a B cell hybridoma it has been described that expresssion of CD44s correlates with aggregation and metastasis formation (HAWLEY et al. 1993).

High levels of CD44s expression have also been noted on solid tumors, e.g., in melanomas (GUO et al. 1994; EAST et al. 1993; THOMAS et al. 1993; HART et al. 1991), gastric cancer (WASHINGTON et al. 1994), mesothelioma (ASPLUND and HELDIN 1994), (breast carcinoma (JOENSUU 1993), and glioblastoma and meningioma (MERZAK et al. 1994; KUPPNER et al. 1992). It was supposed that CD44 plays a role in forming a leading lamella, which is required for efficient locomotion, and that the chondroitin sulfate portion of CD44 is the critical component for the increased motility by interaction with type I collagen (FAASEN et al. 1992, 1993). In line with these notions was the observation that in ovarian tumors a loss in tumorigenicity apparently correlated with loss of CD44 (TEYSSIER et al. 1992). Regarding the underlying mechanism, it was again suggested that CD44 may increase motility (WASHINGTON et al. 1994; ASPLUND and HELDIN 1994; THOMAS et al. 1992) or that it may facilitate penetration by HA degradation (FORSTER et al. 1994) or by interaction with the extracellular matrix (MERZAK et al. 1994). The latter assumption was strongly supported by the notion that melanoma metastasis formation could be inhibited by a CD44–Ig fusion protein, which inhibited binding to HA, but not by mutated CD44–Ig fusion protein (LESLEY et al. 1994).

The observation that expression of CD44v initiated lymphatic spread of solid tumors in the rat (GÜNTHERT et al. 1991) has received much attention. In the rat, this phenomenon appears to be of general validity. In a variety of rat tumor lines with paired sublines either growing locally or metastasizing via the lymphatic system, expression of CD44v was detected exclusively in the metastasizing sublines (ZÖLLER 1995). In line with this finding was the notion that transfection of nonmetatasizing rat tumor lines with CD44v transferred the metastasizing phenotype (GÜNTHERT et al. 1991). This was independent of the histology of the primary tumor and of the grade of dedifferentiation. The capacity to form lymph node metastases solely correlated with the intensity of surface expression of CD44v (M. HOFMANN et al., unpublished observation). Interestingly, in a rat colon carcinoma model it has also been noted that tumorigenicity correlates with CD44v6 expression (LABARRIERE et al. 1994). Experiments aimed at defining the important structural equivalent excluded the possibility that any of the variant exons negatively interfered with metastatic progression (J. SLEEMAN et al., unpublished observation). Moreover, no interference of exons from the standard part of the molecule was observed (A. KASUHIRO et al., unpublished observation). Finally, transfection with exons v6 and v7 or with exon v6 as the only variant exon still conferred metastatic behavior (RUDY et al. 1993; A. KASUHIRO., unpublished finding). However, it remains to be explored whether the variant exons by themselves, interactions between the standard exons and exon v6, or conformational changes by the insertion of variant exons were the structural equivalent for metastasis induction.

Irrespective of these open questions, many institutions have now started to screen human tumors for the expression of CD44v and to correlate expression with prognostic parameters. As mentioned earlier, it appears that expression of CD44v in humans is not as strongly restricted to metastasis formation as described in the rat (JACKSON et al. 1993), e.g., there are tumors which do not express CD44v such as neuroblastoma (GROSS et al. 1994), where tumor aggressiveness even correlates with repression of CD44 expression (COMBARET et al. 1995); tumors arising from CD44v$^+$ tissues, especially skin and squamous epithelium including the lung, appear to lose expression upon tumor progression (SALMI et al. 1993; JACKSON et al. 1994; SEITER et al.1996a; HEROLD-MENDE 1996). On the other hand, CD44v is frequently upregulated in tumors infiltrating the skin, but this appears to be associated with tissue injury rather than with tumor progression (PENNEYS et al. 1993). There are other tumors which, in contrast to their nonmalignant counterparts, express CD44v, but even at early stages of malignant transformation and not linked to metastatic progression, e.g., prostate cancer (LIU 1994), gastrinomas (CHAUDHRY et al. 1994). However, there are some tumors in which in humans, too, progression appears to be closely linked to expression of CD44v. This has been described for breast carcinoma (KAUFMANN et al. 1995; LIDA and BOURGIGNON 1995; DALL et al. 1995), bladder carcinoma (MATSUMARA et al. 1994), high-grade NHL and large cell lymphoma (SALLES et al. 1993; KOOPMAN et al. 1993b), kidney carcinoma (TERPE et al. 1994), high-grade glioblastoma and meningioma (HARN et al. 1994a), and hepatocellular carcinoma (HARN et al. 1994b). Yet there are also opposing results in some tumor systems such as the colon, where correlation of tumor progression with expression of CD44v6 was noted by some groups (WIELENGA et al. 1993; HEIDER et al. 1993a; TANABE et al. 1993a; FINN et al. 1994; MULDER et al. 1994), while others did not detect expression of CD44 splice variants or noted it early and independent of progression (ABBASI et al. 1993; KIM et al. 1994; KORETZ et al. 1995). Upregulation of CD44v expression was also described for cervical cancer (DALL et al. 1994), but could not be detected by another group (WOERNER et al. 1995). With gastric cancer, too, published evidence supports either correlation of CD44v expression with progression in the intestinal, more undifferentiated type only (HEIDER et al. 1993b; YOKOZAKI and TAHARA 1994; GUO et al. 1994d) or in general (MAYER et al. 1993; HARN et al. 1995). Different sets of reagents might explain some of the discrepancy.

Interestingly, as far as a correlation between tumor progression and CD44v expression was noted, it was not essentially exon v6, which was upregulated in human malignancies. Instead, expression of other variant exons has been described to be important for metastasis formation, e.g., exon v9 in kidney carcinoma (TERPE et al. 1994), exon v10 in skin metastasis of melanoma (SEITER et al. 1996b), exon v5 or exon v9 in some types of gastric cancer (HEIDER et al. 1993a; MAYER et al. 1993), exon v5 for the settlement of melanoma cells in lymph node tissue (SEITER et al. 1996b), exon v7-v8 in carcinoma of the cervix uteri (DALL et al. 1993), and exon v4 and v5 in hepatocellular carcinoma (HARN et al. 1994b).

Even though CD44v cannot be considered as a general metastasis marker in humans, the notion that metastasis formation and CD44v expression correlates in some systems, and taking into account the generality of the phenomenon in the rat, allows the working hypothesis that CD44v-induced metastasis formation may be based on the recruitment of physiological programs. Since tumor cells which gain metastatic capacity recruit new functions and/or loose other functions of nontransformed cells, it is also tempting to speculate that they may adopt pathways of functional activities, in particular from cells where activation or silencing of genes occurs physiologically (HERRLICH et al. 1993; PALS et al. 1993). This is frequently observed during ontogeny, in stem cell differentiation, and during lymphocyte activation.

5.2 Joint Functions of CD44 in Metastasis Formation and Lymphocyte Maturation and Activation

Based on these considerations and in view of the notion that CD44 splice variants are so far the only molecules which can initiate the metastatic cascade, it would be of great value to unravel the molecular mechanisms underlying the distinct functions of CD44 isoforms. This might enable joint programs of tumor progression, lymphocyte maturation, and lymphocyte activation to be elucidated.

Regarding joint functions of CD44 isoforms in lymphocyte maturation and tumor progression, malignancies of the hematopoietic system should be considered in particular, since it is well documented that these tumors frequently resemble early stages of development. They have been shown to be mostly accompanied by a high expression of CD44, and CD44 has also been proven to facilitate stem cell and progenitor cell expansion. Indeed, as outlined above, many hematological malignancies are accompanied by a high expression of CD44, and progressive states were defined by upregulation of CD44s and CD44v6. So far, however, a possible growth-inhibiting potential of anti-CD44 has not been evaluated with native leukemia or lymphoma. However, it has been shown that a $CD44^-$, Burkitt lymphoma line (Namalwa) transfected with CD44s cDNA displayed increased tumorigenicity and metastatic potential upon intravenous injection (Sy et al. 1991). Furthermore, tumor growth, was inhibited by a CD44s–Ig fusion protein (Sy et al. 1992). Interestingly, a slightly reversed effect was noted with CD44v-transfected Namalwa cells. We can speculate that CD44s confers growth-promoting activities, while CD44v expression may induce differentiation. So far, this assumption is only speculative and remains to be experimentally verified.

As outlined above, adhesion molecules are not only important in lymphocyte maturation, but are essentially required during the activation process, where in addition to establishing cell–cell contacts they are known to be involved in signal transduction. Furthermore, special adhesion molecules are required for the recruitment of activated lymphocytes at the place of injury.

The process of activation takes place within the draining lymph node, and there is preliminary evidence that activation of lymphocytes and growth of tumor cells which have settled in the lymph node may occur by similar mechanisms. When lymphocytes and tumor cells are cultured on antibody (anti-TCR and/or anti-CD44v6)-coated plates, an increase in the proliferation rate is noted. We interpreted this to indicate that cross-linking of the TCR concomitantly with CD44v6 on the lymphocyte or of CD44v6 solely on the tumor cell initiates signals with growth-promoting activity. Furthermore, lymphocytes and $CD44v^+$ tumor cells preferentially adhere to dendritic cells. While antigen-specific activation of purified T cells essentially depends on the presence of antigen-presenting cells, proliferation of tumor cells is also clearly augmented. Adhesion of lymphocytes/tumor cells can be blocked by anti-CD44v6. As a consequence, lymphocytes do not respond to nominal antigen, and the growth advantage for tumor cells supplied by antigen-presenting cells is abolished. Finally, there is preliminary evidence that upon

CD44v–ligand binding secretion of cytokines may be augmented, e.g., upregulation of TNF and IL-1 secretion by CD44-mediated monocyte–tumor cell interactions has been described (WEBB et al. 1990; ZEMBALA et al. 1994). It thus appears that via CD44v on tumor cells as well as on lymphocytes signals can be transduced in CD44v$^+$ and CD44v ligand-bearing cells, which in addition to proliferation also initiate cytokine production.

The last step in the cascade is the infiltration of injured tissue by activated lymphocytes and the organ specificity of metastasis formation. Rolling, tethering, and extravasation have been linked to selectins and integrins (LAWRENCE and SPRINGER 1991; ALON 1994). There are no reports which associate CD44v with these processes. However, tumor cells, have been described to modify adhesion molecules at each of the involved organs (DEDHAR 1990; MCCARTHY 1991; RUSCIANO and BURGER 1992; EVANS 1992; LESTER and MCCARTHY 1992; GLINSKY 1993; MIYASAKA and TOYAMA-SORIMACHI 1993; MAREEL et al. 1993; ZETTER 1993; BEHRENS 1993), and it appears that tissue-specific infiltration again may be accompanied by CD44v expression, expecially in the skin. JACKSON et al. (1995) recently reported that a special variant isoform exists spanning v3 and v8–v10 (JACKSON et al. 1995). Exon v3 was found to contain glycosaminoglycan-related sequences, which are known to act as reservoirs for growth factors in many tissues (YANAGISHITA and HASCALL 1992). There was no ligand structure on the endothelial cells. However, as suggested by the authors, cytokine production may be initiated via CD44v3, in particular by keratinocytes, monocytes, and dendritic cells in the skin (BENNETT et al. 1995). Of special interest is also the observation that intracutaneous lymphomas express CD44v10. This particular variant exon is not expressed on lymphocytes during systemic activation, nor is it expressed on hematopoetic precursor cells. It is, however, found on intracutaneous lymphocytes during allergic reactions as well as during inflammation of the skin (S.N. WAGNER et al., submitted). These features of shared expression of CD44v10 on activated as well as malignant lymphocytes related to the homing organ appear to indicate joint features in organ-specific metastasis formation and lymphocyte infiltration in autoimmunity or in response to injury. Taken together, the data strongly suggest that special variant isoforms are active beyond the process of lymphocyte/tumor cell expansion in the draining lymph node and also facilitate organ-specific homing.

Adhesion molecules are important in organogenesis, stem cell differentiation, and lymphocyte activation. Due to their plasticity, they can change their functional programs and adjust repeatedly to differing requirements. This is exemplified for CD44 isoforms, which facilitate processes as divergent as migration, penetration into preformed tissue, and activation of genes, the expression of which may be required for proliferation, recruitment of cytokines, and/or differentiation. Tumor cells, such as lymphocytes, also need to fulfil divergent functions during the process of metastasis formation. Expression of distinct CD44 isoforms apparently can be sufficient to complete the metastatic cascade.

References

Abbasi AM, Chester KA, Talbot IC, Macpherson AS, Boxer G, Forbes A, Malcolm AD, Begent RH (1993) CD44 is associated with proliferation in normal and neoplastic human colorectal epithelial cells. Eur J Cancer 29A: 1995–2002

Abe M, Tominaga K, Wakasa H (1994) Phenotypic characterization of human B-lymphocyte subpopulations, particularly human CD5+ B-lymphocyte subpopulation within the mantle zones of secondary follicles. Leukemia 8: 1039–1044

Aho R, Ekfors T, Haltia M, Kalimo H (1993) Pathogenesis of primary central nervous system lymphoma: invasion of malignant lymphoid cells into and within the brain parenchyme. Acta Neuropathol (Berl) 86: 71–76

Akamatsu M, Takeshita M, Ohshima K, Kikuchi M, Suzumiya J (1994) Analysis of adhesion molecules in Ki-1 anaplastic large-cell lymphoma. Virchows Arch 425: 33–39

Albeda SM, Buck CA (1990) Integrins and other cell adhesion molecules. FASEB J 4: 2868–2880

Alon R, Rossiter H, Wang X, Springer TA, Kupper TS (1994) Distinct cell surface ligands mediate T lymphocyte attachment and rolling on P and E selectin under physiological flow. J Cell Biol 127: 1485–1495

Anderson H (1990) Adhesion molecules and animal development. Experientia 46: 2–13

Antica M, Wu L, Shortman K, Scollay R (1994) Thymic stern cells in mouse bone marrow. Blood 84: 111–117

Arch R, Wirth K, Hofmann M, Ponta H, Matzku S, Herrlich P, Zöller M (1992) Participation in normal immune responses of a metastasis-inducing splice variant of CD44. Science 257: 682–685

Aruffo A, Stamenkovic I, Melnick M, Underhill CB, Seed B (1990) CD44 is the principal cell surface receptor for hyaluronate. Cell 61: 1303–1313

Asplund T, Heldin P (1994) Hyaluronan receptors are expressed on human malignant mesothelioma cells but not on normal mesothelial cells. Cancer Res 15: 4516–4523

Barker HF, Ball J, Drew M, Hamilton MS, Franklin IM (1992) The role of adhesion molecules in multiple myeloma. Leuk Lymphoma 8: 189–196

Bartolazzi A, Peach R, Aruffo A, Stamenkovic I (1994) Interaction between CD44 and hyaluronate is directly implicated in the regulation of tumor development. J Exp Med 180: 53–66

Behrens J (1993) The role of cell adhesion molecules in cancer invasion and metastasis. Breast Cancer Res Treat 24: 175–184

Behrens J, Frixen U, Schipper J, Weidner M, Birchmeier W (1992) Cell adhesion in invasion and metastasis. Semin Cell Biol 3: 169–178

Belitsos PC, Hildreth JE, August JT (1990) Homotypic cell aggregation induced by anti-CD44 (Pgp-1) monoclonal antibodies and related to CD44 (Pgp-1) expression. J Immunol 144: 1661–1670

Bell SE, Zamoyska R (1991) Identification of thymocyte precursors in murine fetal liver. Eur J Immunol 21: 2931–2936

Bennett KL, Jackson DG, Simon JC, Tanczos E, Peach R, Modrell B, Stamenkovic I, Plowman G, Aruffo A (1995) CD44 isoforms containing exon v3 are responsible for the presentation of heparin-binding growth factor. J Cell Biol 128: 687–698

Berg EL, Goldstein LA, Jutila MA, Nakache M, Picker LP, Streeter PR, Wu NW, Zhou D, Butcher EC (1989) Homing receptors and vascular addressins: cell adhesion molecules that direct lymphocyte traffic. Immunol Rev 108: 5–18

Birch M, Mitchell S, Hart IR (1991) Isolation and characterization of human melanoma cell variants expressing high and low levels of CD44. Cancer Res 51: 6660–6667

Birchmeier W, Behrens J, Weidner KM, Frixen UH, Schipper J (1991) Dominant and recessive genes involved in tumor cell invasion. Curr Opin Cell Biol 3: 832–840

Birchmeier W, Behrens J (1994) Cadherin expression in carcinomas: role in the formation of cell junctions and the prevention of invasiveness. Biochem Biophys Acta 1198: 11–26

Böhrer C, Berlin C, Jablonski-Westrich D, Holzmann B, Thiele HG, Hamann A (1992) Lymphocyte activation and regulation of three adhesion molecules with supposed function in homing: LECAM-1 (MEL-14 antigen), LPAM-1/2 (4-integrin) and CD44 (Pgp-1). Scand J Immunol 35: 107–120

Bosman FT (1993) Integrins: cell adhesives and modulators of cell function. Histochem J 25: 469–477

Bosworth BT, St John T, Gallatin WM, Harp JA (1991) Sequence of the bovine CD44 cDNA: comparison with human and mouse sequences. Mol Immunol 28: 1131–1135

Bourguignon LY, Lokeshwar VB, He H, Chen X, Bourguignon GJ (1992) A CD44-like endothelial cell transmembrane glycoprotein (GP116) interacts with extracellular matrix and ankyrin. Mol Cell Biol 12: 4464–4471
Bourguignon LY, Lokeshwar VB, Chen X, Kerrick WG (1993) Hyaluronic acid-induced lymphocyte signal transduction and HA receptor (GP85/CD44)-cytoskeleton interaction. J Immunol 151: 6634–6644
Brown TA, Bouchard T, St John T, Wayner E, Carter WG (1991) Human keratinocytes express a new CD44 core protein (CD44E) as a heparan-sulfate intrinsic membrane proteoglycan with additional exons. J Cell Biol 113: 207–221
Bruynzeel I, Koopman G, van der Raaij LM, Pals ST, Willemze R (1993) CD44 antibody stimulates adhesion of peripheral blood T cells to keratinocytes through the leukocyte function-associated antigen-1/intercellular adhesion molecule-1 pathway. J Invest Dermatol 100: 424–428
Budd RC, Cerottini JC, Mac Donald HR (1987) Selectively increased production of interferon-γ by subsets of Lyt-2^+ and L3T4^+ T cells identified by expression of Pgp-1. J Immunol 138: 3586
Camp RL, Kraus TA, Pure E (1991a) Variations in the cytoskeletal interaction and posttranslational modification of the CD44 homing receptor in macrophages. J Cell Biol 115: 1283–1292
Camp RL, Kraus TA, Birkeland ML, Pure E (1991b) High levels of CD44 expression distinguish virgin from antigen-primed B cells. J Exp Med 173: 763–766
Camp RL, Scheynius A, Johansson C, Pure E (1993) CD44 is necessary for optimal contact allergic responses but is not required for normal leukocyte extravasation. J Exp Med 178: 497–507
Cannistra SA, Kansas GS, Niloff J, DeFranzo B, Kim Y, Ottensmeier C (1993) Binding of ovarian cancer cells to peritoneal mesothelium in vitro is partly mediated by CD44H. Cancer Res 53: 3830–3838
Carlos TM, Harlan JM (1994) Leukocyte-endothelial adhesion molecules. Blood 84: 2068–2101
Carter WG, Wayner EA (1988) Characterization of the class III collagen receptor, a phosphorylated, transmembrane glycoprotein expressed in nucleated human cells. J Biol Chem 263: 4193–4201
Chaudry A, Gobl A, Eriksson B, Skogseid B, Oberg K (1994) Different splice variants of CD44 are expressed in gastrinomas but not in other subtypes of endocrine pancreatic tumors. Cancer Res 54: 981–986
Chong ASF, Boussy IA, Graf LH, Scuderi P (1992) Stimulation of IFN-γ, TNF-α and TNF-β secretion in IL-2-activated T cells: costimulatory roles for LFA-1, LFA-2, CD44 and CD45 molecules. Cell Immunol 144: 69–79
Chong AS, Jiang XL, Scuderi P, Lamas M, Graf LH Jr (1994) ICAM-1 and LFA-3 enhance the ability of anti-CD3 mAb to stimulate interferon gamma production in interleukin-2-activated T cells. Cancer Immunol Immunother 39: 127–134
Clark BR, Gallagher JT, Dexter TM (1992) Cell adhesion in the stromal regulation of haemopoiesis. Baillieres Clin Haematol 5: 619–652
Collado A, de Andres A, Canadas E, Ruiz-Cabello F, Gomez O, Pedrinaci S, Garrido F (1991) Characterization of CD44 antigen during lymphoid ontogeny. Immunobiology 183: 1–11
Combaret V, Lasset C, Bouvier R, Frappaz D, Thiesse P, Rebillard AC, Philip T, Favrot MC (1995) Expression de CD44: un nouveau facteur prognostique pour le neuroblastome. Bull Cancer 82: 131–136
Conrad P, Rothman BL, Kelley KA, Blue ML (1992) Mechanism of peripheral T cell activation by coengagement of CD44 and CD2. J Immunol 149: 1833–1839
Cooper DL, Dougherty G, Harn HJ, Jackson S, Baptist EW, Byers J, Datta A, Phillips G, Isola NR (1992) The complex CD44 transcriptional unit; alternative splicing of three internal exons generates the epithelial form of CD44. Biochem Biophys Res Commun 182: 569–578
Csanaky G, Kalasz V, Kelenyi G, Losonczy H, Baliko Z, Toth A (1993) Expression of an adhesion molecule and homing in B-cell chronic lymphocytic leukaemia. I. Application of the HEV-binding assay to a clinical series. Med Oncol Tumor Pharmacother 10: 125–130
Culty M, Miyake K, Kincade PW, Silorski E, Butcher EC, Underhill C, Sikorski E (1990) The hyaluronate receptor is a member of the CD44 (H-CAM) family of cell surface glycoproteins. J Cell Biol 111: 2765–2774
Culty M, Nguyen HA, Underhill CB (1992) The hyaluronan receptor (CD44) participates in the uptake and degradation of hyaluronan. J Cell Biol 116: 1055–1062
Culty M, O'Mara TE, Underhill CB, Yeager H Jr, Swartz RP (1994) Hyaluronan receptor (CD44) expression and function in human peripheral blood monocytes and alveolar macrophages. J Leukoc Biol 56: 605–611
Culty M, Shizari M, Thompson EW, Underhill CB (1994) Binding and degradation of hyaluronan by human breast cancer cell lines expressing different forms of CD44 correlation with invasive potential. J Cell Physiol 160: 275–286

Dall P, Heider KH, Hekele A, von Minckwitz G, Kaufmann M, Ponta H, Herrlich P (1994) Surface protein expression and messenger RNA-splicing analysis of CD44 in uterine cervical cancer and normal cervical epithelium. Cancer Res 54: 3337–3341

Dall P, Heider KH, Sinn HP, Skroch-Angel P, Adolf G, Kaufmann M, Herrlich P, Ponta H (1995) Comparison of immunohistochemistry and RT-PCR for detection of CD44v expression, a new prognostic factor in human breast cancer. Int J Cancer 60: 471–477

Dedhar S (1990) Integrins and tumor invasion. Bioessays 12: 583–590

Degrassi A, Hilbert DM, Rudikoff S, Anderson AO, Potter M, Coon HG (1993) In vitro culture of primary plasmacytomas requires stromal cell feeder layers. Proc Natl Acad Sci USA 90: 2060–2064

Delfino DV, Patrene KD, DeLeo AB, Deleo R, Herberman RB, Boggs SS (1994) Role of CD44 in the development of natural killer cells from precursors in long-term cultures of mouse bone marrow. J Immunol 152: 5171–5179

Denning SM, Le PT, Singer KH, Haynes BF (1990) Antibodies against the CD44 p80, lymphocyte homing receptor molecule augment human peripheral blood T cell activation. J Immunol 144: 7–15

Dougherty GJ, Landorp PM, Cooper DL, Humphries RK (1991) Molecular cloning of CD44R1 and CD44R2, two novel isoforms of the human CD44 lymphocyte "homing" receptor expressed by hemopoietic cells. J Exp Med 174: 1–5

Dustin ML, Springer TA (1991) Role of lymphocyte adhesion receptors in transient interactions and cell locomotion. Annu Rev Immunol 9: 27–66

East JA, Hart IR (1993) CD44 and its role in tumour progression and metastasis. Eur J Cancer 29A: 1921–1922

East JA, Mitchell SD, Hart IR (1993) Expression and function of the CD44 glycoprotein in melanoma cell lines. Melanoma Res 3: 341–346

Edelman GM (1983) Cell adhesion molecules. Science 219: 450–457

Edelman GM (1986) Cell adhesion molecules in the regulation of normal form and tissue pattern. Annu Rev Cell Biol 2: 81–116

Edwards JC, Wilkinson LS, Jones HM, Soothill P, Henderson KJ, Worrall JG, Pitsillides AA (1994) The formation of human synovial joint cavities: a possible role for hyaluronan and CD44 in altered interzone cohesion. J Anat 185: 355–367

Evans CW (1992) Cell adhesion and metastasis. Cell Biol Intern Rep 16: 1–10

Faasen AE, Schrager JA, Klein DJ, Oegema TR, Couchman JR, McCarthy J (1992) A cell surface chondroitin sulfate proteoglycan, immunologically related to CD44, is involved in type I collagen-mediated melanoma cell motility and invasion. J Cell Biol 116: 521–531

Faasen AE, Mooradian DL, Tranquillo RT, Dickinson RB, Letourneau PC, Oegema TR, McCarthy JB (1993) Cell surface CD44-related chondroitin sulfate proteoglycan is required for transforming growth factor-beta-stimulated mouse melanoma cell motility and invasive behavior on type I collagen. J Cell Sci 105: 501–511

Fenderson BA, Stamenkovic I, Aruffo A (1993) Localization of hyaluronan in mouse embryos during implantation, gastrulation and organogenesis. Differentiation 54: 85–98

Finn L, Dougherty G, Finley G, Meisler A, Becich M, Cooper DL (1994) Alternative splicing of CD44 pre-mRNA in human colorectal tumors. Biochem Biophys Res Commun 200: 1015–1022

Foets BJ, van den Oord JJ, Volpes R, Missotten L (1992) In situ immunohistochemical analysis of cell adhesion molecules on human corneal endothelial cells. Br J Ophthalmol 76: 205–209

Forster R, Emrich T, Kremmer E, Lipp M (1994) Expression of the G-protein-coupled receptor BLR1 defines mature, recirculating B cells and a subset of T-helper memory cells. Blood 84: 830–840

Fox SB, Gatter KC, Jackson DG, Screaton GR, Bell MV, Bell JI Harris AL, Simmons D, Fawcett J (1993) CD44 and cancer screening. Lancet 342: 548–549

Fox SB, Fawcett J, Jackson DG, Collins I, Gatter KC, Harris AL, Gearing A, Simmons DL (1994) Normal human tissues, in addition to some tumors, express multiple different CD44 isoforms. Cancer Res 54: 4539–4546

Frogner FJ, O'Neill HC (1992) Lymphocyte recirculation: the need for site-specific receptors to dictate T-lymphocyte localization into different tissue sites. Scand J Immunol 35: 627–632

Funaro A, Spagnoli GC, Momo M, Knapp W, Malavasi F (1994) Stimulation of T cells via CD44 requires leukocyte-function-associated antigen interactions and interleukin-2 production. Hum Immunol 40: 267–278

Gabrilovich DI, Woods GM, Patterson S, Harvey JJ, Knight SC (1994) Retrovirus-induced immunosuppression via blocking of dendritic cell migration and down-regulation of adhesion molecules. Immunology 82: 82–87

Galandrini R, Albi N, Tripodi G, Zarcone D, Terenzi A, Moretta A, Grossi CE, Velardi A (1993) Antibodies to CD44 trigger effector functions of human T cell clones. J Immunol 150: 4225–4335

Galandrini R, deMaria R, Piccoli M, Frati L, Santoni A (1994) CD44 triggering enhances human NK cell cytotoxic functions. J Immunol 153: 4399–4407
Galandrini R, Galluzzo E, Albi N, Grossi CE, Velardi A (1994) Hyaluronate is costimulatory for human T cell effector functions and binds to CD44 and activated T cells. J Immunol 153: 21–31
Gallagher JT (1992) The protein and proteoglycan guises of Hermes/CD44. Glycobiology 2: 93–94
Gallatin WM, Wayner EA, Hoffman PA, St. John T, Butcher EC, Carter WG (1989) Structural homology between lymphocyte receptors for high endothelium and class III extracellular matrix receptor. Proc Natl Acad Sci USA 86: 4654–4658
Geiger B, Ayalon O (1992) Cadherins. Annu Rev Cell Biol 8: 307–332
Geppert TD, Lipsky PE (1991) Association of various T cell-surface molecules with the cytoskeleton. Effect of cross-linking and activation. J Immunol 146: 3298–3305
Giancotti FG, Mainiero F (1994) Integrin-mediated adhesion and signaling in tumorigenesis. Biochim Biophys Acta 1198: 47–64
Girgrah N, Letarte M, Becker LE, Cruz TF, Theriault E, Moscarello MA (1991) Localization of the CD44 glycoprotein to fibrous astrocytes in normal white matter and to reactive astrocytes in active lesions in multiple sclerosis. J Neuropathol Exp Neurol 50: 779–792
Gjierset RA, Arya J, Volkman S, Haas M (1992) Association of induction of a fully tumorigenic phenotype in murine radiation-induced T-lymphoma cells with loss of differentiation antigens, gain of CD44, and alterations in p53 protein levels. Mol Carcinog 5: 190–198
Glinsky GV (1993) Cell adhesion and metastasis: is the site specificity of cancer metastasis determined by leukocyte-endothelial cell recognition and adhesion? Crit Rev Oncol Hematol 14: 229–277
Goldstein LA, Zhou DFH, Picker LJ, Minty CN, Bargatze RF, Ding JF, Butcher EC (1989) A human lymphocyte homing receptor, the hermes antigen, is related to cartilage proteoglycan core and link proteins. Cell 56: 1063–1072
Goldstein LA, Butcher EC (1990) Identification of mRNA that encodes an alternative form of H-CAM (CD44) in lymphoid and nonlymphoid tissues. Immunogenetics 32: 389–397
Goridis C, Brunet JF (1992) NCAM: structural diversity, function and regulation of expression. Semin Cell Biol 3: 189–197
Gross N, Beretta C, Peruisseau G, Jackson D, Simmons D, Beck D (1994) CD44H expression by human neuroblastoma cells: relation to mycN amplification and lineage differentiation. Cancer Res 54: 4238–4242
Grossi CE, Zarcone D, Tenca C, De Rossi G, Mandelli F (1992) Expression of adhesion molecules in lymphoproliferative disorders. Leukemia 6 [Suppl 4]: 35–37
Gruber MF, Webb DS, Gerrard TL (1992) Stimulation of human monocytes via CD45, CD44, and LFA-3 triggers macrophage-colony-stimulating factor production. Synergism with lipopolysaccharide and IL-1 beta. J Immunol 148: 1113–1118
Günthert U (1993) CD44: a multitude of isoforms with diverse functions. Curr Top Microbiol Immunol 184: 47–73
Günthert U, Hofmann M, Rudy W, Reber S, Zöller M, Haussmann I, Matzku S, Wenzel A, Ponta H, Herrilich P (1991) A new variant of glycoprotein CD44 confers metastatic potential to rat carcinoma cells. Cell 65: 13–24
Guo YJ, Ma J, Wang JH, Lin SC, Chang HC, Bigby M, Sy MS (1993) Monoclonal anti-CD44 antibody acts in synergy with anti-CD2 but inhibits anti-CD3 or T cell receptor-mediated signaling in murine T cell hybridomas. Immunology 152: 186–199
Guo YJ, Lin SC, Wang JH, Bigby M, Sy MS (1994a) Palmitoylation of CD44 interferes with CD3-mediated signalling in human T lymphocytes. Int Immunol 6: 213–221
Guo YJ, Wong JH, Lin SC, Aruffo A, Stamenkovic I, Sy MS (1994b) Disruption of T lymphocyte reappearance in anti-Thy-1-treated animals in vivo with soluble CD44 and L-selectin molecules. Cell Immunol 154: 202–218
Guo Y, Ma J, Wang J, Che, X, Narula J, Bigby M, Wu M, Sy M-S (1994c) Inhibition of human melanoma growth and metastasis in vivo by anti-CD44 monoclonal antibody. Cancer Res 54: 1561–1565
Guo YJ, Liu G, Wang X, Jin D, Wu M, Ma J, Sy MS (1994d) Potential use of soluble CD44 in serum as indicator of tumor burden and metastasis in patients with gastric or colon cancer. Cancer Res 54: 422–426
Haeğel H, Tolg C, Hofmann M, Ceredig R (1993) Activated mouse astrocytes and T cells express similar CD44 variants. Role of CD44 in astrocyte/ T cell binding. J Cell Biol 122: 1067–1077
Haggerty JG, Bretton RH, Milstone LM (1992) Identification and characterization of a cell surface proteoglycan on keratinocytes. J Invest Dermatol 99: 374–380
Haig DM, Thomson J, Percival A (1992) Purification and adhesion receptor phenotype of bovine bone marrow-derived haemopoietic colony-forming cells. Vet Immunol Immunopathol 33: 223–236

Hale LP, Haynes BF (1992) Bromelain treatment of human T cells removes CD44, CD45RA, E2/MIC2, CD6, CD7, CD8, and Leu 8/LAM1 surface molecules and markedly enhances CD2-mediated T cell activation. J Immunol 149: 3809–3816
Hamilton MS, Ball J, Bromidge E, Franklin IM (1991) Surface antigen expression of human neoplastic plasma cells includes molecules associated with lymphocyte recirculation and adhesion. Br J Haematol 78: 60–65
Hardingham TE, Fosang AJ (1992) Proteoglycans: many forms and many functions. FASEB J 6: 861–70
Harn HJ, Lee HS, Ho LI, Lee WH, Ding JH (1994a) Selective expression of CD44 messenger RNA splice variants in four high grade human brain tumour cell lines. Biochem Mol Biol Int 33: 743–749
Harn HJ, Ho LI, Yu CP, Wang MW, Lee HS, Lin JJ, Lee WH, Isola NR,Cooper DL (1994b) The variant mRNA isoform of human metastasis gene (CD44V) detected in the cell lines of human hepatocellular carcinoma. Biochem Mol Biol Int 32: 233–8
Harn HJ, Ho LI, Chang JY, Wu CW, Jiang SY, Lee HS, Lee WH (1995) Differential expression of the human metastasis adhesion molecule CD44v in normal and carcinomatous stomach mucosa of chinese subjects. Cancer 75: 1065–1071
Hart IR, Birch M, Marshall JF (1991) Cell adhesion receptor expression during melanoma progression and metastasis. Cancer Metastasis Rev 10: 115 – 128
Hathcock KS, Hirano H, Murakami S, Hodes RJ (1993) CD44 expression on activated B cells. Differential capacity for CD44-dependent binding to hyaluronic acid. J Immunol 151: 6712–6722
Hawley RG, Wang MH, Fong AZ, Hawley TS (1993) Association between ICAM-1 expression and metastatic capacity of murine B-cell hybridomas. Clin Exp Metastasis 11: 213–226
Haynes BF, Liao HX, Patton KL (1991a) The transmembrane hyaluronate receptor (CD44): multiple functions, multiple forms. Cancer Cells 3: 347–350
Haynes BF, Hale LP, Patton KL, Martin ME, McCallum RM (1991b) Measurement of an adhesion molecule as an indicator of inflammatory disease activity. Up-regulation of the receptor for hyaluronate (CD44) in rheumatoid arthritis. Arthritis Rheum 34: 1434–1443
He Q, Lesley J, Hyman R, Ishihara K, Kincade PW (1992) Molecular isoforms of murine CD44 and evidence that the membrane proximal domain is not critical for hyaluronate recognition. J Cell Biol 119: 1711–1719
Heider KH, Hofmann M, Hors E, van den Berg F, Ponta H, Herrlich P, Pals ST (1993a) A human homologue of the rat metastasis-associated variant of CD44 is expressed in colorectal carcinomas and adenomatous polyps. J Cell Biol 120: 227–233
Heider KH, Dämmrich J, Skroch-Angel P, Müller-Hermelink HK, Vollmers HP, Herrlich P, Ponta H (1993b) Differential expression of CD44 splice variants in intestinal- and diffuse-type human gastric carcinomas and normal gastric mucosa. Cancer Res 53: 4197–4203
Hemler ME (1990) VLA proteins in the integrin family: structures, functions, and their role on leukocytes. Annu Rev Immunol 8: 365–400
Herold-Mende C, Seiter S, Jourdan U, Herrmann K, Born AI, Patzelt E, Zöller J, Bosch FX, Zöller M (1996) Expression of splice variants of CD44 on squamous epithelia and squamous cell carcinoma of head and neck. J Pathol (in press)
Herrlich P, Zöller M, Pals ST, Ponta H (1993) CD44 splice variants: metastases meet lymphocytes. Immunol Today 14: 395–399
Heufelder AE, Bahn RS, Boergen KP, Scriba PC (1993) Detection, localization and modulation of hyaluronic acid/CD44 receptor expression in patients with endocrine orbitopathy. Med Klin 88: 181–184, 277
Hirano H, Screaton GR, Bell MV, Jackson DG, Bell JI, Hodes RJ (1994) CD44 isoform expression mediated by alternative splicing: tissue-specific regulation in mice. Int Immunol 6: 49–59
Hoffman S, Edelman GM (1983) Kinetics of homophilic binding by embryonic and adult forms of the neural cell adhesion molecule. Proc Natl Acad Sci USA 80: 5762–5766
Hofmann M, Rudy W, Zöller M, Tölg C, Ponta H, Herrlich P, Günthert U (1991) CD44 splice variants confer metastatic behavior in rats: Homologous sequences are expressed in human tumor cell lines. Cancer Res 51: 5292–5297
Hofmann M, Rudy W, Güntert U, Zimmer SG, Zawadzki V, Zöller M, Lichtner RB, Herrlich P, Ponta H (1993) A link between ras and metastatic behavior of tumor cells: ras induces CD44 promoter activity and leads to low-level expression of metastasis-specific variants of CD44 in CREF cells. Cancer Res 53: 1516–1521
Honn KV, Tang DG (1992) Adhesion molecules and tumor cell interaction with endothelium and subendothelial matrix. Cancer Metastasis Rev 11: 353–375

Horgan KJ, Van Seventer GA, Shimizu Y, Shaw S (1990) Hyporesponsiveness of "naive" (CD45RA+) human T cells to multiple receptor-mediated stimuli but augmentation of responses by co-stimuli. Eur J Immunol 20: 1111–1118
Horst E, Meijer CJ, Duijvestijn AM, Hartwig N, Van der Harten HJ, Pals ST (1990a) The ontogeny of human lymphocyte recirculation: high endothelial cell antigen (HECA-452) and CD44 homing receptor expression in the development of the immune system. Eur J Immunol 20: 1483–1489
Horst E, Meijer CJ, Radaszkiewicz T, Ossekoppele GJ, Van Krieken JH, Pals ST (1990b) Adhesion molecules in the prognosis of diffuse large-cell lymphoma: expression of a lymphocyte homing receptor (CD44), LFA-1 (CD11a/18), and ICAM-1 (CD54). Leukemia 4: 595–599
Horst E, Meijer CJ, Radaskiewicz T, van Dongen JJ, Pieters R, Figdor CG, Hooftman A, Pals ST (1990c) Expression of human homing receptor (CD44) in lymphoid malignancies and related stages of lymphoid development. Leukemia 4: 383–389
Hua Q, Knudson CB, Knudson W (1993) Internalization of hyaluronan by chondrocytes occurs via receptor-mediated endocytosis. J Cell Sci 106: 365–375
Huet S, Groux H, Caillou B, Valentin H, Prieur AM, Bernard A (1989) CD44 contributes to T cell activation J Immunol 143: 798–801
Hughes EN, Colombatti A, August JT (1981) Murine cell surface glycoproteins. J Biol Chem 256: 1014–1021
Hughes DE, Salter DM, Simpson R (1994) CD44 expression in human bone: a novel marker of osteocytic differentiation. J Bone Miner Res 9: 39–44
Humphries MJ, Mould AP, Tuckwell DS (1993) Dynamic aspects of adhesion receptor function – integrins both twist and shout. Bioessays 15: 391–397
Hynes RO (1992a) Integrins: versatility, modulation and signaling in cell adhesion. Cell 69: 11–25.
Hynes RO (1992b) Specificity of cell adhesion in development: the cadherin superfamily. Curr Opin Genet Dev 2: 621–624
Idzerda RL, Carter WG, Nottenburg C, Wayner EA, Gallatin WM, St John T (1989) Isolation and DNA sequence of a cDNA clone encoding a lymphocyte adhesion receptor for high endothelium. Proc Natl Acad Sci USA 86: 4659–4663
Iida N, Bourguignon YW (1995) New CD44 splice variants associated with human breast cancers. J Cell Physiol 162: 127–133
Ishii S, Ford R, Thomas P, Nachman A, Steele G Jr, Jessup JM (1993) CD44 participates in the adhesion of human colorectal carcinoma cells to laminin and type IV collagen. Surg Oncol 2: 255–264
Isola NR, Harn HJ, Cooper DL (1991) Screening recombinant DNA libraries: a rapid and efficient method for isolating cDNA clones utilizing the PCR. Biotechniques 11: 580–582
Jackson DG, Buckley J, Bell JI (1992) Multiple variants of the human lymphocyte homing receptor CD44 generated by insertions at a single site in the extracellular domain. J Biol Chem 267: 4732–4739
Jackson DG, Screaton GR, Bell MV, Bell JI (1993) CD44 and cancer. Lancet 341: 252–253
Jackson DG, Schenker T, Waibel R, Bell JI, Stahel RA (1994) Expression of alternatively spliced forms of the CD44 extracellular-matrix receptor on human lung carcinomas. Int J Cancer Suppl 8: 110–115
Jackson DG, Bell JI, Dickinson R, Timans J, Shields J, Whittle N (1995) Proteoglycan forms of the lymphocyte homing receptor CD44 are alternatively spliced variants containing the v3 exon. J Cell Biol 128: 673–685
Jalkanen S, Bargatze RF, de los Toyos J, Butche EC (1986) A lymphoid cell surface protein involved in endothelial cell recognition and lymphocyte homing in man. Eur J Immunol 16: 1195–1202
Jalkanen S, Jalkanen M, Bargatze R, Tammi M, Butcher EC (1988) Biochemical properties of glycoproteins involved in lymphocyte recognition of high endothelial venules in man. J Immunol 141: 1615–1623
Jalkanen S, Joensuu H, Soderstrom KO, Klemi P (1991) Lymphocyte homing and clinical behavior of non-Hodgkin's lymphoma. J Clin Invest 87: 1835–1840
Jalkanen S, Jalkanen M (1992) Lymphocyte CD44 binds the COOH-terminal heparin-binding domain of fibronectin. J Cell Biol 116: 817–825
Joensuu H, Klemi PJ, Toikkanen S, Jalkanen S (1993a) Glycoprotein CD44 expression and its association with survival in breast cancer. Am J Pathol 143: 867–874
Joensuu H, Ristamaki R, Klemi PJ, Jalkanen S (1993b) Lymphocyte homing receptor (CD44) expression is associated with poor prognosis in gastrointestinal lymphoma. Br J Cancer 68: 428–432
Johnson JP (1991) Cell adhesion molecules of the immunoglobulin supergene family and their role in malignant transformation and progression to metastatic disease. Cancer Metastasis Rev 10: 11–22
Johnson BA, Haines GK, Harlow LA, Koch AE (1993) Adhesion molecule expression in human synovial tissue. Arthritis Rheum 36: 137–146

Jothy S, McClure D, LeDuy L, Günthert U (1993) Selective expression of CD44 messenger RNA splice variants in human colon carcinoma. Proc Am Assoc Cancer Res 34: 553
Juliano RL, Haskill S (1993) Signal transduction from the extracellular matrix. J Cell Biol 120: 577–585
Jutila MA, Bargatze RF, Kurk S, Warnock RA, Ehsanie N, Watson SR, Walcheck B (1994) Cell surface P- and E-selectin support shear-dependent rolling of bovine gamma/delta T cells. J Immunol 153: 3917–3928
Kalomiris EL Bourguignon LYW (1988) Mouse T lymphoma cells contain a transmembrane glycoprotein (GP85) that binds ankyrin. J Cell Biol 106: 319–327
Kansas GS, Wood GS, Daily MO (1989) A family of cell-surface glycoproteins defined by a putative anti-endothelial cell receptor antibody in man. J Immunol 142: 3050–3057
Kansas GS, Muirhead MJ, Dailey MO (1990) Expression of the CD11/CD18, leukocyte adhesion molecule 1, and CD44 adhesion molecules during normal myeloid and erythroid differentiation in humans. Blood 76: 2483–2492
Kaufmann M, Heider KH, Sinn HP, von Minckwitz G, Ponta H, Herrlich P (1995) CD44 variant exon epitopes in primary breast cancer and length of survival. Lancet 345: 615–619
Kee BL, Dadi HK, Tran-Paterson R, Quackenbush EJ, Andrulis IL, Letarte M (1991) CD10 and CD44 genes of leukemic cells and malignant cell lines show no evidence of transformation-related alterations. J Cell Physiol 148: 414–420
Kennel SJ, Lankford TK, Foote LJ, Shinpock SG, Stringer C (1993) CD44 expression on murine tissues. J Cell Sci 104: 373–382
Kim JC Ishii S, Ford R, Thomas P, Steele GJr, Jessup JM (1993) Epitopes for homotypic binding of human carcinoembryonic antigen also participate in adhesion to extracellular matrix proteins. Proc Am Assoc Cancer Res 34: 197
Kim H, Yang XL, Rosada C, Hamilton SR (1994) CD44 expression on colorectal adenomas is an early event occuring prior to K-ras and p53 gene mutation. Arch Biochem Biophys 310: 504–507
Kincade PW (1991) Molecular interactions between stromal cells and B lymphocyte precursors. Semin Immunol 3: 379–390
Kincade PW (1992) Cell interaction molecules and cytokines which participate in B lymphopoieses. Baillieres Clin Haematol 5: 575–598
Kincade, PW, He Q, Ishihara K, Miyake K, Lesley J, Hyman R (1993) CD44 and other cell interaction molecules contributing to B lymphopoieses. Curr Top Microbiol Immunol 184: 215–222
Knudson CB, Knudson W (1993) Hyaluronan-binding proteins in development, tissue homeostasis, and disease. FASEB J 7: 1233–1241
Knudson W, Bartnik E, Knudson GB (1993) Assembly of pericellular matrices by COS-7 cells transfected with CD44 lymphocyte-homing receptor genes. Proc Natl Acad Sci USA 90: 4003–4007
Kobayashi M, Imamura M, Uede T, Sakurada K, Maeda S, Iwasaki H, Tsuda Y, Musashi M, Miyazaki T (1994) Expression of adhesion molecules on human hematopoietic progenitor cells at different maturational stages. Stem Cells (Dayt) 12: 316–321
Koopman G, van Kooyk-Y, de Graaff M, Meyer CJ, Figdor CD, Pals ST (1990) Triggering of the CD44 antigen on T lymphocytes promotes T cell adhesion through the LFA-1 pathway. J Immunol 145: 3589–3593
Koopman G, Griffioen AW, Ponta H, Herrlich P, van den Berg F, Manten-Horst E, Pals ST (1993a) CD44 splice variants; expression on lymphocytes and in neoplasia. Res Immunol 144: 750–754
Koopman G, Heider KH, Horst E, Adolf GR, van den Berg F, Ponta H, Herrlich P, Pals ST (1993b) Activated human lymphocytes and aggressive non-Hodgkin's lymphomas express a homologue of the rat metastasis-associated variant of CD44. J Exp Med 177: 897–904
Koretz K, Möller P, Lehnert T, Hinz U, Otto HF, Herfarth C (1995) Effect of CD44v6 on survival in colorectal carcinoma. Lancet 345: 327–328
Koshiishi I, Shizari M, Underhill CB (1994) CD44 can mediate the adhesion of platelets to hyaluronan. Blood 84: 390–396
Krakauer T (1994) Cell adhesion molecules are co-receptors for staphylococcal enterotoxin B-induced T-cell activation and cytokine production. Immunol Lett 39: 121–125
Kuppner MC, Van Meir E, Gauthier T, Hamou MF, De Tribolet N (1992) Differential expression of the CD44 molecule in human brain tumours. Int J Cancer 50: 572–577
Labarriere N, Piau JP, Otry C, Denis M, Lustenberger P, Meflah K, Le Pendu J (1994) H blood group antigen carried by CD44V modulates tumorigenicity of rat colon carcinoma cells. Cancer Res 54: 6275–6281
Lasky LA (1992) Selectins: interpreters of cell-specific carbohydrate information during inflammation. Science 258: 964–969
Lawrence MB, Springer TA (1991) Leukocytes roll on a selectin at physiologic flow rates: distinction from and prerequisite for adhesion through integrins. Cell 65: 859–873

Lepesant H, Reggio H, Pierres M, Naquet P (1990) Mouse thymic epithelial cell lines interact with and select a CD3lowCD4+CD8+ thymocyte subset through an LFA-1-dependent adhesion–de-adhesion mechanism. Int Immunol 2: 1021–1032
Lesley J, Schulte R, Hyman R (1990) Binding of hyaluronic acid to lymphoid cell lines is inhibited by monoclonal antibodies against Pgp-1. Exp Cell Res 187: 224–233
Lesley J, Hyman R (1992) CD44 can be activated to function as an hyaluronic acid receptor in normal murine T cells. Eur J Immunol 22: 2719–2723
Lesley J, He Q, Miyake K, Hamann A, Hyman R, Kincade PW (1992) Requirements for hyaluronic acid binding by CD44: a role for the cytoplasmic domain and activation by antibody. J Exp Med 175: 257–266
Lesley J, Hyman R, Kincade PW (1993a) CD44 and its interaction with extracellular matrix. Adv Immunol 54: 271–335
Lesley J, Kincade PW, Hyman R (1993b) Antiboby-induced activation of the hyaluronan receptor function of CD44 requires multivalent binding by antibody. Eur J Immunol 23: 1902–1909
Lesley J,Howes N, Perschl A, Hyman R (1994) Hyaluronan binding function of CD44 is transiently activated on T cells during an in vivo immune response. J Exp Med 180: 383–387
Lester BR, McCarthy JB(1992) Tumor cell adhesion to the extracellular matrix and signal transduction mechanisms implicated in tumor cell motility, invasion and metastasis. Cancer Metastasis Rev 11: 31–44
Lewinsohn DM, Nagler A, Ginzton N, Greenberg P, Butcher EC (1990) Hematopoietic progenitor cell expression of the H-CAM (CD44) homing-associated adhesion molecule. Blood 75: 589–595
Li H, Hamou MF, de Tribolet N, Jaufeerally R, Hofmann M, Diserens AC, Van Meir EG (1993) Variant CD44 adhesion molecules are expressed in human brain metastases but not in glioblastomas. Cancer Res 53: 5345–5349
Liao HX, Levesque MC, Patton K, Bergamo B, Jones D, Moody MA, Telen MJ, Haynes BF (1993) Regulation of human CD44H and CD44E isoform binding to hyaluronan by phorbol myristate acetate and anti-CD44 monoclonal and polyclonal antibodies. J Immunol 151: 6490–6499
Liu AY (1994) Expression of CD44 in prostate cancer cells. Cancer Lett 76: 63–69
Lokeshwar VB, Bourguignon LY (1991) Post-translational protein modification and expression of ankyrin-binding site(s) in GP85 (Pgp-1/CD44) and its biosynthetic precursors during T-lymphoma membrane biosynthesis. J Biol Chem 266: 17983–17989
Lokeshwar VB, Bourguignon LY (1992) The lymphoma transmembrane glycoprotein GP85 (CD44) is a novel guanine nucleotide-binding protein which regulates GP85 (CD44)-ankyrin interaction. J Biol Chem 267: 22073–22078
Lokeshwar VB, Fregien N, Bourguignon LY (1994) Ankyrin-binding domain of CD44(Gp85) is required for the expression of hyaluronic acid-mediated adhesion function. J Cell Biol 126: 1099–1109
Lund-Johansen F, Terstappen LW (1993) Differential surface expression of cell adhesion molecules during granulocyte maturation. J Leukoc Biol 54: 47–55
Mackay CR, Terpe HJ, Stauder R, Marston WL, Stark H, Gunthert U (1994) Expression and modulation of CD44 variant isoforms in humans. J Cell Biol 124: 71–82
Mareel MM, Van Roy FM, Bracke ME (1993) How and when do tumor cells metastasize? Crit Rev Oncog 4: 559–594
Matsumura Y, Tarin D (1992) Significance of CD44 gene products for cancer diagnosis and disease evaluation. Lancet 340: 1053–1058
Matsumura Y, Hanbury D, Smith J, Tarin D (1994) Non-invasive detection of malignancy by identification of unusual CD44 gene activity in exfoliated cancer cells. BMJ 308: 619–624
Matzku S, Komitowski D, Miltenberger M, Zöller M (1983) Characterization of BSp73, a spontaneous rat tumor and its in vivo selected variants showing different metastasizing capacities. Invasion Metast 3: 109–123
Matzku S, Wenzel A, Liu S, Zöller M (1989) Antigenic differences between metastatic and non-metastatic BSp73 rat tumor variants characterized by monoclonal antibodies. Cancer Res 49: 1294–1299
Mayer B, Jauch KW, Günthert U, Figdor CG, Schildberg FW, Funke I, Johnson JP (1993) De-novo expression of CD44 and survival in gastric cancer. Lancet 342: 1019–1022
McCarthy JB, Skubitz AP, Lika J, Mooradian DL, Wilke MS, Furcht LT (1991) Tumor cell adhesive mechanisms and their relationship to metastasis. Semin Cancer Biol 2: 155–167
Merzak A, Koocheckpour S, Pilkington GJ (1994) CD44 mediates human glioma cell adhesion and invasion in vitro. Cancer Res 54: 3988–92
Miyake K, Kincade PW (1990) A new adhesion mechanism involving hyaluronate and CD44. Curr Top Microbiol Immunol 166: 87–90

Miyake K, Medina KL, Hayashi SL, Ono S, Hamaoka T, Kincade PW (1990a) Monoclonal antibodies to Pgp-1/CD44 block lympho-hemopoiesis in long-term bone marrow cultures. J Exp Med 171: 477–488

Miyake K, Underhill CB, Lesley J, Kincade PW (1990b) Hyaluronate can function as a cell adhesion molecule and CD44 participates in hyaluronate recognition. J Exp Med 172: 69–75

Miyasaka M, Toyama-Sorimachi N (1993) Tumor metastasis and adhesion molecules. Gan To Kagaku Ryoho 20: 359–362

Moll J, Schmidt A, Arch R, van der Putten H, Plug R, Ponta H, Herrlich P, Zöller M (1996) Accelerated T cell responsiveness in variant CD44 expressing transgenic mice. J Immunol (in press)

Morimoto K, Robin E, Le Bousse Kerdiles MC, Li Y, Clay D, Jasmin C, Smadja-Joffe F (1994) CD44 mediates hyaluronan binding by human myeloid KG1A and KG1 cells. Blood 83: 657–662

Mulder JW, Kruyt PM, Sewnath M, Oosting J, Seldenrijk CA, Weidema WF, Offerhaus GJ, Pals ST (1994) Colorectal cancer prognosis and expression of exon-v6-containing CD44 proteins. Lancet 344: 1470–1472

Munzig E, Eckert K, Harrach T, Graf H, Maurer HR (1994) Bromelain protease F9 reduces the CD44 mediated adhesion of human peripheral blood lymphocytes to human umbilical vein endothelial cells. FEBS Lett 351: 215–8

Murakami S, Miyake K, June CH, Kincade PW, Hodes RJ (1990) IL-5 induces a Pgp-1 (CD44) bright B cell subpopulation that is highly enriched in proliferative and Ig secretory activity and binds to hyaluronate. J Immunol 145: 3618–3627

Murakami S, Miyake K, Abe R, Kincade PW, Hodes RJ (1991) Characterization of autoantibody-secreting B cells in mice undergoing stimulatory (chronic) graft-versus-host reactions.Identification of a CD44hi population that binds specifically to hyaluronate. J Immunol 146: 1422–1427

Murakami S, Shimabukuro Y, Miki Y, Saho T, Hino E, Kasai D, Nozaki T, Kusumoto Y, Okada H (1994) Inducible binding of human lymphocytes hyaluronate via CD44 does not require cytoskeleton association but does require new protein synthesis. J Immunol 152: 467–477

Nathke IS, Linck LE, Nelson WJ (1991) Epithelial cell adhesion and development of cell surface polarity: possible mechanisms for modulation of cadherin function, organization and distribution. J Cell Sci Suppl 17: 139–145

Naujokas MF, Morin M, Anderson MS, Peterson M, Miller J (1993) The chondroitin sulfate form of invariant chain can enhance stimulation of T cell responses through interaction with CD44. Cell 74: 257–268

Neame SJ, Isacke CM (1992) Phosphorylation of CD44 in vivo requires both Ser323 and Ser325, but does not regulate membrane localization or cytoskeletal interaction in epithelial cells. EMBO J 11: 4733–4738

Neame PJ, Barry FP (1993) The link proteins. Experientia 49: 393–402

Neame SJ, Isacke CM (1993) The cytoplasmic tail of CD44 is required for basolateral localization in ephitelial MDCK cells but does not mediate association with the detergent-insoluble cytoskeleton of fibroblasts. J Cell Biol 121: 1299–1310

Nigam AK,Pignatelli M, Boulos PB(1994) Current concepts in metastasis. Gut 35: 996–1000

Noble PW, Lake FR, Henson PM, Riches DW (1993) Hyaluronate activation of CD44 induces insulin-like growth factor-1 expression by a tumor necrosis factor-alpha-dependent mechanism in murine macrophages. J Clin Invest 91: 2368–2377

O'Sullivan NL,Skandera CA,Chin YH,Montgomery PC (1994) In vitro adhesive interactions between rat lymphocytes and lacrimal gland acinar epithelium. Phenotype of adherent lymphocytes and involvement of adhesion molecules. J Immunol 152: 1684–1692

Omary MB, Trowbridge IS, Letarte M, Kagnoff MF, Isacke CM (1988) Structural heterogeneity of human Pgp-1 and its relationship with p85. Immunogenetics 27: 460–464

Oppenheimer Marks N,Davis LS, Lipsky PE (1990) Human T lymphocyte adhesion to endothelial cells and transendothelial migration. Alteration of receptor use relates to the activation status of both the T cell and the endothelial cell. J Immunol 145: 140–148

Pals ST, Koopman G, Heider KH, Griffioen A, Adolf GR, Van den Berg F, Ponta H, Herrlich P, Horst E (1993) CD44 splice variants: expression during lymphocyte activation and tumor progression. Behring-Inst Mitt 92: 273–277

Pardi R, Inverardi L, Bender JR (1992) Regulatory mechanisms in leukocyte adhesion: flexible receptors for sophisticated travelers. Immunol Today 13: 224–230

Peach RJ, Hollenbaugh D, Stamenkovic I, Aruffo A (1993) Identification of hyaluronic acid binding sites in the extracellular domain of CD44. J Cell Biol 122: 257–264

Penneys NS (1993) CD44 expression in normal and inflamed skin. J Cutan Pathol 20: 250–253

Penneys NS, Kaiser M (1993) Cylindroma expresses immunohistochemical markers linking it to eccrine coil. J Cutan Pathol 20: 40–43
Picker LJ, Nakache M, Butcher EC (1989) Monoclonal antibodies to human lymphocyte homing receptors define a novel class of adhesion molecules on diverse cell types. J Cell Biol 109: 927–937
Pierres A, Lipcey C, Mawas C, Olive D (1992) A unique CD44 monoclonal antibody identifies a new T cell activation pathway. Eur J Immunol 22: 413–417
Pilkington GJ, Akinwunmi J, Ognjenovic N, Rogers JP (1993) Differential binding of anti-CD44 on human gliomas in vitro. Neuroreport 4: 259–262
Postigo AA, Sanchez-Madrid F (1993) Adhesion and homing molecules. Transpl Proc 25:65–69
Prieto J, Eklund A, Patarroyo M (1994) Regulated expression of integrins and other adhesion molecules during differentiation of monocytes into macrophages. Cell Immunol 156: 191–211
Prince HE, York J, Jensen ER (1992) Phenotypic comparison of the three populations of human lymphocytes defined by CD45RO and CD45RA expression. Cell Immunol 145: 254–262
Quackenbush EJ, Vera S, Greaves A, Letarte A (1990) Confirmation by peptide sequence and co-expression on various cell types of the identity of CD44 and P85 glycoprotein. Mol Immunol 27: 947–955
Ratner S (1992) Lymphocyte migration through extracellular matrix. Invasion Metastasis 12: 82–100
Reber S, Matzku S, Günthert U, Ponta H, Herrlich P, Zöller M(1990) Retardation of metastatic tumor growth after immunization with metastasis-specific monoclonal antibodies. int J Cancer 46: 919–927
Reuss-Borst MA, Buhring HJ, Klein G, Muller CA (1992) Adhesion molecules on CD34+ hematopoietic cells in normal human bone marrow and leukemia. Ann Hematol 65: 169–174
Ristamaki R, Joensuu H, Salmi M, Jalkanen S (1994) Serum CD44 in malignant lymphoma: an association with treatment response. Blood 84: 238–243
Roos E (1991) Adhesion molecules in lymphoma metastasis. Cancer Metastasis Rev 10: 33–48
Rosales C, Juliano RL(1995) Signal transduction by cell adhesion receptors in leukocytes. J Leukoc Biol 57: 189–198
Rosen SD (1993) Robert Feulgen Lecture 1993. L-selectin and its biological ligands. Histochemistry 100: 185–191
Rosenman SJ, Ganji AA, Tedder TF, Gallatin WM (1993) Syn-capping of human T lymphocyte adhesion/activation molecules and their redistribution during interaction with endothelial cells. J Leukoc Biol 53: 1–10
Rothman BL, Blue ML, Kelley KA, Wunderlich D, Mierz DV, Aune TM (1991) Human T cell activation by OKT3 is inhibited by a monoclonal antibody to CD44. J Immunol 147: 2493–2499
Rudy W, Hofmann M, Schwartz-Albiez R, Zöller M, Heider K-H, Ponta H, Herrlich P(1993) The two major CD44 proteins expressed on a metastatic rat tumor cell line are derived from different splice variants: each one individually suffices to confer metastatic behavior. Cancer Res 53:1262–1268
Ruoslathi E (1991) Integrins. J Clin Invest 87: 1–5
Rusciano D, Burger MM(1992) Why do cancer cells metastasize into particular organs? Bioessays 14: 185–194
Rutishauser U, Acheson A, Hall AK, Sunshine J(1988) NCAM as a regulator of cell-cell interactions. Science 240: 53–57.
Ryan DH (1993) Adherence of normal and neoplastic human B cell precursors to the bone marrow microenvironment. Blood Cells 19: 225–241
Ryan DH, Nuccie BL, Abboud CN, Liesveld JL (1990) Maturation-dependent adhesion of human B cell precursors to the bone marrow microenvironment. J Immunol 145: 477–484
Salles G, Zain M, Jiang EM, Boussiotis VA, Shipp MA (1993) Alternatively spliced CD44 transcripts in diffuse large-cell lymphomas: characterization and comparison with normal activated B cells and epithelial malignancies. Blood 82: 3539–3547
Sallusto F, Lanzavecchia A (1994) Efficient presentation of soluble antigen by cultured human dendritic cells is maintained by granulocyte/macrophage colony-stimulating factor plus interleukin 4 and downregulated by tumor necrosis factor alpha. J Exp Med 179: 1109–1118
Salmi M, Gron-Virta K, Sointu P, Grenman R, Kalimo H, Jalkanen S (1993) Regulated expression of exon v6 containing isoforms of CD44 in man: downregulation during malignant transformation of tumors of squamocellular origin. J Cell Biol 122: 431–422
Sampson PM, Rochester CL, Freundlich B, Elias JA (1992) Cytokine regulation of human lung fibroblast hyaluronan (hyaluronic acid) production. Evidence for cytokine-regulated hyaluronan (hyaluronic acid) degradation and human lung fibroblast-derived hyaluronidase. J Clin Invest 90: 1492–1503
Scheeren RA, Koopman G, Van der Baan S,Meijer CJ,Pals ST (1991) Adhesion receptors involved in clustering of blood dendritic cells and T lymphocytes. Eur J Immunol 21: 1101–1105

Sconocchia G, Titus JA, Segal DM (1994) CD44 is a cytotoxic triggering molecule in human peripheral blood NK cells. J Immunol 153: 5473–5481
Screaton GR, Bell MV, Jackson DG, Cornelis FB, Gerth U, Bell JI (1992) Genomic structure of DNA encoding the lymphocyte homing receptor CD44 reveals at least 12 alternatively spliced exons. Proc Natl Acad Sci USA 89: 12160–12164
Screaton GR, Bell MV, Bell JL, Jackson DG (1993) The identification of a new alternative exon with highly restricted tissue expression in transcripts encoding the mouse Pgp-1 (CD44) homing receptor. Comparison of all 10 variable exons between mouse, human, and rat. J Biol Chem 268: 12235–12238
Seiter S, Arch R, Reber S, Komitowski D, Hofmann M, Ponta H, Herrlich P, Matzku S, Zöller M (1993) Prevention of tumor metastasis formation by anti-variant CD44. J Exp Med 177: 443–445
Seiter S, Herrmann K, Möller P, Patzelt E, Tilgen W, Zöller M (1996a) Expression of splice variant of CD44 on human skin and epidermal tumours. Virchows Arch Pathol (in press)
Seiter S, Schadendorf D, Herrmann K, Schneider M, Rösel M, Arch R, Tilgen W, Zöller M (1996b) Expression of splice variants of CD44 during melanoma progression. Clin Cancer Res (in press)
Seth A,Gote L, Nagarkatti M, Nagarkatti PS (1991) T-cell-receptor-independent activation of cytolytic activity of cytotoxic T lymphocytes mediated through CD44 and gp90MEL-14. Proc Natl Acad Sci USA 88: 7877–7881
Shimizu Y, van Seventer GA, Siraganian R, Wahl L, Shaw S (1989) Dual role of the CD44 molecule in T-cell adhesion and activation. J Immunol 143: 2457–2463
Shimizu Y, Shaw S (1991) Lymphocyte interactions with extracellular matrix. FASEB J 5: 2292–2299
Shimizu Y, Newman W, Tanaka Y, Shaw S (1992) Lymphocyte interactions with endothelial cells. Immunol Today 13: 106–112
Shtivelman E, Bishop JM (1991) Expression of CD44 is repressed in neuroblastoma cells. Mol Cell Biol 11: 5446–5453
Si Z, Hersey P (1994) Immunohistological examination of the relationship between metastatic potential and expression of adhesion molecules and 'selectins' on melanoma cells. Pathology 26: 6–15
Simmons PJ, Zannettino A, Gronthos S, Leavesley D (1994) Potential adhesion mechanisms for localisation of haemopoietic progenitors to bone marrow stroma. Leuk Lymphoma 12: 353–363
Sonnenberg A (1993) Integrins and their ligands. Curr Top Microbiol Immunol 184: 7–35
Springer TA (1990) Adhesion receptors of the immune system. Nature 346: 425–433
Sretavan DW, Feng L, Pure E, Reichardt LF (1994) Embryonic neurons of the developing optic chiasm express L1 and CD44 cell surface molecules with opposing effects on retinal axon growth. Neuron 12: 957–975
StJohn T, Meyer J, Idzerda R, Gallatin WM (1990) Expression of CD44 confers a new adhesive phenotype on transfected cells. Cell 60: 45–52
Stamenkovic I, Amiot M, Pesando JM, Seed B (1989) A lymphocyte molecule implicated in lymph node homing is a member of the cartilage link protein family. Cell 56: 1057–1062
Stamenkovic L, Aruffo A, Amiot M, Seed B (1991) The hematopoietic and epithelial forms of CD44 are distinct polypeptides with different adhesion potentials for hyaluronate-bearing cells. EMBO J 10: 343–348
Sugimoto K, Tsurumaki Y, Hoshi H, Kadowaki S, LeBousse-Kerdiles MC, Smadja-Joffe F, Mori KJ (1994) Effects of anti-CD44 monoclonal antibody on adhesion of erythroid leukemic cells (ELM-I–1) to hematopoietic supportive cells (MS-5): CD44 but not hyaluronate-mediated, cell-cell adhesion. Exp Hematol 22: 488–494
Sy MS, Guo YJ, Stamenkovic I (1991) Distinct effects of two CD44 isoforms on tumor growth in vivo. J Exp Med 174: 859–866
Sy MS Guo YJ, Stamenkovic I (1992) Inhibition of tumor growth in vivo with a soluble CD44-immunoglobulin fusion protein. J Exp Med 176: 623–627
Takahashi H, Soderstrom K, Nilsson E, Kiessling R, Patarroyo M (1992) Integrins and other adhesion molecules on lymphocytes from synovial fluid and peripheral blood of rheumatoid arthritis patients. Eur J Immunol 22: 2879–2885
Tan PH, Santos EB, Rossbach HC, Sandmaier BM (1993) Enhancement of natural killer activity by an antibody to CD44. J Immunol 150: 812–820
Tanabe KK, Nischi T, Saya H (1993a) Expression of CD44R1 adhesion molecule in colon carcinomas and metastases. Lancet 341: 725–726
Tanabe KK, Nischi T, Saya H (1993b) Novel variants of CD44 arising from alternative splicing: changes in the CD44 alternative splicing pattern of MCF-7 breast carcinoma cells treated with hyaluronidase. Mol Carcinog 7: 212–220

Tanaka Y, Adams DH, Hubscher S, Hirano H, Siebenlist U, Shaw S (1993) T-cell adhesion induced by proteoglycan-immobilized cytokine MIP-1 beta. Nature 361: 79–82
Tarin D, Matsumura Y (1993a) Deranged activity of the CD44 gene and other loci as biomarkers for progression to metastatic malignancy. J Cell Biochem Suppl 17G:173–85
Tarin D, Matsumara Y (1993b) Deranged CD44 gene activity in malignancy. J Pathol 171: 249–250
Terpe A, Franke F, Stark H, Gustmann C, Mackay C, Marston W, Gunthert U (1993) Occurrence of CD44 and its isoforms under orthological and pathological conditions. Verh Dtsch Ges Pathol 77: 276–281
Terpe HJ, Stark H, Prehm P, Gunthert U (1994) CD44 variant isoforms are preferentially expressed in basal epithelial of non-malignant human fetal and adult tissues. Histochemistry 101:79–89
Teyssier JR, Couillin P, Benard J, Ravisse N, Ulrich E, Cornillet P (1992) A multidrug-resistant ovarian carcinoma cell line with a malignant suppressed phenotype is a CD44 gene expression defective mutant. Cancer Genet Cytogenet 60: 14–19
Thomas L, Byers HR, Vink J, Stamenkovic I (1992) CD44H regulates tumor cell migration on hyaluronate-coated substrate. J Cell Biol 118: 971–977
Thomas L (1993) CD44, the hyaluronic acid cell receptor. Its role in neoplastic invasion and metastatic dissemination. Bull Cancer (Paris) 80: 833–44
Thomas L, Etoh T, Stamenkovic I, Mihm MC Jr, Byers HR (1993) Migration of human melanoma cells on hyaluronate is related to CD44 expression. J Invest Dermatol 100: 115–120.
Tölg C, Hofmann M, Herrlich P, Ponta H (1993) Splicing choice from ten variant exons establishes CD44 variability. Nucleic Acids Res 21: 1225–1229
Toyama-Sorimachi N, Miyake K, Miyasaka M (1993) Activation of CD44 induces ICAM-1/LFA-1-independent, Ca^{2+}-, Mg^{2+}-independent adhesion pathway in lymphocyte-endothelial cell interaction. Eur J Immunol 23: 439–446
Toyama-Sorimachi N, Miyasaka M (1994) A novel ligand for CD44 is sulfated proteoglycan. Int Immunol 6: 655–660
Tsukita S, Oishi K, Sato N, Sagara J, Kawai A, Tsukita S (1994) ERM family members as molecular linkers between the cell surface glycoprotein CD44 and actin-based cytoskeletons. J Cell Biol 126: 391–401
Tuckwell DS, Weston SA, Humphries MJ (1993) Integrins: a review of their structure and mechanisms of ligand binding. Symp Soc Exp Biol 47: 107–136
Underhill C (1992) CD44: the hyaluronan receptor. J Cell Sci 103: 293–298
Underhill CB (1993) Hyaluronan is inversely correlated with the expression of CD44 in the dermal condensation of the embryonic hair follicle. J Invest Dermatol 101: 820–826
Underhill C, Nguyen HA, Shizari M, Culty M (1993) CD44 positive macrophages take up hyaluronan during lung development. Dev Biol 155: 324–336
Valent P, Majdic O, Maurer D, Bodger M, Muhm M, Bettelheim P (1990) Further characterization of surface membrane structures expressed on human basophils and mast cells. Int Arch Allergy Appl Immunol 91: 198–203
Verfaillie CM, Benis A, Lida J, McGlave PB, McCarthy JB (1994) Adhesion of committed human hematopoietic progenitors to synthetic peptides from the C-terminal heparin-binding domain of fibronectin: cooperation between the integrin alpha 4 beta 1 and the CD44 adhesion receptor. Blood 84: 1802–1811
Vestweber D (1992) Selectins: cell surface lectins which mediate the binding of leukocytes to endothelial cells. Semin Cell Biol 3: 211–220
Walter J, Wolf J, Pawlita M, Schirrmacher V, Moldenhauer G, Moller P (1991) Invasive, metastatic growth of lymphoblastoid B cells in immunodeficient SCID mice. Verh Dtsch Ges Pathol 75: 224–228
Wang W, Kobayashi S, Uede T (1992) Dysregulated expression of cellular adhesion molecules in autoimmune-prone mice. J Dermatol 19: 831–835
Wang W (1993) Functional studies of adhesion molecules on CD4-CD8-double negative T cells of autoimmune MRL/Mp-lpr-mice. Hokkaido Igaku Zasschi 68: 755–766
Washington K, Gottfried MR, Telen MJ (1994) Expression of the cell adhesion molecule CD44 in gastric adenocarcinomas. Hum Pathol 25: 1043–9
Wayner E, Carter W (1987) Identification of multiple cell adhesion receptors for collagen and fibronectin in human fibrosarcoma cells possessing unique α and common β submits. J Cell Biol 105: 1873–1884
Webb DS, Shimizu Y, Van Seventer GA, Shaw S, Gerrard TL (1990) LFA-3, CD44, and CD45: physiologic triggers of human monocyte TNF and IL-1 release. Science 249: 1295–1297
Weber B, Rösel M, Arch R, Möller P, Zöller M (1996) Expression of variant isoforms of CD44 during ontogeny of the rat: evidence for divergent functions of distinct exon combinations. Differentiation (in press)

Westphal JR, de Waal RMW (1992) The role of adhesion molecules in endothelial cell accessory function. Mol Biol Rep 17: 47–59

Wheatley SC, Isacke CM, Crossley PH (1993) Restricted expression of the hyaluronan receptor, CD44, during postimplantation mouse embryogenesis suggests key roles in tissue formation and patterning. Development 119: 295–306

Wielenga VJM, Heider KH, Offerhaus JA, Adolf GR, van den Berg FM, Ponta H, Herrlich P. Pals ST (1993) Expression of CD44 variant proteins in human colorectal cancer is related to tumor progression. Cancer Res 53: 4754–4756

Wilkinson JM, McDonald G, Smith S, Galea-Lauri J, Lewthwaite J, Henderson B, Revell PA (1993) Immunohistochemical identification of leucocyte populations in normal tissue and inflamed synovium of the rabbit. J Pathol 170: 315–320

Williams AF, Barclay AN (1988) The immunoglobulin superfamily-domains for cell surface recognition. Annu Rev Immunol 6: 381–405

Wirth K, Arch R, Somasundaram C, Hofmann M, Weber B, Herrlich P, Matzku S, Zöller M (1993) Expression of CD44 isoforms carrying metastasis-associated sequences in newborn and adult rats. Eur J Cancer 29A: 1172–1177

Woerner SM, Givehchian M, Dürst M, Schneider A, Costa S, Melsheimer P, Lacroix J, Zöller M, von Knebel-Doeberitz M (1995) Expression of CD44 splice variants in normal, dysplastic and neoplastic cervical epithelium. Clin Cancer Res 1: 1125–1132

Wolffe EJ, Gause WC, Pelfrey CM, Holland SM, Steinberg AD, August JT (1990) The cDNA sequence of mouse Pgp-1 and homology to human CD44 cell surface antigen and proteoglycan core/link proteins. J Biol Chem 265: 341–347

Wu L, Kincade PW, Shortman K (1993) The CD44 expressed on the earliest intrathymic precursor population functions as a thymus homing molecule but does not bind to hyaluronate. Immunol Lett 38: 69–75

Wylie C (ed)(1993) Signals, polarity and adhesion in development. Proc Br Soc Dev Bio Suppl 1: 1–306

Yanagishita M, Hascall VC (1992) Cell surface heparan sulfate proteoglycans. J Biol Chem 267: 9451–9454

Yang H, Binns RM (1993) CD44 is involved in porcine natural cytotoxicity. Cell Immunol 149: 227–236

Yang B, Yang BL, Savani RC, Turley EA (1994) Identification of a common hyaluronan binding motif in the hyaluronan binding proteins RHAMM, CD44 and link protein. EMBO J 13: 286–296

Yap UJ, Stokes AN, Pearson GJ (1994) Concepts of adhesion – a review. N Z Dent J 90: 91–97

Yokozaki H, Tahara E (1994) Metastasis related genes. Gan To Kagaku Ryoho 21: 2541–2548

Yong K, Khwaja A (1990) Leucocyte cellular adhesion molecules. Blood Rev 4: 211–225

Zeine R, Owens T (1992) Direct demonstration of the infiltration of murine central nervous system by Pgp-1/CD44high CD45RB(low) $CD4^+$ T cells that induce experimental allergic encephalomyelitis. J Neuroimmunol 40: 57–69

Zembala M, Siedlar M, Ruggiero I, Wieckiewiez J, Mytar B, Mattei M, Colizzi V (1994) The MHC class-II and CD44 molecules are involved in the induction of tumour necrosis factor (TNF) gene expression by human monocytes stimulated with tumour cells. Int J Cancer 56: 269–274

Zetter BR (1993) Adhesion molecules in tumor metastasis. Semin Cancer Biol 4: 219–229

Zhou DFH, Ding JF, Picker LF, Bargatze RF, Butcher EC, Goeddel DV (1989) Molecular cloning and expression of Pgp-1 – the mouse homolog of the human H-CAM (Hermes) lymphocyte homing receptor. J Immunol 143: 3390–3395

Zöller M (1995) CD44 variant expression, lymphocyte homing and lymphogenic metastasis. In: Siess W, Lorenz R, Weber PC (eds) Adhesion molecules and cell signaling: biology and clinical applications, vol 1. Raven, New York, pp 201–218

The CD44 Proteins in Embryonic Development and in Cancer

L. SHERMAN, J. SLEEMAN, P. DALL, A. HEKELE, J. MOLL, H. PONTA, and P. HERRLICH

1 Introduction

During the past decade, considerable advances have been made in our understanding of the cellular and molecular biology of carcinogenesis. The analysis of the metastatic process, however, which is the chief contributor to cancer patient morbidity, has met with various obstacles due to its complexity. As a result, we still know very little about how cancer cells spread. It is generally assumed that cancer cells randomly diversify during tumor progression and that clones of such cells with metastatic properties are selected. The process by which these cells then metastasize is believed to occur via several steps, each of which can be linked to the aquisition of new cellular functions (see, e.g., reviews by HART et al. 1989; KERBEL 1990). For example, a stationary epithelial cell must first be converted into a mobile, nonadherent cancer cell which can invade subepithelial tissue. The cell must then move towards lymphatic and blood vessels, enter these vessels, and survive during their transport within the fluid medium. It must also colonize draining lymphatic tissue and further disseminate through efferent lymphatics into the blood stream, extravasate from blood vessels and, finally, preferentially settle in various distant locations.

Forschungszentrum Karlsruhe, Institut für Genetik, PO Box 3640, 76021 Karlsruhe, Germany

A further complication for the analysis of the metastatic process is that any proliferative advantage along the metastatic pathway will expand the pool of diversifying cells. Such expansion would, consequently, increase the chance of generating new cells that can pass subsequent obstacles in the metastatic route. Specifically invasive and migratory properties, therefore, are not the only contributors to metastasis. Nonetheless, understanding how cancer cells gain and utilize their invasive and migratory abilities will be crucial to finding ways to control the pathogenesis of cancer.

Invasion and migration are also important processes in embryonic development and in several normal adult cellular functions. During gastrulation, for example, there are massive cellular movements which establish the primary germ layers. Later in development, in a process that is perhaps analogous to at least part of the metastasic process, neural crest cells dislodge from the dorsal neural tube, invade somitic tissue, migrate over long distances, then localize in various places giving rise to the peripheral nervous system, the adrenal medulla, and numerous other structures (Le Douarin 1984). Axons in the brain and in the periphery also invade and migrate across much of the organism to specifically connect to their targets. In the adult, leukocytes with migratory potential are generated throughout the life of an organism. It is reasonable to assume, therefore, that cancer cells make use of behavior patterns that are genetically preformed and which occur in normal cells during their progression to the metastatic state. If this is true, then molecules participating in normal migration, especially during embryogenesis, will likely be part of a cancer cell's repertoire.

2 The CD44 Protein Family

The CD44 epitope, originally discovered on the surface of leukocytes (Trowbridge et al. 1982), is now known to be carried by a family of transmembrane proteins which are widely distributed on various mesodermal and ectodermal derivatives, including hematopoietic cells, keratinocytes, and numerous other cell types in both the adult and the embryo (Lesley et al. 1993). Particular interest in CD44 arose from the detection of splice variants in a metastatic rat pancreatic carcinoma (Günthert et al. 1991). Artificial overexpression of these CD44 molecules promoted metastatic spreading of a related nonmetastatic carcinoma cell line. This effect could be blocked by an antibody against a variant epitope (encoded by exon v6, for nomenclature see Tölg et al. 1993, and see Fig. 1) but not by antibodies against the NH_2-terminal portion of the protein (Reber et al. 1990; Seiter et al. 1993), suggesting that certain CD44 splice variants could play a crucial role in promoting metastasis. Consistent with this idea, splice variants of CD44, containing sequences encoded by variant exons, have also been detected in certain other animal tumor cells (Günthert et al. 1991; Hofmann et al. 1993).

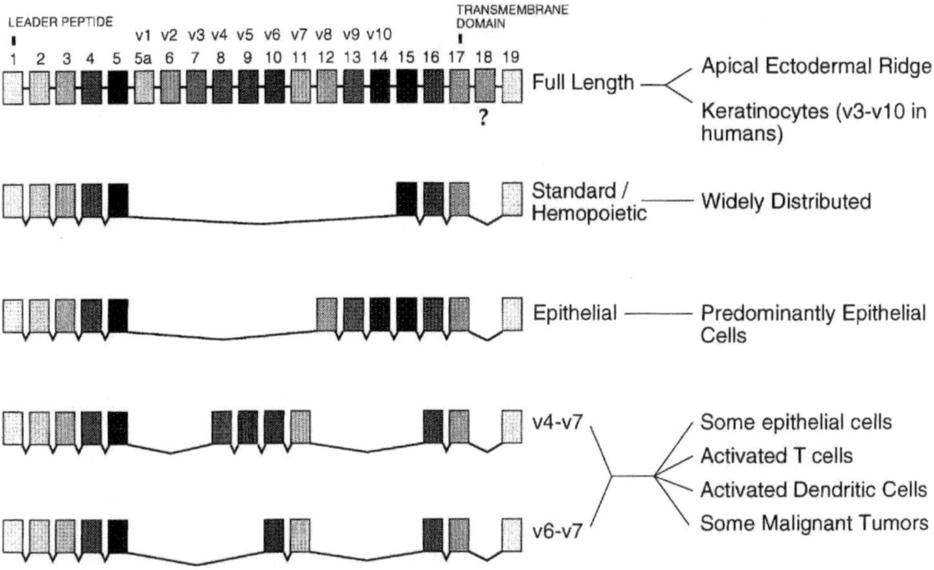

Fig. 1. Localization of some CD44 splice variants. The ten variant exons (representing exons 5a-14) are referred to as *v1-v10*. Note that in humans, exon 5a (v1) is not utilized. The *question mark* under exon 18 indicates that it is not clear whether this exon is spliced out of all transcripts

Examination of the CD44 gene has indicated that at least ten adjacent variant (v) exons may be utilized during its transcription (SCREATON et al. 1992; TÖLG et al. 1993). Sequences encoded by these exons can be found in the middle of the CD44 proteins, just outside the transmembrane region. The metastasis-associated variants initially found in the rat metastatic pancreatic carcinoma mentioned above exhibit the structures CD44v4-v7 and CD44v6-v7 (GÜNTHERT et al. 1991; RUDY et al. 1993) as shown in Fig. 1. As discussed below, it is now clear that tumor cells frequently express many different CD44 splice variants. Theoretically, greater than 100 splice variants could be formed using combinations of the variant exons. While it is likely that certain splice variants will not occur, the variability of primary structure is expected to be tremendous. Furthermore, CD44 proteins are differentially glycosylated. In the metastatic rat cancer cells in which CD44v4-v7 was first discovered, glycosylation accounts for more than three fourths of the molecular mass. Both N and O linkages exist and part of these modifications are negatively charged (GÜNTHERT et al. 1991; RUDY et al. 1993). While most such modifications concern the NH_2-terminal half of the protein, specific glycosaminoglycan (GAG) moieties have been found on variant epitope sequences, including a heparin sulfate addition site encoded by exon v3 (BENNETT et al. 1995; JACKSON et al. 1995).

Experimental studies on the role of CD44 during tumor progression have indicated that CD44 proteins are only likely to have a metastasis-promoting function in cells that already have certain tumorigenic properties (HOFMANN et al. 1993). An important question, therefore, is what functions individual CD44 proteins contribute to tumor behavior. This question has been addressed using lymphoma and melanoma cells which are negative for CD44. Such tumor cell lines are rare, since almost all tumors express at least the smallest known CD44 isoform, referred to as the "standard" form of the protein (CD44s). CD44s was shown to enhance the growth and metastatic abilities of Namalwa cells injected subcutaneously or intravenously into SCID mice (SY et al. 1991, 1992). A larger CD44 protein, CD44v8-v10, however, failed to confer this increased growth rate. Growth enhancement was also achieved in CD44-negative melanoma cells upon transfection with CD44s (BARTOLAZZI et al. 1994). Interestingly, CD44s mutants defective in hyaluronate binding, a function attributed to the NH_2-terminal portion of CD44 proteins (ARUFFO et al. 1990; MIYAKE et al. 1990; PEACH *et al. 1993;* YANG et al. 1994), were unable to promote melanoma cell growth (BARTOLAZZI et al. 1994). It is possible, therefore, that the induction of metastatic spread by CD44s in these cases is due to their growth-promoting effects and that their effects on cell growth are somehow hyaluronate-dependent. In the case of the nonmetastatic pancreatic carcinoma mentioned above, however, the cells already carried CD44s and did not profit from further increases of CD44s expression (HERRLICH et al. 1993a). These cells only became metastatic when larger CD44 variants (CD44v4-v7, CD44v6/v7) were introduced (GÜNTHERT et al. 1991; RUDY et al. 1993). These data indicate that individual larger splice variants provide tumor cells with properties that are different from those induced by CD44s and suggest that different cell types can utilize the same CD44 proteins in different ways.

3 Molecular Properties of CD44 Proteins

In view of the above mentioned complexities of metastasis formation and the heterogeneous expression of CD44 in normal cells, it is not surprising that only a few molecular functions have been unequivocally assigned to individual CD44 proteins. One of the first reported differences between CD44s and a variant of CD44 (CD44v8-v10) was in the ability of these proteins to bind hyaluronic acid (HA). CD44s, when expressed on human Namalwa cells, could bind to HA while the CD44v8-v10 isoform could not (STAMENKOVIC et al. 1991). Data from other laboratories, however, contradict this observation (DOUGHERTY et al. 1994; SLEEMAN et al. 1995), and it seems more likely that modifications (e.g., by glycosylation) influence the HA-binding properties and not the primary structure of the different isoforms (SLEEMAN et al., unpublished observations; see also review by SHERMAN et al. 1994).

The binding of HA to CD44 is conferred by two regions in the extracellular, NH_2-terminal part of CD44 proteins which are present on all known CD44 isoforms (PEACH et al. 1993; YANG et al. 1994). This binding can be induced by activating antibodies (LESLEY et al. 1992) and can be up-regulated upon T cell activation by antigenic stimulation (LESLEY and HYMAN et al. 1992) or treatment with phorbol myristate acetate (LIAO et al. 1993). Interestingly, transfection of a CD44v4-v7 variant into a rat tumor cell led to increased HA binding (SLEEMAN et al. 1995), whereas expression of the same transgene on T cells of transgenic mice did not induce HA binding (MOLL et al. 1994; MOLL et al., unpublished), again supporting the notion that posttranslational modifications, and not the primary structure of the protein, are important for HA binding.

Binding to HA cannot, therefore, explain the functional differences between CD44s and variant isoforms in promoting metastases. Although HA binding could be a feature that is required for more specific molecular functions provided by different CD44 isoforms, this does not seem to be the case in the immune system. The activation of lymphocytes by antigenic stimulation is accompanied by the expression of CD44 variant isoforms (ARCH et al. 1992; KOOPMAN et al. 1993). This expression is functionally relevant, since antibodies that specifically recognize the variant portion of the protein abolish activation of lymphocytes when administered at the time when the immune response is raised (ARCH et al. 1992). These activated lymphocytes are unable to bind to HA as judged by FACS analysis using biotinylated hyaluronate (MOLL et al., unpublished). In agreement with this observation are the results obtained with transgenic mice, in which the expression of the rat CD44v4-v7 isoform is targeted to T lymphocytes by means of the Thy 1 promoter. The expression of this isoform does not provide the T cell with HA binding ability (see also above) but has a pronounced effect in the immune response. T cell-dependent antigenic or polyclonal stimulation in vivo or in vitro using isolated lymphocytes from transgenic mice is enhanced and accelerated by at least 24 h (MOLL et al. 1995). This accelerated response is abolished when the activity of the transgene is blocked by antibodies which specifically recognize the variant sequence of the rat transgene product.

In addition to HA, a number of other components of the extracellular matrix have been implicated as CD44 ligands. These include fibronectin (JALKANEN and JALKANEN 1992), collagen types I and IV (WAYNER and CARTER 1987; CARTER and WAYNER 1988), and the chondroitin sulfate-modified invariant chain (NAUJOKAS et al. 1993). A recent study by TOYAMA-SORIMACHI and MIYASAKA (1994) has identified serglycin to be another CD44 ligand, interacting with chondroitin-4-sulfate-modified residues on CD44s. Serglycin is a small proteoglycan that is stored in intracellular secretory granules of lymphoid, myeloid and some tumor cells (for review see STEVENS et al. 1988). Although the functions of this protein are not known, it has been shown to be exocytosed by cytotoxic T cells (CTLs) when they come into contact with target tumor cells (MACDERMOTT et al. 1985), suggesting that it plays a role in cell-mediated cytotoxicity. Interestingly, when serglycin interacts with CD44, it augments the CD3-dependent degranulation of CD44 positive CTL clones, an observation that is consistent with the notion that CD44 and serglycin regulate lymphoid cell adherence and activation (TOYAMA-SORIMACHI and MIYASAKA 1994).

As mentioned above, CD44 splice variants can be modified by specific sugars and GAGs, including heparin sulfate. The role for these modifications in the normal and tumor-promoting functions of CD44 is, however, not yet understood.

Interestingly, the specific modification of a CD44 variant by α (1-2) fucosyltransferase has recently been correlated with increased malignancy of a colon carcinoma cell line (LABARRIÈRE et al. 1994). This enzyme catalyzes the synthesis of H antigens from β-galactoside precursors. CD44 splice variants containing sequences encoded by exon v6 were found to be modified by H blood group antigens in highly tumorigenic clones of PROb tumor cells. When these cells were transfected with antisense fragments of the fucosyltransferase sequence, the expression of cell surface H antigen was reduced as was tumorigenicity. It is possible, therefore, that such modifications of certain CD44 splice variants may play a direct role in tumor-promoting capacity.

While we are learning a great deal about the putative functions of the extracellular portion of CD44, comparatively little is understood about the protein's intracellular domain (for review see ISACKE 1994). The cytoplasmic tail of CD44 is believed to interact with components of the actin cytoskeleton. A group of studies have suggested that part of this interaction occurs via a 72 kDa ankyrin-like protein (KALOMIRES and BOURGUIGNON 1988; BOURGUIGNON et al. 1986, 1992). The ezrin-radixin-moesin (ERM) group of actin-binding proteins have also been shown to interact with CD44 (TSUKITA et al. 1994). These proteins belong to a family of actin-binding proteins which include talin, band 4.1, and the neurofibromatosis type 2 (NF2) gene product merlin (for review see ARPIN et al. 1994). It is possible that some of these proteins link CD44 to a signal transduction pathway. These proteins are known, for example, to be phosphorylated on tyrosine in response to growth factors (BRETSCHER 1989; FAZIOLI et al. 1993) and, in T lymphocytes, upon stimulation through the CD3 antigen (EGERTON et al. 1992). Furthermore, tyrosine phosphorylation of ezrin has been shown in cells carrying CD3 or CD4 ligated to ezrin (THUILLIER et al. 1994). This phosphorylation is apparently achieved by the p56-lck tyrosine kinase and modulated by the tyrosine phosphatase CD45R. The ERM proteins have been shown to be involved in cell-cell and cell-matrix adhesion and in microvilli formation, probably by regulating actin filament-plasma membrane interactions (TSUKITA et al. 1992). The association between these and other cytoskeleton-associated proteins and CD44 might be mediated by phosphorylation of one or more of the four serines in the sequence of the CD44 cytoplasmic tail, although no data have been presented so far which directly support this idea. These serines have, however, been suggested to be necessary for binding to hyaluronate and CD44-mediated adhesion of T cells to smooth muscle cells (LAZAAR et al. 1994; PURÈ et al. 1995).

Another hint for the functional relevance of the interaction between the cytoplasmic tail of CD44 and ERM family members comes from recent preliminary studies in our laboratory on the relationship between merlin and CD44. In primary Schwann cell and smooth muscle cell cultures, there is at least a weak interaction between CD44 and merlin (SHERMAN et al., unpublished). Interestingly, in humans, mutations in the NF2 gene typically result in either solitary or multiple benign Schwann cell tumors in which the merlin protein is nonfunctional. We have shown that these tumors as well as experimentally induced rat schwannomas express multiple splice variants of CD44 (SHERMAN et al. 1995). Furhtermore, in these rat

Schwann cell tumor lines, wild-type human merlin can significantly reduce CD44 expression. Although these observations do not prove that merlin can regulate CD44 expression by binding to CD44, it does indicate that a functional link between these proteins is likely to exist. As one possibility, therefore, we are pursuing the idea that merlin may mediate an autoregulatory function of CD44. Perturbing such a function could conceivably lead to aberrant CD44 expression, such as that seen on schwannomas and numerous other tumors.

4 CD44 Proteins in Development

As mentioned above, CD44 proteins are found in a variety of normal tissues throughout life (reviewed by Ruiz et al. 1995). A number of recent immunohistochemical and RT-PCR data have shown that both standard and variant CD44 proteins are widely expressed during mouse (Wheatley et al. 1993), rat (Wirth et al. 1993), and human embryonic development (Terpe et al. 1994a; Campbell et al. 1995). Collectively, these data indicate that CD44 proteins are expressed from the onset of embryogenesis in both embryonic and extraembryonic (e.g., placental stromal cells) tissues, that certain splice variants have restricted patterns of expression, that this expression often persists through adulthood, and that CD44 expression does not necessarily correlate with sites of HA-dependent morphogenesis (Wheatley et al. 1993; Terpe et al. 1994a). Furthermore, although it has been suggested that CD44 expression is somehow associated with the functions of instructive epithelia (Wheatley et al. 1993), numerous tissues which do not fall into this category also express CD44 isoforms. It is likely, therefore, that the functions of particular CD44 proteins will depend highly on where they are expressed.

Given the complexity of the pattern of CD44 expression during development, the mechanisms employed by embryos to regulate CD44 expression and splicing will be difficult to elucidate. One way to study this question, however, is to examine CD44 expression during the differentiation of embryonic stem (ES) cells. Using the totipotent D3 mouse ES cell line, Haegel and coworkers (1993) have found that these cells initially do not express detectable amounts of CD44 proteins, although they do express a limited number of CD44 transcripts. Following differentiation, however, a subpopulation of CD44-positive cells were identified, and additional RNA transcripts could be found, including some splice variants. Since ES cells can be induced to differentiate into various combinations of phenotypes, this system should provide a means by which to study the molecular basis of developmentally regulated CD44 promoter activity and splicing.

So far, it has been possible to relate specific CD44 isoforms to defined developmental processes in only a limited number of cases. One example can be found in the mammalian optic chiasm, an X-shaped pattern of axons formed by projections from retinal neurons as they enter the brain laterally from opposite sides of the ventral diencephalon. While some of these retinal axons cross the midline of the chiasm, others turn away and grow into the ipsilateral tract. A population of neuronal cells in the optic chiasm appear to be decisive in directing the paths of these axons. In mice, these cells carry both L1, an axon growth-promoting molecule, and CD44s (Sretavan et al. 1994). In fact, the CD44s protein defines the midline of the optic chiasm during ipsilateral and contralateral axon pathfinding. When retinal explants were grown on collagen gels seeded with CD44-positive fibroblast membrane fragments, they extended neurites poorly compared to explants grown in the presence of CD44-negative membranes. This effect could be partially blocked by CD44 antibodies (Sretavan et al. 1994). These data are consistent with the notion that CD44 has an axon growth-inhibiting function, which might influence the decision of retinal axons to cross or not to cross the chiasm midline. These data also raise the possibility that CD44 may somehow mediate axon growth that is stimulated by L1.

Interestingly, we have recently found that Schwann cell precursors express splice variants of CD44 as they differentiate from multipotential neural crest cells (Sherman et al., unpublished results). Schwann cells come into close contact with neurons during peripheral nerve development and provide the myelin sheaths that wrap around axons. Although the functional consequences of CD44 variant expression by these cells are not known, we have found that late Schwann cell precursors and mature Schwann cells express only standard CD44 (Sherman et al. 1995). Like the neurons in the optic chiasm, these more differentiated Schwann cells also express L1. It is intriging to speculate, therefore, that CD44 may function similarly in the midline of the optic chiasm and in Schwann cell-neuron or Schwann cell-Schwann cell contact during peripheral nerve morphogenesis.

Another place where we have found a functional role for a CD44 protein is during limb development. By RT-PCR and whole mount immunohistochemistry, we have determined that the full-length form of CD44 (CD44vl-v10) is expressed in the apical ectodermal ridge (AER) of rat embryos (Sherman et al., unpublished results). This structure keeps the underlying mesenchymal cells, which express only CD44s, in a state of proliferation and is necessary for limb outgrowth. In a rat embryo culture system, we found that we could retard limb outgrowth by removing the AER, treating it with an antibody that recognizes an epitope encoded by exon v6, and then grafting it back onto the limb mesenchyme. Limb mesenchyme receiving AER tissue that had been treated with isotype-matched control antibodies recognizing a different CD44 epitope grew normally. These data suggest that the full-length form of CD44 plays a role in the growth-promoting function of the AER.

Members of the fibroblast growth factor (FGF) family can completely substitute for the AER to induce limb development in the underlying mesenchyme (Niswander et al. 1993). Interestingly, one member of this family, FGF-8, is expressed throughout the AER (Crossley and Martin 1995) in a pattern that is identical to CD44. As mentioned above, CD44 proteins containing exon v3 sequences can be modified with heparan sulfate (Jackson et al. 1995), enabling CD44 to bind to heparin-binding growth factors, including FGFs (Bennett et al.

1995). Since the CD44 variant expressed on the AER contains exon v3 sequences, one possible function of CD44 in the AER might be the presentation of growth factors to the mesenchyme. As mentioned above, however, CD44 may also be involved in signaling pathways, and such functions should also be considered in the case of limb development.

5 CD44 Epitopes in Human Cancer

It is, as mentioned in the Introduction, highly likely that cancer cells acquire invasive and migratory properties that normally play major roles in the embryo and, in a more restricted fashion, in certain specialized cells of the adult. We have already demonstrated that the expression of certain CD44 proteins is one such putative property in animal models. We will now review data on the expression of CD44 in human cancers. With the enormous structural and functional diversity of the CD44 proteins, it is hard to predict which one may confer an advantage in the metastatic process of specific human tumors. In addition, although various metastatic pathways seem similar, human cancers result from a variety of different heterogeneous populations of cells and it is not obvious that common traits are to be found in all cases. Nevertheless, CD44 epitopes have been followed through carcinogenesis and tumor progression in several cancers. We will review examples of tumor types for which interesting correlations have been revealed. We will first consider tumors originating from tissues that normally express CD44 proteins.

5.1 Cancer of Tissues That Normally Express CD44 Variant Proteins

5.1.1 Non-Hodgkins Lymphoma

CD44 is constantly expressed in circulating lymphocytes. These cells apparently perform the steps of recirculation without the need for large CD44 splice variants, since, in their resting state, they express only CD44s. After antigenic stimulation, however, several larger splice variants are transiently expressed on the lymphocyte surface suggesting that these variants mediate a specific step different from CD44s (ARCH et al. 1992; KOOPMAN et al. 1993; MACKAY et al. 1994; HIRANO et al. 1994). Since CD44s seems to catalyze, at least indirectly, steps of recirculation (JALKANEN et al. 1987), one question is whether the malignant counterparts of normal lymphocytes, non-Hodgkin lymphomas (NHLs), utilize CD44 for their dissemination. Indeed, it has been shown that the more aggressive NHLs carry elevated levels of the CD44 NH_2-terminal epitopes recognized by the antibodies Hermes-3, NKI-P1, and NKI-P2 (PICKER et al. 1988; HORST et al. 1990; JOENSUU et al.

1993; Stauder et al. 1995). The current histological definitions of low grade and high grade NHLs very coarsely predict the clinical features of the disease. Low grade NHLs disseminate and invade tissues less than high grade NHLs that aggressively infiltrate various organs and terminate life within 1–2 years. It would be desirable to have criteria available that could better predict clinical outcome (see review by Shipp 1994). Despite the presence of CD44 NH_2-terminal epitopes (present on all known CD44 proteins) on normal lymphocytes, their increased levels paralleled poor prognosis (Horst et al. 1990; Stauder et al. 1995). This relationship was considered insignificant in another study (Fujiwara et al. 1993). The reason for this discrepancy could be that it is not the overall level of CD44 proteins but rather an altered splice pattern which determines NHL behavior.

Using antibodies specific for epitopes encoded by variant exons, a more persuasive relationship between CD44 expression and NHL prognosis has been revealed. High grade NHLs carry larger CD44 variants, similar to those found in activated lymphocytes (Koopman et al. 1993; Salles et al. 1993; Terpe et al. 1994b; Stauder et al. 1995). Epitopes encoded by both exon v6 and exon v3 were most consistently related to aggressive NHL behavior (Koopman et al. 1993; Stauder et al. 1995). In a recent extensive study of 138 patients, the expression level of total CD44 and the expression of exon v3 and v6 correlated with poor overall survival (Stauder et al. 1995). CD44v6 proved to represent an independent prognostic factor (Stauder et al. 1995). Interestingly, exons v3 and v6 are also coexpressed after lymphocyte activation (Hirano et al. 1994; Stauder et al. 1995). Aggressive NHLs may share properties with activated lymphocytes and this notion has led to various hypotheses on the molecular action of these CD44 isoforms. It is likely that activated lymphocytes as well as CD44 variant expressing tumor cells such as NHLs enter lymphatic tissues through the same route, the afferent lymphatic vessels, and CD44v3/v6 exon containing variants may promote entry into and/or expansion in lymphatic tissue (Herrlich et al. 1993b). In any case, these data suggest that CD44 immunohistochemistry may provide valuable information in determining the diagnosis and prognosis of NHLs.

5.1.2 Cervical Cancer

Normal keratinizing epithelia such as external skin, oral mucosa and cervical mucosa express large CD44 splice variants. Analysis of several cell lines derived from skin keratinocytes have suggested that the predominant CD44 protein carries sequences of exons v3 to v10 (Hoffmann et al. 1991). Uterine cervical mucosa, however, does not express CD44v3-v10. By RT-PCR no product of the appropriate size was detected (Dall et al. 1994); rather, smaller mRNAs have been observed, corresponding to CD44 v3-v7 and CD44v8-v10. This finding matches the absence of an epitope that is jointly encoded by exons v7 and v8 (and recognized by mAB VFF17). Interestingly, during carcinogenesis the expression pattern apparently changes. RT-PCR analysis demonstrates that, in tumor material, a larger CD44 RNA is expressed, presumably comprising exons v3 to v10. Furthermore, these cells become positive for the v7/v8 epitope. While only four of 21 samples of low grade squamous intraepithelial lesions were positive for the VFF17 epitope, the frequency increased to 17 of 35 high grade squamous intraepithelial lesions (DALL et al., unpublished results). All samples of carcinoma-in-situ and all samples of invasive carcinoma were positive for the v7/v8 epitope. These data suggest that the VFF17 antibody may be a useful marker for precancerous stages of cervical cancer that is suitable for early diagnostic screening.

5.1.3 Neuroectodermal Tumors

Neuroblastomas are tumors of neural crest origin that develop during childhood. A number of recent studies have shown that most high grade neuroblastomas do not express any CD44 proteins, that low grade tumors tend to express CD44s but no splice variants, and that the expression of CD44s predicts an increased survival probability (FAVROT et al. 1993; COMBARET et al. 1995; GROSS et al. 1995; CRISTIANSEN et al. 1995). The expression of CD44 by these tumors seems to correlate with the differentiated state of the tumors: S-type neuroblastomas, which have Schwann cell-like characteristics, express CD44 while N-type, which have more neuronal characteristics, do not (GROSS et al. 1994). Interestingly, neither premigratory nor migratory neural crest cells express CD44 proteins (WHEATLEY et al. 1993; L.SHERMAN, unpublished results). When neural crest cells are allowed to differentiate in vitro, however, a subpopulation becomes positive for both CD44s and at least some splice variants (L. SHERMAN, unpublished results). Neural crest-derived neurons never appear to express CD44 (SHERMAN et al. 1995). It is likely, therefore, that at least some of the cells among this CD44-positive crest-derived subpopulation are precursors of Schwann cells, which, as mentioned above, are known to express CD44. These data are consistent with the notion that the state of differentiation in neuroblastomas and the likely prognosis are reflected by CD44 expression.

A number of other neural crest-derived tumors have also been shown to express CD44 and CD44 splice variants including schwannomas and neurofibromas, in which the significance of the expression has yet to be examined (SHERMAN et al. 1995; SHERMAN et al., unpublished), and melanomas, in which, as mentioned above, CD44 may provide some advantages in tumor cell growth (BIRCH et al. 1991; GUO et al. 1994). Additional neuroectodermal tumors that express at least CD44s include glioblastomas, astrocytomas, and medulloblastomas (NAGASAKA et al. 1955; LI et al. 1993; RADOTRA et al. 1994; EIBL et al. 1995). The expression of CD44s by these tumors may reflect the expression by their normal counterparts, since both astrocytes and some oligodendrocytes typically express CD44s (MORETTO et al. 1993). In gliomas, experimental data indicate that CD44 may mediate tumor cell adhesion and invasiveness (MERZAK et al. 1994; RADOTRA et al. 1994). It remains to be shown whether this is also true for other tumors of the central nervous system. There are conflicting reports as to whether some of these tumors may also express splice variants of CD44 (LI et al. 1993; RADOTRA et al. 1994; EIBL et al. 1995). Although the significance of CD44 splice variant expression by these tumors is also not known, it is interesting to note that metastases into the brain by tumors of other origins have been shown to express CD44 variants (Li et al. 1993).

5.2 Cancer of Tissues That Normally Do Not Express CD44 Variant Proteins

5.2.1 Mammary Carcinoma

Several markers have been described that are thought to predict the course of breast cancers following removal of the primary tumor. These markers include the epidermal growth factor receptor (Sainsbury et al. 1987), erbB2 amplification (Berger et al. 1988), expression of tissue plasminogen activator (Vespaget et al. 1995) and of cathepsin D (Seshadri et al. 1994), reduced levels of thrombospondin (Zabrenetzky et al. 1994), and numerous others. The diagnostic and prognostic value of some of these markers, however, is still being investigated, and it is clear that additional markers are needed. Several laboratories have reported on the occurrence of CD44 in breast cancer. It is difficult, however, to compare these data since methodologies and materials differed substantially in each case. The results of these studies, therefore, will remain controversial until larger series of patients have been screened using standardized procedures.

Careful examination of ductal epithelium has revealed that normal epithelium does not express CD44. Only myoepithelium surrounding the ductal epithelium carries larger CD44 isoforms comprising exons v3 through v10 (Fig. 2). The initial studies done by RT-PCR could not resolve this point (Matsumura and Tarin 1992). The data are, however, consistent. Using sections from tumor patients and 30 cycles of RT-PCR with primers from outside the v exon region, one report found CD44s RNA expression in normal breast tissue of a magnitude resembling that in peripheral blood leukocytes (Matsumura and Tarin 1992). It is, of course, not possible to discern which cells express CD44s using this method. Under conditions in which tumor samples were strongly positive for material hybridizing to exon v5 and v6 sequences, normal breast tissue showed no v5 or v6 expression. This result has been confirmed using 25 cycles of RT-PCR (Dall et al. 1995). Samples from normal breast tissue, however, contained mRNA hybridizing to exon sequences v8, v9 and v10. A more informative analysis can be performed with immunohistochemistry since the cell types can be defined. Prominent surface staining with CD44v6 epitope specific antibodies was found in fresh samples of normal myoepithelial cells (Fox et al. 1993; Terpe et al. 1994a; Dall et al. 1995; Fig. 2). This expression has obviously escaped the RT-PCR analysis. Normal epithelial cells and benign tumors could be stained neither with polyclonal variant exon region antibodies nor monoclonals recognizing exon v6 epitopes (Fox et al. 1993; Sinn et al. 1995; Dall et al. 1995). An example is shown in Fig. 2. In the middle of the picture there is a normal mammary duct showing strong staining of the myoepithelial cells and no staining of the underlying epithelial cells with a CD44 exon v6 specific antibody. On the upper side there are invasive tumor cells that also stain with the antibody. As in many sections of human tissues, stromal

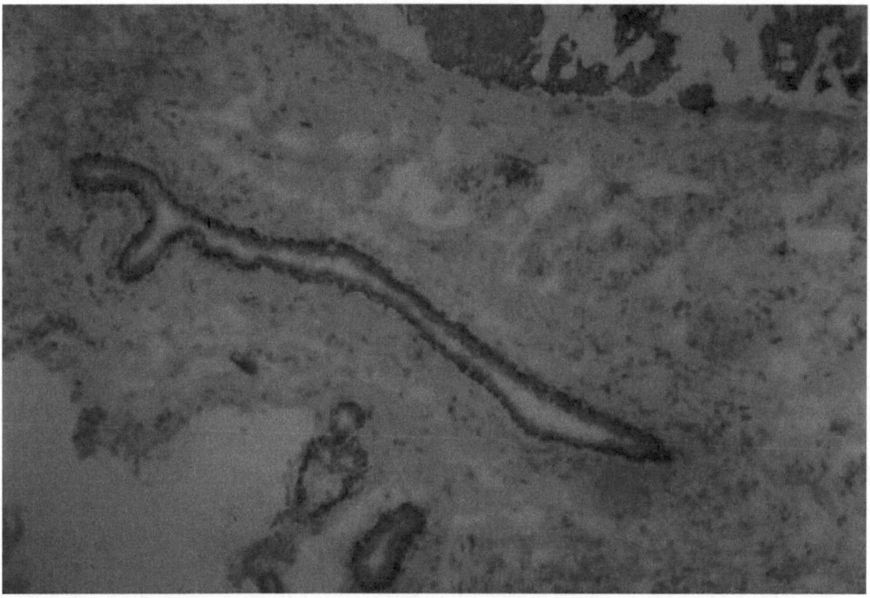

Fig. 2. Immunohistochemical staining of normal mammary ducts and tumor cells with CD44 exon v6-specific monoclonal antibodies. A frozen section of a mammary tumor was fixed in ice-cold methanol followed by ice-cold acetone treatment, washed in PBS and preincubated with goat serum (10% in PBS). The section was then washed another three times followed by incubation with the CD44 exon v6 specific antibody VFF18 (Bender and Co. GmBH, Vienna, Austria). Secondary antibody treatment and peroxidase staining was performed according to standard protocol (see, e.g., HEIDER et al. 1993)

cells are strongly positive for NH_2-terminal epitopes, supporting the conclusion that CD44s is expressed in stroma. Immunocytochemical analyses are semi-quantitative and need, of course, to be done under strictly defined conditions. This may make comparisons between laboratories difficult (MACKAY et al. 1994). Immunohistochemistry has, however, the great advantage of demonstrating in which cells an epitope is expressed.

Mammary malignancies were found to react with a pan-CD44 antibody at higher proportions with increasing degrees of tumor progression (invasiveness, estrogen receptor negativity) (JOENSUU et al. 1993). Immune reactivity to CD44 variant exon sequences including CD44v6 epitopes was convincingly associated with breast cancer metastases (100%) and some of the primary tumors (70%) (SINN et al. 1995; KAUFMANN et al. 1995; DALL et al. 1995). Furthermore, immune reactivity to CD44 exon v6 epitopes matches observations obtained by RT-PCR (MATSUMURA and TARIN 1992).

In a study of 136 patients with primary breast cancer, the presence of CD44v5 and v6 epitopes (but not v7/8; v8 and v10) correlated with poor overall survival (p=0.05). Multivariate analysis revealed significant independence of CD44v6 from the progesterone receptor (PR) status (PONTA et al. 1995; KAUFMANN et al. 1995). Unpublished data from another laboratory (FRIEDRICHS et al., cited in GÜTHERT et al. 1995), however, have not revealed significant correlations between CD44 and breast cancer patient survival. The different findings of these studies may be due to the tumor samples, since the former collection contained relatively advanced tumors (average diameter 3.5 cm) while FRIEDRICHS and coworkers may have examined earlier stages. These discrepancies need to be settled by examining tumors in a larger and more standardized patient series.

5.2.2 Colorectal Cancer

Enormous progress has been made during the past few years in understanding colorectal carcinogenesis. Part of the reason for this advance is due to the characterization of human mutants with elevated cancer risk: hereditary polyposis coli and nonpolyposis colorectal cancer syndromes (VOGELSTEIN and KINZLER 1993). The colorectal mucosa is thought to be exposed to numerous carcinogenic agents whose effects accumulate over time. Mutant cells that arise under these conditions have no chance of surviving unless they escape the rapid shedding from the tips of the villi. Such "escapees" make themselves known in the form of intramucosal hypertrophies, adenomas (polyps) of various thickness, and as colorectal carcinoma. Cancer often originates from the polyps, and histological features permit one to order precancerous and cancerous lesions in a scheme of increasing malignancy. These stages in colorectal carcinogenesis are molecularly characterized by the frequency of mutations found in APC, Ki-*ras*, p53 and DCC (FEARON and VOGELSTEIN 1990; VOGELSTEIN and KINZLER 1993).

Normal colorectal mucosa carries little CD44 protein. Using an antibody that recognizes the CD44 NH_2-terminal, strong staining of submucosal tissues was contrasted by little immunofluorescence on cells at the base of the crypts (HEIDER et al. 1993a). Using a different monoclonal NH_2-terminal antibody, a similar correlation with the proliferative capacity of crypt cells was seen (ABBASI et al. 1993). Because of the high CD44 levels in submucosal tissue, it is impossible to decide which cell type gave rise to the CD44s RNA detected by PCR of normal mucosa (MATSUMURA and TARIN 1992). A polyclonal variant exon-specific antibody (HEIDER et al. 1993a) and a v9 exon-specific monoclonal (mAB FW 11.24; TERPE et al. 1994a) stained the base of the crypts only very weakly.

CD44 variant exon epitopes are found on mucosal cells of lesions ranging from early adenomas to invasive cançer and metastases. There is, however, an interesting change in epitope pattern suggesting selection for certain CD44 variant proteins. Epitopes for exons v8, v9 and v10 are observed in early adenomas (WIELENGA et al. 1993; FINKE et al., cited in GÜNTHERT et al. 1995). Also, exon v5 appears to be present in about 70% of early adenomas (WIELENGA et al. 1993). The most interesting selective accumulation was seen for epitopes encoded by exon v6: the frequency increased from essentially zero in early adenomas to about 55% in Dukes C/D invasive cancer (WIELENGA et al. 1993). It appears that epitopes that arise early (e.g., v5) are retained throughout carcinogenesis. PCR data obtained from colorectal tumors are compatible with these immunochemical findings (MATSUMURA and TARIN 1992; TANABE et al. 1993; WIELENGA et al. 1993; FINN et al. 1994).

Since only about half of the invasive carcinomas carry epitopes of exon v6, a comparison of the expression of exon v6 with the clinical outcome has been performed. Some 68 patients with primary colorectal carcinoma who had been operated on between 1983 and 1986 entered this first study. Patients with tumors of the category zero expression or low (below 10% of the cells) had significantly better overall survival probability (p=0.002) than those with tumors with strong expression levels. Patients with tumors of an intermediate category survived with a probability between the two extremes (SLEEMAN et al. 1995; MULDER et al. 1994).

5.2.3 Other Cancers

For renal cell cancer and gastric carcinoma we refer to the review by GÜNTHERT and coworkers (1995). Many other types of cancer are currently being studied. Unfortunately, most of the published data often address findings with tumor cell lines or report only very few cases. While these studies may suggest interesting correlations, large-scale examinations will be necessary before any general conclusions about CD44's role in a particular type of cancer can be drawn. A selected bibliography of some of these studies may guide the interested reader: MACKAY et al. 1994; EAST et al. 1993; PENNO et al. 1994; JACKSON et al. 1994; CHAUDHRY et al. 1994; CULTY et al. 1994; MAYER et al. 1993; SALMI et al. 1993; HEIDER et al. 1993b; ROSENBERG et al. 1995; WASHINGTON et al. 1994.

6 Concluding Remarks

The CD44 proteins appear to be part of the program of tumor progression in several animal and human cancers. Their detection, especially by immunohistochemistry, will likely become part of the clinical evaluation of many tumors. The data we have reviewed here, for example, strongly suggest that certain CD44 epitopes can serve as early diagnostic markers in colorectal (v5) and cervical (v7/v8) carcinoma. Although the consequences of CD44 expression by tumor cells is not clear in most cases, at least some CD44 proteins seem to share a proliferation-promoting function. These and other isoforms, however, are also expressed in various normal tissues and apparently can promote tumor metastasis via additional mechanisms (e.g.,CD44v6-v7 and CD44v4-v7 vs CD44s). Furthermore, various biochemical differences have been detected between these proteins. It is, therefore, likely that the various functions of CD44 proteins are exerted in strict dependence on ligands in the microenvironment. This would be in agreement with the specific roles detected in mature lymphocytes, in neuronal cells of the optic chiasm, in neural crest cell differentiation and in the AER of the developing limb. Considering the vast diversity of putative functions attributed to this protein family, the continued study of CD44 in both embryogenesis and cancer will likely lead to advances in our understanding of the process by which cells differentiate, migrate, and metastasize.

References

Abbasi AM, Chester KA, Talbot IC, Macpherson AS, Boxer G, Forbes A, Malcolm ADB, Begent RHJ (1993) CD44 is associated with proliferation in normal and neoplastic human colorectal epithelial cells. Eur J Cancer 29A: 1995–2002

Arch R, Wirth K, Hofmann M, Ponta H, Matzku S, Herrlich P, Zöller M (1992) Participation in normal immune responses of a splice variant of CD44 that encodes a metastasis-inducing domain. Science 257: 682–685

Arpin M, Algrain M, Louvard D (1994) Membrane-actin microfilament connections: an increasing diversity of players related to band 4.1. Curr Opin Cell Biol 6: 136–141

Aruffo A, Stamenkovic I, Melnick M, Underhill CB, Seed B (1990) CD44 is the principal cell surface receptor for hyaluronate. Cell 61: 1303–1313

Bartolazzi A, Peach R, Aruffo A, Stamenkovic I (1994) Interaction between CD44 and hyaluronate is directly implicated in the regulation of tumor development. J Exp Med 180: 53–66

Bennett KL, Jackson DG, Simon JC, Tanczos E, Peach R, Modrell B, Stamenkovic I, Plowman G, Aruffo A (1995) CD44 isoforms containing exon v3 are responsible for the presentation of heparin-binding growth factor. J Cell Biol 128: 687–698

Berger MS, Locher GW, Saurer S, Gullick WJ, Waterfield MD, Groner B, Hynes NE (1988) Correlation of c-erbB-2 gene amplification and protein expression in human breast carcinoma with nodal status and nuclear grading. Cancer Res 48: 1238–1243

Birch M, Mitchell S, Hart IR (1991) Isolation and characterization of human melanoma cell variants expressing high and low levels of CD44. Cancer Res 51: 6660–6667

Bourguignon LYW, Walker G, Suchard SJ, Balazovich K (1986) A lymphoma plasma membrane-associated protein with ankyrin-like properties. J Cell Biol 102: 2115–2124

Bourguignon LYW, Lokeshwar VB, He J, Chen X, Bourguignon GJ (1992) A CD44-like endothelial cell transmembrane glycoprotein (GP 116) interacts with extracellular matrix and ankyrin. Mol Cell Biol 12: 4464–4471

Bretscher A (1989) Rapid phosphorylation and reorganization of ezrin and spectrin accompany morphological changes induced in A-431 cells by epidermal growth factor. J Cell Biol 108: 921–930

Carter WG, Wayner EA (1988) Characterization of the class III collagen receptor, a phosphorylated, transmembrane glycoprotein expressed in nucleated human cells. J Biol Chem 263: 4193–4201

Campbell S, Swann HR, Aplin JD, Seif MW, Kimber SJ, Elstein M (1995) CD44 is expressed throughout pre-implantation human embryo development. Hum Reprod 10: 425–430

Chaudhry A, Gobl A, Eriksson B, Skogseid B, Oberg K (1994) Different splice variants of CD44 are expressed in gastrinomas but not in other subtypes of endocrine pancreatic tumors. Cancer Res 54: 981–986

Christiansen H, Sahin K, Berthold F, Hero B, Terpe HJ, Lampert F (1995) Comparison of DNA aneuploidy, chromosome 1 abnormalities, MYCN amplification and CD44 expression as prognostic factors in neuroblastoma. Eur J Cancer 31A: 541–544

Combaret V, Lasset C, Frappaz D, Bouvier R, Thiesse P, Rebillard AC, Philip T, Favrot MC (1995) Evaluation of CD44 prognostic value in neuroblastoma: comparison with the other prognostic factors. Eur J Cancer 31A: 545–549

Crossley PH, Martin GR (1995) The mouse FGF8 gene encodes a family of polypeptides and is expressed in regions that direct outgrowth and patterning in the developing embryo. Development 121: 439–451

Culty M, Shizari M, Thompson EW, Underhill CB (1994) Binding and degradation of hyaluronan by human breast cancer cell lines expressing different forms of CD44: correlation with invasive potential. J Cell Physiol 160: 275–286

Dall P, Heider K-H, Hekele A, von Minckwitz G, Kaufmann M, Ponta H, Herrlich P (1994) Surface protein expression and messenger RNA-splicing analysis of CD44 in uterine cervical cancer and normal cervical epithelium. Cancer Res 54: 3337–3341

Dall P, Heider K-H, Sinn H-P, Skroch-Angel P, Adolf G, Kaufmann M, Herrlich P, Ponta H (1995) Comparison of immunohistochemistry and RT-PCR for detection of CD44v-expression, a new prognostic factor in human breast cancer. Int J Cancer 60: 471–477

Dougherty GJ, Cooper DL, Memory JF, Chiu RK (1994) Ligand binding specificity of alternatively spliced CD44 isoforms. Recognition and binding of hyaluronan by CD44R1. J Biol Chem 269: 9074–9078

East JA, Mitchell SD, Hart IR (1993) Expression and function of the CD44 glycoprotein in melanoma cell lines. Melanoma Res 3: 341–346

Egerton M, Burgess WH, Chen D, Druker BJ, Bretscher A, Samelson LE (1992) Identification of ezrin as a 81-kDa tyrosine-phosphorylated protein in T cells. J Immunol 149: 1847–1852

Eibl RH, Pietsch T, Moll J, Skroch-Angel P, Heider K-H, Ammon vK, Wiestler OD, Ponta H, Kleihues P, Herrlich P (1995) Expression of variant CD44 epitopes in human astrocytic brain tumors. J Neurooncol (in press)

Favrot MC, Combaret V, Lasset C, Bérard CL (1993) CD44 – a new prognostic marker for neuroblastoma. Lancet 329: 1965

Fazioli F, Wong WT, Ullrich SJ, Sakaguchi K, Appella E, Di Fiore PP (1993) The ezrin-like family of tyrosine kinase substrates: receptor-specific pattern of tyrosine phosphorylation and relationship to malignant transformation. Oncogene 8: 1335–1345

Fearon ER, Vogelstein B (1990) A genetic model for colorectal tumorigenesis. Cell 61: 759–767

Finn L, Dougherty G, Finley G, Meisler A, Becich M, Cooper DL (1994) Alternative splicing of CD44 pre-mRNA in human colorectal tumors. Biochem Biophys Res Commun 200: 1015–1022

Fox SB, Gatter KC, Jackson DG, Screaton GR, Bell MV, Bell JI, Harris AL, Simmons D, Fawcett J (1993) CD44 and cancer screening. Lancet 342: 548–549

Fujiwara K, Yoshino T, Miyake K, Ohara N, Akagi T (1993) Expression of a lymphocyte adhesion molecule (CD44) in malignant lymphomas: relevance to primary site, histological subtype and clinical stage. Acta Med Okayama 47: 215

Gross N, Beretta C, Peruisseau G, Jackson D, Simmons D, Beck D (1994) CD44H expression by human neuroblastoma cells: relation to MYCN amplification and lineage differentiation. Cancer Res 54: 4238–4242

Gross N, Beck D, Beretta C, Jackson D, Perruisseau G (1995) CD44 expression and modulation on human neuroblastoma tumors and cell lines. Eur J Cancer 31A: 471–475

Günthert U, Hofmann M, Rudy W, Reber S, Zöller M, Haußmann I, Matzku S, Wenzel A, Ponta H, Herrlich P (1991) A new variant of glycoprotein CD44 confers metastatic potential to rat carcinoma cells. Cell 65: 13–24

Günthert U, Stauder R, Mayer B, Terpe H-J, Finke L, Friedrichs K (1995) Are CD44 variant isoforms involved in human tumour progression? Cancer Surv 24: 19–42

Guo Y, Ma J, Wang J, Che X, Narula J, Bigby M, Wu M, Sy MS (1994) Inhibition of human melanoma growth and metastasis in vivo by anti-CD44 monoclonal antibody. Cancer Res 54: 1561–1565

Haegel H, Tolg C, Hofmann M, Ceredig R (1993) Activated mouse astrocytes and T cells express similar CD44 variants. Role of CD44 in astrocyte/T cell binding. J Cell Biol 122: 1067–77

Hart IR, Goode NT, Wilson RE (1989) Molecular aspects of the metastatic cascade. Biochim Biophys Acta 989: 65–84

Heider K-H, Hofmann M, Horst E, van den Berg F, Ponta H, Herrlich P, Pals ST (1993a) A human homologue of the rat metastasis-associated variant of CD44 is expressed in colorectal carcinomas and adenomatous polyps. J Cell Biol 120: 227–233

Heider K-H, Dämmrich J, Skroch-Angel P, Müller-Hermelink H-K, Vollmers HP, Herrlich P, Ponta H (1993b) Differential expression of CD44 splice variants in intestinal and diffuse type human gastric carcinoma and normal gastric mucosa. Cancer Res 53: 4197–4203

Herrlich P, Rudy P, Hofmann M, Arch R, Zöller M, Zawadzki V, Tölg C, Hekele A, Koopman G, Pals S, Heider K-H, Sleeman J, Ponta H (1993a) CD44 and splice variants of CD44 in normal differentiation and tumor progression. In: Hemler ME, Mihich E (eds) Cell adhesion molecules. Plenum, New York, pp 265–288

Herrlich P, Zöller M, Pals ST, Ponta H (1993b) CD44 splice variants: metastases meet lymphocytes. Immunol Today 14: 395–399

Hirano H, Screaton GR, Bell MV, Jackson DG, Bell JI, Hodes RJ (1994) CD44 isoform expression mediated by alternative splicing: tissue-specific regulation in mice. Int Immunol 6: 49–59

Hofmann M, Rudy W, Zöller M, Tölg C, Ponta H, Herrlich P, Günthert U (1991) CD44 splice variants confer metastatic behavior in rats: homologous sequences are expressed in human tumor cell lines. Cancer Res 51: 5292–5297

Hofmann M, Rudy W, Günthert U, Zimmer SG, Zawadzki V, Zöller M, Lichtner RB, Herrlich P, Ponta H (1993) A link between RAS and metastatic behavior of tumor cells: ras induces CD44 promoter

activity and leads to low-level expression of metastasis-specific variants of CD44 in CREF cells. Cancer Res 53: 1516–1521
Horst E, Meijer CJLM, Radaszkiewicz T, van Dongen JJM, Pieters R, Figdor CG, Hooftmann A, Pals ST (1990) Expression of a human homing receptor (CD44) in lymphoid malignancies and related stages of lymphoid development. Leukemia 4: 383–389
Isacke CM (1994) The role of the cytoplasmic domain in regulating CD44 function. J Cell Sci 107: 2353–2359
Jackson DG, Schenker T, Waibel R, Bell JI, Stahel RA (1994) Expression of alternatively spliced forms of the CD44 extracellular-matrix receptor on human lung carcinomas. Int J Cancer [Suppl] 8: 110–115
Jackson DG, Bell JI, Dickinson R, Timans J, Shields J, Whittle N (1995) Proteoglycan forms on the lymphocyte homing receptor CD44 are alternatively spliced variants containing the v3 exon. J Cell Biol 128: 673–685
Jalkanen S, Jalkanen M (1992) Lymphocyte CD44 binds the COOH-terminal heparin-binding domain of fibronectin. J Cell Biol 116: 817–825
Jalkanen S, Bargatze RF, de los Toyos J, Butcher EC (1987) Lymphocyte recognition of high endothelium: antibodies to distinct epitopes of an 85–95 kD glycoprotein antigen differentially inhibit lymphocyte binding to lymph node, mucosal, or synovial endothelial cells. J Cell Biol 105: 983–990
Joensuu H, Klemi PJ, Toikkanen S, Jalkanen S (1993) Glycoprotein CD44 expression and its association with survival in breast cancer. Am J Pathol 143: 867–874
Kalomires EL, Bourguignon LYW (1988) Mouse T lymphoma cells contain a transmembrane glycoprotein (GP85) that binds ankyrin. J Cell Biol 106: 319–327
Kaufmann M, Heider K-H, Sinn H-P, von Minckwitz G, Ponta H, Herrlich P (1995) CD44 variant exon epitopes in primary breast cancer and length of survival. Lancet 345: 615–619
Kerbel RS (1990) Growth dominance of the metastatic cancer cell: cellular and molecular aspects. Adv Cancer Res 55: 87–132
Koopman G, Heider K-H, Horst E, Adolf GR, van den Berg F, Ponta H, Herrlich P, Pals ST (1993) Activated human lymphocytes and aggressive Non-Hodgkin lymphomas express a homologue of the rat metastasis-associated variant of CD44. J Exp Med 177: 897–904
Labarrière N, Piau JP, Otry C, Denis M, Lustenberger P, Le Pendu J (1994) H-blood group antigen carried by CD44v modulates tumorigenicity of rat colon carcinoma cells. Cancer Res 54: 6275–6281
Lazaar AL, Albelda SM, Pilewski JM, Brennan B, Pure E, Panettieri RA Jr (1994) T lymphocytes adhere to airway smooth muscle cells via integrins and CD44 and induce smooth muscle cell DNA synthesis. J Exp Med 180: 807–816
Le Douarin NM (1984) Cell migrations in embryos. Cell 38: 353–360
Lesley J, Hyman R (1992) CD44 can be activated to function as an hyaluronic acid receptor in normal murine T cells. Eur J Immunol 22: 2719–2723
Lesley J, He Q, Miyake K, Hamann A, Hyman R, Kincade PW (1992) Requirements for hyaluronic acid binding by CD44: a role for the cytoplasmic domain and activation by antibody. J Exp Med 175: 257–266
Lesley J, Hyman R, Kincade PW (1993) CD44 and its interaction with the extracellular matrix. Adv Immunol 54: 271–335
Li H, Hamou M-F, de Tribolet N, Jaufeerally R, Hofmann M, Diserens M-C, Van Meir EG (1993) Variant CD44 adhesion molecules are expressed in human brain metastases but not in glioblastomas. Cancer Res 53: 5345–5349
Liao H-X, Levesque MC, Patton K, Bergamo B, Jones D, Moody MA, Telen MJ, Haynes BF (1993) Regulation of human CD44H and CD44E isoform binding to hyaluronan by phorbol myristate acetate and anti-CD44. J Immunol 151: 6490–6499
Mac Dermott RP, Schmidt RE, Caulfield JP, Hein A, Vartley GT, Ritz J, Schlossman SF, Austen KF, Stevens RL (1985) Proteoglycans in cell-mediated cytotoxicity. Identification, localization, and exocytosis of a chondroitin sulfate proteoglycan from human cloned natural killer cells during target cell lysis. J Exp Med 162: 1771–1787
Mackay CR, Terpe H-J, Stauder R, Marston WL, Stark H, Günthert U (1994) Expression and modulation of CD44 variant isoforms in humans. J Cell Biol 124: 71–82
Matsumura Y, Tarin D (1992) Significance of CD44 gene products for cancer diagnosis and disease evaluation. Lancet 340: 1053–1058
Mayer B, Jauch KW, Günthert U, Figdor CG, Schildberg FW, Funke I, Johnson JP (1993) De-novo expression of CD44 and survival in gastric cancer. Lancet 342: 1019–1022
Merzak A, Koocheckpour S, Pilkington GJ (1994) CD44 mediates human glioma cell adhesion and invasion in vitro. Cancer Res 54: 3988–3992
Miyake K, Underhill CB, Lesley J, Kincade PW (1990) Hyaluronate can function as a cell adhesion molecule and CD44 participates in hyaluronate recognition. J Exp Med 172: 69–75

Moll J, Sleeman J, Kondo K, Hekele A, Plug R, Sherman L, Schmidt A, Zöller M, Horst E, Ponta H, Herrlich P, Pals ST (1994) Analysis of molecular functions of the tumor metastasis promoting surface molecule CD44v4-v7 using transgenic mice. In: Hirohashi S, Moses HL, Ruoslahti E, Sugimura T, Takeichi M, Terada M (eds) Molecular and cellular basis for cell to cell interaction: its significance in cancer. Princeton Scientific, Princeton, pp 142–151

Moll J, Schmidt A, van der Putten H, Plug R, Ponta H, Herrlich P, Zöller M (1995) Accelerated immune response in transgenic mice expressing rat CD44v4-v7 on T cells. J Immunol (in press)

Moretto G, Xu RY, Kim SU (1993) CD44 expression in human astrocytes and oligodendrocytes in culture. J Neuropathol Exp Neurol 52: 419–423

Mulder J-WR, Kruyt PM, Sewnath M, Oosting J, Seldenrijk CA, Weidema WF, Offerhaus GJA, Pals ST (1994) Colorectal cancer prognosis and expression of exon-v6-containing CD44 proteins. Lancet 344: 1470–1472

Nagasaka S, Tanabe KK, Bruner JM, Saya H, Sawaya RE, Morrison RS (1995) Alternative RNA splicing of the hyaluronic acid receptor CD44 in the normal human brain and in brain tumors. J Neurosurg 82: 858–863

Naujokas MF, Morin M, Anderson MS, Peterson M, Miller J (1993) The chondroitin sulfate form of invariant chain can enhance stimulation of T cell responses through interaction with CD44. Cell 74: 257–268

Niswander L, Tickel C, Vogel A, Booth I, Martin GR (1993) FGF-4 replaces the apical ectodermal ridge and directs outgrowth and patterning of the limb. Cell 75: 579–587

Peach RJ, Hollenbaugh D, Stamenkovic I, Aruffo A (1993) Identification of hyaluronic acid binding sites in the extracellular domain of CD44. J Cell Biol 122: 257–264

Penno MB, August JT, Baylin SB, Mabry M, Linnoila RI, Lee VS, Croteau D, Yang XL, Rosada C (1994) Expression of CD44 in human lung tumors. Cancer Res 54: 1381–1387

Picker LJ, Medieros LJ, Weiss LM, Warnke RA, Butcher EC (1988) Expression of lymphocyte homing receptor antigen in nonHodgkin's lymphoma. Am J Pathol 130: 496–504

Ponta H, Hekele A, Heider K-H, Kaufmann M, von Minckwitz G, Sinn H-P, Pals ST, Wielenga VJM, Herrlich P (1995) Relevance of CD44 variant epitopes for diagnosis and therapy of human breast cancer. In: Contribution to oncology. Karger, Basel (in press)

Puré E, Camp RL, Peritt D, Panettieri RA Jr, Lazaar AL, Nayak S (1995) Defective phosphorylation and hyaluronate binding of CD44 with point mutations in the cytoplasmic domain. J Exp Med 181: 55–62

Radotra B, McCormick D, Crockard A (1994) CD44 plays a role in adhesive interactions between glioma cells and extracellular matrix components. Neuropathol Appl Neurobiol 20: 399–405

Rosenberg WMC, Prince C, Kaklamanis L, Fox SB, Jackson DG, Simmons DL, Chapman RW, Trowell JM, Jewell DP, Bell JI (1995) Increased expression of CD44v6 and CD44v3 in ulcerative colitis but not colonic Crohn's disease. Lancet 345: 1205–1209

Rudy W, Hofmann M, Schwartz-Albiez R, Zöller M, Heider K-H, Ponta H, Herrlich P (1993) The two major CD44 proteins expressed on a metastatic rat tumor cell line are derived from different splice variants: each one individually suffices to confer metastatic behavior. Cancer Res 53: 1262–1268

Reber S, Matzku S, Ginthert U, Ponta H, Herrlich P, Zöller M (1990) Retardation of metastatic tumor growth after immunization with metastasis-specific monoclonal antibodies. Int J Cancer 46: 919–927

Ruiz P, Schwärzler C, Günthert U (1995) CD44 isoforms during differentiation and development. Bioessays 17: 17–24

Sainsbury JRC, Farndon JR, Needham GK, Malcalm AJ, Harris AL (1987) Epidermal-growth-factor receptor status as predictor of early recurrence of and death from breast cancer. Lancet 1: 1398–1402

Salles G, Zain M, Jiang WM, Boussiotis VA, Shipp MA (1993) Alternatively spliced CD44 transcripts in diffuse large-cell lymphomas: characterization and comparison with normal activated B cells and epithelial malignancies. Blood 82: 3539–3547

Salmi M, Grön-Virta K, Sointu P, Grenman R, Kalimo H, Jalkanen S (1993) Regulation expression of exon v6 containing isoforms of CD44 in man: downregulation during malignant transformation of tumors of squamocellular origin. J Cell Biol 122: 431–442

Screaton GR, Bell MV, Jackson DG, Cornelis FB, Gerth U, Bell JI (1992) Genomic structure of DNA encoding the lymphocyte homing receptor CD44 reveals at least 12 alternatively spliced exons. Proc Natl Acad Sci USA 89: 12160–12164

Seiter S, Arch R, Komitowski D, Hofmann, Ponta H, Herrlich P, Matzku S, Zöller M (1993) Prevention of tumor metastasis formation by anti-variant CD44. J Exp Med 177: 443–455

Seshadri R, Horsfall DJ, Firgaira F, McGaul K, Setlur V, Chalmers AH, Yeo R, Ingram D, Dawkins H, Hahnel R (1994) The relative prognostic significance or total Cathepsin D and HER-2/neu oncogene amplification in breast cancer. Int J Cancer 56: 61–65

Sherman L, Sleeman J, Herrlich P, Ponta H (1994) Hyaluronate receptors: key players in growth, differentiation, migration and tumor progression. Curr Opin Cell Biol 6: 726–733

Sherman L, Skroch-Angel P, Moll J, Schwecheimer K, Ponta H, Herrlich P, Hofmann M (1995) Schwann cell tumors express characteristic patterns of CD44 splice variants. J Neurooncology (in press)

Shipp A (1994) Prognostic factors in aggressive non-Hodgin's lymphoma: who has "high-risk" disease? Blood 83: 1165–1173

Shtivelman E, Bishop M (1991) Expression of CD44 is repressed in neuroblastoma cells. Mol Cell Biol 11: 5446–5453

Sinn H-P, Heider K-H, Skroch-Angel P, Minckwitz Gv, Kaufmann M, Herrlich P, Ponta H (1995) Human mammary carcinomas express homologues of metastasis-associated variants of CD44. Breast Cancer Res Treat 36: 307–313

Sleeman J, Moll J, Sherman L, Dall P, Pals ST, Ponta H, Herrlich P (1995) The role of CD44 splice variants in human metastatic cancer. Ciba Found Symp 189: 142–156

Sretavan DW, Feng L, Pure E, Reichardt LF (1994) Embryonic neurons of the developing optic chiasm express L1 and CD44, cell surface molecules with opposing effects in retinal axon growth. Neuron 12: 957–975

Stamenkovic I, Aruffo A, Amiot M, Seed B (1991) The hematopoietic and epithelial forms of CD44 are distinct polypeptides with different adhesion potentials for hyaluronate-bearing cells. EMBO J 10: 343–348

Stauder R, Eisterer W, Thaler J, Günthert U (1995) CD44 variant isoforms in Non-Hodgin's lymphoma: a new independent prognostic factor. Blood 85: 2885–2899

Stevens RL, Kamada MM, Serafin WE (1988) Structure and function of the family of proteoglycans that reside in the secretory granules of natural killer cells and other effector cells of the immune response. Curr Top Microbiol Immunol 140: 93

Sy MS, Guo Y-J, Stamenkovic I (1991) Distinct effects of two CD44 isoforms on tumor growth in vivo. J Exp Med 174: 859–866

Sy M-S, Guo Y-J, Stamenkovic I (1992) Inhibition of tumor growth in vivo with a soluble CD44-immunoglobulin fusion protein. J Exp Med 176: 623–627

Tanabe KK, Ellis LM, Saya H (1993) Expression of CD44R1 adhesion molecule in colon carcinomas and metastases. Lancet 341: 725–726

Terpe H-J, Stark H, Prehm P, Günthert U (1994a) CD44 variant isoforms are preferentially expressed in basal epithelia of non-malignant human fetal and adult tissues. Histochemistry 101: 79–89

Terpe HJ, Koopman R, Imhof B, Günthert U (1994b) Expression of integrins and CD44 isoforms in non-Hodgkin's lymphomas: CD44 isoforms are preferentially expressed in high-grade malignant lymphomas. J Pathol 174: 89–100

Thuillier L, Hivroz C, Fagard R, Andreoli C, Mangeat P (1994) Ligation of CD4 surface antigen induces rapid tyrosine phosphorylation of the cytoskeletal protein ezrin. Cell Immunol 156: 322–331

Toyama-Sorimachi N, Miyasaka M (1994) A novel ligand for CD44 is sulfated proteoglycan. Int Immunol 6: 655–660

Tölg C, Hofmann M, Herrlich P, Ponta H (1993) Splicing choice from ten variant exons establishes CD44 variability. Nucleic Acids Res 21: 1225–1229

Trowbridge IS, Lesley J, Schulte R, Hyman R, Trotter J (1982) Biochemical characterization and cellular distribution of a polymorphic, murine cell-surface glycoprotein expressed in lymphoid tissues. Immunogenetics 15: 299–312

Tsukita S, Tsukita S, Naagafuchi A, Yonemura S (1992) Molecular linkage between cadherins and actin filaments in cell-cell adherens junctions. Curr Opin Cell Biol 4: 834–839

Tsukita S, Oishi K, Sato N, Sagara J, Kawai A, Tsukita S (1994) ERM family members as molecular linkers between the cell surface glycoprotein CD44 and actin-based cytoskeletons. J Cell Biol 126: 391–401

Verspaget HW, Sier CFM, Ganesh S, Griffioen G, Lamers CBHW (1995) Prognostic value of plasminogen activators and their inhibitors in colorectal cancer. Eur J Cancer (in press)

Vogelstein B, Kinzler KW (1993) The multistep nature of cancer. TIG 9: 138–141

Washington K, Gottfried MR, Telen MJ (1994) Expression of the cell adhesion molecule CD44 in gastric adenocarcinomas. Hum Pathol 25: 1043–1049

Wayner EA, Carter WG (1987) Identification of multiple cell adhesion receptors for collagen and fibronectin in human fibrosarcoma cells possessing unique alpha and common beta subunits. J Cell Biol 105: 1873–1884

Wheatley SC, Isacke CM, Crossley PH (1993) Restricted expression of the hyaluronan receptor, CD44, during postimplantation mouse embryogenesis suggests key roles in tissue formation and patterning. Development 199: 295–306

Wielenga VJM, Heider K-H, Offerhaus GJA, Adolf GR, van den Berg FM, Ponta H, Herrlich P, Pals ST (1993) Expression of CD44 variant proteins in human colorectal cancer is related to tumor progression. Cancer Res 53: 4754–4756

Wirth K, Arch R, Somasundaram C, Hofmann M, Weber B, Herrlich P, Matzku S, Zöller M (1993) Expression of CD44 isoforms carrying metastasis-associated sequences in newborn and adult rats. Eur J Cancer 29A: 1177–1183

Yang B, Yang BL, Savani RC, Turley EA (1994) Identification of a common hyaluronan binding motif in the hyaluronan binding proteins RHAMM, CD44 and link protein. EMBO J 13: 286–296

Zabrenetzky V, Haris CC, Steeg PS, Roberts DD (1994) Expression of the extracellular matrix molecule thrombospondin inversely correlates with malignant pogression in melanoma, lung and breast carcinoma cell lines. Int J Cancer 59: 191–195

CD44 in Maligant Disorders

U. GÜNTHERT

1 Introduction

CD44 has been implicated to act in processes such as lymphocyte homing, hematopoiesis, tumor dissemination, lymphocyte activation, pattern formation in embryogenesis, and inflammatory reactions. Possible explanations for the supposedly multipurpose nature of CD44 include the existence of an enormous number of isoforms and/or the potential likelihood that the role of CD44 in all these diverse processes is not as manifold as it sounds, because we still lack essential understanding of the function of the variant isoforms.

Many review articles and commentaries have appeared recently on CD44 (HAYNES et al. 1991a; UNDERHILL 1992; GALLAGHER 1992; GÜNTHERT 1993; LESLEY et al. 1993; KINCADE 1993; KINCADE et al. 1993; HERRLICH et al. 1993a, b; KOOPMAN et al. 1993a; EAST and HART 1993; PALS et al. 1993 a, b; ZÖLLER and KAUFMANN 1994; SHERMAN et al. 1994; TANABE and SAYA 1994; TARIN and MATSUMURA 1994; ISACKE 1994; PONTA et al. 1994, 1995; MOLL et al. 1994; LAZAR and PURÉ 1995 GÜNTHERT et al. 1995; RUIZ et al. 1995; STAUDER and GÜNTHERT 1995; COOPER and DOUGHERTY 1995; ZÖLLER 1995; SLEEMAN et al. 1995; including those from the current series: SHERMAN et al. 1996; ZÖLLER 1996; SY et al. 1996; RISTAMÄKI et al. 1996). The present review focuses on additional aspects: Do the recent studies on correlation between CD44 and tumor progression or inflammation provide evidence for a functional involvement of CD44 in these processes? How does CD44 fit into the

Basel Institute for Immunology, Grenzacherstrasse 487, 4005 Basel, Switzerland

current view of the network of metastasis-related molecules (see the articles in the present CTMI series and RUIZ and GÜNTHERT 1995)?

2 The Structure

Although encoded by only one gene, CD44 represents a large family of molecules which differ in primary structure (summarized by GÜNTHERT 1993). Not only ten so-called variant exons can be joined in multiple combinations, but also the two flanking standard exons (5 and 16) share in the alternative splicing process, as well as exons 19 and 20, encoding the cytoplasmic domain (Fig. 1). In addition, two cryptic splice sites in exons 3 and 10 of the variant region (exons 8 and 15, respectively, counting contiguously) may still enlarge the variability of the isoforms. In theory the sum of the possible exon combinations easily exceeds 1000,

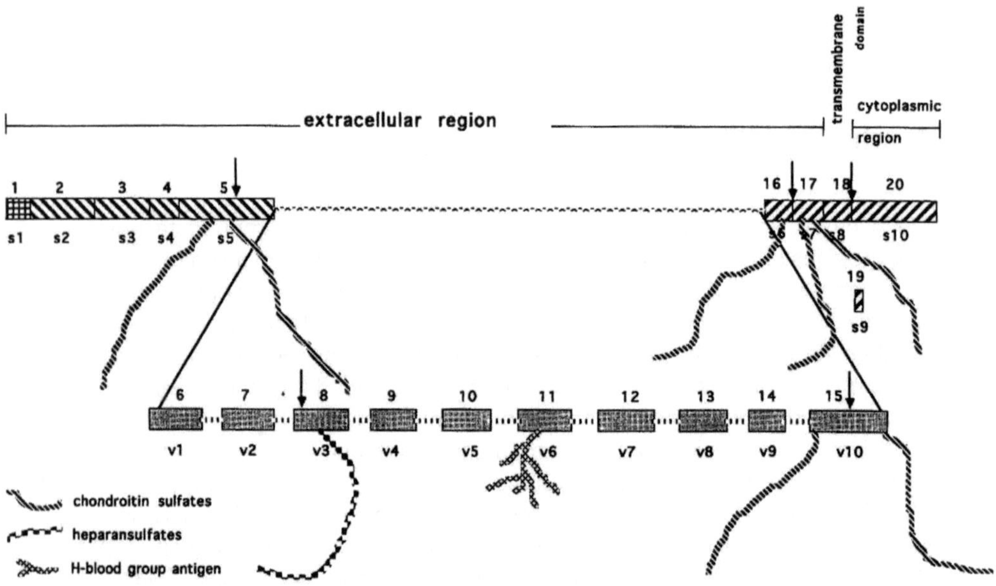

Fig. 1. Organization of the CD44 exons. The nomenclature above the exons represents the contiguous numbering from 1 to 20, while the numbers below the exons indicate the difference between the standard (s1–10) and the variant (v1–v10) region. *Striped boxes*, encoding of the standard or hematopoietic form of CD44; *gray boxes*, the alternatively spliced isoforms. *Square pattern*, leader peptide. *Arrowheads*, additional splice sites in the standard region and some cryptic splice sites; *Striped lines*, potential chondroitin sulfate side chains; *small squares*, heparan sulfate side chain in exon v3. The attachment sites for chondroitin sulfates in exon s5 (5') and in s7 (3') are conserved in humans, rats, and mice, as well as the site for attachment of heparan sulfate in exon v3. The other sites are present only in humans (s5:3', s6, and v10:3') or in mice and rats (v10:5' and s7:5'). The H-blood group antigen in exon v6 is shown as a tree structure; presence has been demonstrated only in rats. Exon v1 (or 6 when contiguously counting) is not expressed in humans due to a stop codon, but is in rats and mice

surpassing in complexity even the potential variations of the integrin family (DIAMOND and SPRINGER 1994). While the substrate specificity of the integrins is mediated by dimerization between one of the already identified 15 α subunits with any one of the 8 β subunits, all of which are encoded on different genes, the CD44 complexity is apparently generated more efficiently by alternative splicing of not more than 13 variably expressed exons. As with the integrins, however, only a limited number of combinations have been identified for the CD44 isoforms, not more than 100 at present. Analysis of exon combinations with reverse transcriptase polymerase chain reaction (RT-PCR) has revealed distinct combinations in correlation to specific physiological states of the cells analyzed (RUIZ et al. 1995; STAUDER and GÜNTHERT 1995; SHERMAN et al. 1996; ZÖLLER 1996).

To increase the complexity of the CD44 family even more, the primary structure is modified with various glycans, such as N- and O-linked glycosylations, chondroitin, and heparan sulfate side chains (COOPER and DOUGHERTY 1995). Cell type specific differences in modification have been observed (BROWN et al. 1991; HOFMANN et al. 1991). Variable glycosylation of the standard region regulates the interaction with hyaluronan such that the presence of complex sugars can inhibit ligand binding (KATOH et al. 1995; LESLEY et al. 1995). In addition to several more N-and O-linked sugars, the variant region has been shown to carry H-blood group antigens on v6, heparan sulfate side chains on v3, and presumably chondroitin sulfate side chains on v10 (GÜNTHERT 1993; LABARRIERE et al. 1994; JACKSON et al. 1995). The association of growth-and angiogenesis-promoting factors such as basic fibroblast growth factor (b-FGF) and heparan-binding epithelial growth factor (HB-EGF) with heparan sulfate side chains of v3-containing isoforms (BENNETT et al. 1995) clearly indicates functional implications of the variant region. In clear cell renal cell carcinomas the presence of v3-containing isoforms (CD44v3,v8–v10) is highly correlated with advanced grading stages (TERPE et al. 1995b). In line with these data is the observation that an increased amount of bFGF in the serum of patients with renal cell tumors is correlated well with tumor stage and poor survival (FUJIMOTO et al. 1991; NANUS et al. 1993; NGUYEN et al. 1994). Secondary modifications of the primary structure and their ability to capture factors and present them locally at a high concentration (TANAKA et al. 1993) is definitely an important feature of some of the CD44 isoforms. Depending on the nature of the factors the presenting isoforms may promote growth and angiogenesis (in tumor progression and embryogenesis), activation of integrins (in lymphocyte homing), and repair processes (in tissue regeneration and inflammation).

Recent cell mixing studies of transfected murine lymphoma cells suggest that homotypic aggregations are possible between cells expressing CD44v8–v10 containing isoforms (DROLL et al.1995). The region involved probably spans exon v10 and parts of the standard sequence, which is distinct from the hyaluronan binding region (DROLL et al. 1995). Furthermore, human keratinocytes are able to adhere to rat pancreatic carcinoma cells transfected with CD44v4–v7 forming adherens junctions between them (HUDSON et al. 1995). Since keratinocytes express predominantly CD44v3–v10 (HOFMANN et al. 1991), homotypic interactions between 4v–v7 regions may also be possible. Why this presumable interaction between variant regions can be blocked with hyaluronan and hyaluronidase (Hudson et al. 1995) although the hyaluronan binding regions are confined to the standard sequence (Yang et al. 1994; Peach et al. 1993) needs further evaluation.

Deletions of the cytoplasmic domain abrogate binding to hyaluronan (Perschl et al. 1995). When the transmembrane region of CD44 (without the cytoplasmic domain) is replaced by that of the CD3ζ chain, which mediates homodimerization via intermolecular disulfide bonds, hyaluronan binding is restored (Perschl et al. 1995). Thus the cytoplasmic tail may be necessary for clustering CD44 on the surface of cells, thereby increasing the ability to bind hyaluronan or any other ligand.

In summary, the CD44 glycoprotein family represents many isoforms differing not only in primary structure but also in their posttranslational modifications. These differences are most likely responsible for the heterogeneity in function.

3 CD44 in Tumor Progression

Most of the studies on CD44 isoform expression have been carried out with cell lines and tumor material without clinical follow-up (summarized in this series by Zöller 1996; Sherman et al. 1996). The following deals only with those studies in which expression of CD44 isoforms is correlated with the survival data of patients (see also Günthert et al. 1995).

G*astric carcinoma* was the first tumor entity to be analyzed for the expression of CD44 variant isoforms, in particular for v9-containing variants (Mayer et al. 1993). While normal mucosa does not express CD44, the presence of CD44v9 is significantly correlated with distant metastases and, in curatively resected patients, with tumor recurrence and increased mortality. Remarkably, in atrophic gastritis and intestinal metaplasia (precancerous lesions), an intense leukocyte infiltrate is strongly associated with CD44v9 positivity (Mayer et al. 1993). The following section returns to these observations. The two major histological subtypes, intestinal- and the diffuse-type carcinomas, display distinct expression patterns for CD44; significantly more of the intestinal- type than of the diffuse-type carcinomas are positive for CD44s and CD44v6 (Hong et al. 1995a). However, the presence of CD44 in intestinal-type carcinomas is not correlated with adverse prognosis for the patients. In diffuse-type carcinomas, although the presence of CD44v6-containing isoforms indicates a strong incidence for infiltrative tumor growth and lymph node involvement, it unexpectedly is not correlated with poor survival (Hong et al. 1995a). Chronic inflammation and intestinal metaplasia are often found in intestinal-type gastric carcinomas, in conjunction with elevated CD44 expression levels (Mayer et al. 1993). Although v6-containing isoforms were not analyzed in the earlier study, it is likely that most of the v9-positive carcinomas are also positive for v6. Why prognosis is correlated only with the presence of v9-containing isoforms and not with those containing v6 still requires clarification.

Since CD44 isoforms are involved in leukocyte development and activation, it was reasonable to analyze CD44 expression in *non-Hodgkin's lymphomas* (NHL) because these represent the neoplastic image of leukocytes arrested at certain stages of activation and maturation (STAUDER and GÜNTHERT 1995). Immunohistochemical analyses reveal a poor prognosis for patients expressing elevated levels of CD44s as well as de novo expression of CD44v3 and CD44v6 (STAUDER et al. 1995). By RT-PCR presence of v6-containing isoforms in conjunction with other variant exons are detected exclusively in aggressive high-grade malignant NHL and are shown to be associated with a shorter overall survival of the patients. Multivariate analyses, moreover, establish v6 positivity as a new independent prognostic parameter in high-grade NHL compared to the risk groups defined by the INTERNATIONAL NON-HODGKIN'S LYMPHOMA PROGNOSTIC FACTORS PROJECT (1993). While in 97 cases of low-grade NHL only one was weakly positive for CD44v6 (STAUDER et al. 1995), another study described preferential expression of v6 in low-grade lymphomas (RISTAMÄKI et al. 1995). The v6-positive cases similarly have a poorer prognosis than the negative ones, but survival data are not stratified into histological subclassification to indicate whether high- or low-grade v6-positive cases have a worse prognosis (RISTAMÄKI et al. 1995). Two other, nonclinical studies demonstrate v6 positivity exclusively in high-grade NHL (KOOPMAN et al. 1993b; TERPE et al. 1994b). The reason for the differences in assessment of CD44v6-positive cases may be a bias in the RISTAMÄKI study (1995) concerning the classification into the subgroups of low-and high-grade NHL.

CD44 variant isoforms have also been detected in patients with multiple myeloma (STAUDER et al., submitted). The presence of larger isoforms exhibiting combinations of various exons from the variant region and the upregulation of v9-containing isoforms is correlated with unfavorable clinical presentation (STAUDER et al., submitted).

Conflicting data have emerged in the study of *colorectal cancer*. In an analysis of 68 patients with carcinoma of Dukes' B and C stages presence of CD44v6 is associated with poorer outcome of the patients (MULDER et al. 1994). Using the same antibody, another group found no correlation between v6 expression and prognosis in 180 patients (KORETZ et al. 1995). Furthermore, no correlation between v6 positivity and the presence of metastases was demonstrated; most of the distant metastases analyzed exhibited downregulation of the CD44v6 expression in comparison to the primary tumor (FINKE et al. 1995; FINKE et al., submitted). However, the downregulated expression in distant metastases does not imply that CD44v6-containing isoforms are not involved in invasion and metastatic spread of cells from the primary lesion. Upregulation of CD44v6 is an early event in colorectal carcinogenesis, detectable as soon as the first genetic changes become evident (FINKE et al., submitted). Thus, increased expression of mutant p53 protein is correlated significantly with the upregulation of CD44v6 isoforms during tumor progression (MULDER et al. 1995). While in early, preinvasive stages CD44v6 expression is upregulated, the functional significance of these isoforms may possibly be achieved only when other metastasis-related molecules are either gained or lost in function.

Evaluation of CD44v6 expression in primary *breast cancer* has also revealed controversial data. While KAUFMANN and colleagues (1995) described reactivity for CD44v6 isoforms as an indicator for adverse prognosis in 91 patients, but another study on 227 cases detected no correlation of v6 reactivity to survival (FRIEDRICHS et al. 1995a, b). The latter data are further substantiated by other clinical studies on 270 more cases (FRIEDRICHS et al. 1995a). Although for better comparison identical antibodies were used, the data remain contradictory. These inconsistent data are not restricted to the tumor material but extend to the evaluation of CD44v6 expression in normal breast epithelium. This ranges from lack of expression (KAUFMANN et al. 1995) to positivity only in the myoepithelia (DALL et al. 1995; SHERMAN et al. 1996) to expression in myoepithelia and ductal epithelia, although in the latter only focally (FRIEDRICHS et al. 1995a; FRANKE et al., in preparation). Ductal epithelia are generally the origin of neoplasia in the breast.

The patient populations evaluated may have been influenced by a bias concerning the distribution of risk factors, such as an overpresentation of younger patients with advanced stages of disease (KAUFMANN et al. 1995). Such a bias can have a strong effect on the outcome of the analysis and can be overcome only by evaluating much larger cohorts, representing a standard distribution of clinical parameters.

Although not substantiated by clinical data, the analysis of *renal cell tumors* (clear cell carcinomas, chromophilic and chromophobe cell carcinomas, oncocytomas) is included in this survey because the histopathological evaluation according to WHO standards (THOENESS et al. 1986) is correlated very well with clinical prognosis (STÖRKEL et al. 1989). A significant increase in expression of CD44v6 and CD44v9-containing isoforms in the course of tumor progression from G1 to G3 stages has been observed in clear cell and chromophilic cell carcinomas (TERPE et al. 1995b). While these tumors arise from the proximal ducts, oncocytomas and chromophobe cell carcinomas originate from the collecting ducts (THOENES et al. 1986). The benign oncocytomas are negative for all CD44 isoforms, whereas the invasive chromphobe cell carcinomas are strongly positive for v6- and v9-containing isoforms (TERPE et al. 1995b). An analysis of exon compositions expressed in G3 clear cell carcinomas by RT-PCR indicated preferential presence of v8-v10 and v3, v8-v10, and v6-containing isoforms, as opposed to G1 tumors (TERPE et al. 1995b).

Increased levels of bFGF in the serum of patients with renal tumors are correlated with poor prognosis (FUJIMOTO et al. 1991; NANUS et al. 1993; NGUYEN et al. 1994). Interestingly, the recently described presentation of the angiogenesis-inducing factor bFGF on CD44v3-containing isoforms (BENNETT et al. 1995) is correlated well with the increased propensity of G3-stage renal tumors to invade and metastasize. Another noteworthy observation is that most of the G1 cases, which are positive for v6 RNA but do not express v6 protein on the tumor cells show a strong leukocyte infiltrate. The positive reaction for v6 in RT-PCR thus reflects v6-positive activated leukocytes and not tumor cells. The correlation between inflammation and the presence of v3- and v6-positive leukocytes and their possible functional implications for tumor progression is discussed in the following section.

CD44v6 expression has also been studied in *urothelial cancer* tissues to explore the potential prognostic value for patients' survival (HONG et al. 1995b). Noteworthy, the expression of CD44v6 and E-cadherin, an established tumor suppresssor (BRACKE et al. 1996), which have been evaluated concomitantly, is similar in most of the tumors. While strong expression of both molecules is associated with well-differentiated histology, predominantly weak to negative expression is found in poorly differentiated, invasive tumors (HONG et al. 1995b). Correlations between the expression of either molecule and lymph node involvement or the presence of distant metastases do not reveal statistical significance. Likewise, survival and tumor recurrence cannot be predicted in transitional cell carcinomas when applying CD44v6 and E-cadherin specific antibodies for diagnosis (HONG et al. 1995b).

Studies of *neuroblastoma* tumors for CD44 expression have described an inverse correlation between the classical neuroblastoma marker, MYCN, and CD44s (COMBARET et al. 1995; CHRISTIANSEN et al. 1995; TERPE et al. 1995a). Variant isoforms have not been detected at any stage of the disease (TERPE et al. 1995a). Both sets of studies indicate the presence of CD44s and the absence of MYCN as favorable prognostic factors. However, whereas one of the studies, involving 52 tumor samples, concludes that CD44s positivity is an independent prognostic indicator for patients' survival (COMBARET et al. 1995), in the other study group, analyzing 377 patients, MYCN emerged as the only relevant independent prognostic factor in multivariate analyses (CHRISTIANSEN et al. 1995).

In conclusion, analysis of CD44 isoform expression in human tumor progression has not revealed concordant data so far. There is a certain trend to observe that tumors originating from tissues which express variant isoforms before the onset of malignant transformation, such as stratified epithelia (TERPE et al. 1994a; FOX et al. 1994), show a downregulation of these isoforms during tumor progression. Conversely, tumors originating from variant-negative or variant-low tissues, such as lymphomas and renal cell tumors, exhibit upregulation of variant isoforms. Nevertheless, changes in the expression of CD44 variant isoforms are not necessarily independently involved in human tumor progression but may become evident for dysplastic alterations only when other factors are concomitantly up- or downregulated.

Not only studies on the involvement of CD44 isoforms in human tumor progression have revealed controversial data; analyses in *animal tumor models* have also led to divergent results. The initial observation that nonmetastatic rat pancreas adenocarcinoma cells transfected with CD44v4–v7, but not with CD44s, gain metastatic potential was obtained with a spontaneous metastasis assay in syngeneic animals (GÜNTHERT et al. 1991). The decisive influence of the v6 region has been further substantiated by metastasis assays using CD44v6–v7 transfectants (RUDY et al. 1993) and moreover by preventing metastasis formation with monoclonal antibodies against an epitope located in exon v6 (SEITER et al. 1993).

However, using a different animal model, CD44s transfected human Burkitt lymphoma Namalwa cells showed enhanced local and metastatic tumor formation when injected intravenously into immune deficient nude mice (Sy et al. 1991). This tumor formation was prevented by applying a soluble CD44s-immunoglobulin fusion protein (Sy et al. 1992). CD44v8–v10 transfected Namalwa cells do not promote local tumor growth and even exhibit a reduced rate of metastasis formation (Sy et al. 1991).

This tumor model has recently been expanded by using transfectants which express several more variant isoforms (Bartolazzi et al. 1995). While Namalwa cells expressing v6–v10, v7–v10, and v8–v10 containing isoforms show delayed tumor formation when injected intravenously or subcutaneously, v3–v10 and v3, v8–v10 transfectants exhibit extremely slow tumor growth after subcutaneous injection but rapid development of tumors in the bone marrow following intravenous injection (Bartolazzi et al. 1995). The authors explain these variations by different degrees of shedding of the isoforms into the circulation of the mice and by different abilities to bind to hyaluronan. The v3-containing isoforms are shed into the medium supernatant much less than are the non-v3 isoforms and exhibit reduced binding to hyaluronan, suggesting that the v3 region inhibits shedding and also abrogates binding to hyaluronan (Bartolazzi et al. 1995). The v3 region is a target for heparan sulfate side chains to which growth and angiogenesis-promoting factors such as bFGF are bound (Bennett et al. 1995). These novel properties for CD44v3-containing isoforms may be pivotal in promoting bone marrow metastasis formation.

The two animal models used to analyze the metastatic potential of CD44 variant isoforms differ substantially. While in the rat model syngeneic rat carcinoma cells have been transfected with rat CD44 sequences, the nude mouse model has used lymphoma cells transfected with human CD44 constructs (Günthert et al. 1991; Sy et al. 1991; Bartolazzi et al. 1995). Additionally, analogous sequences in the expression constructs have not been used in either tumor model. Although the CD44 gene is conserved between 50% and 80% in humans, rats, and mice (Günthert 1993), species-specific differences cannot be excluded in the interaction of the isoforms with their potential ligands. Further analyses in syngeneic animal model systems preferably employing distinct transfectants should be performed. Regions of importance may be subjected to refined analyses by in vitro mutagenesis.

4 CD44 in Inflammatory Processes

Evidence is increasing that CD44 variant isoforms are functionally involved in inflammatory processes. Inflamed tissue is infiltrated by activacted lymphocytes, which upregulate the expression of many cell surface molecules, among them CD44 variant isoforms (Stauder and Günthert 1995; Stauder et al. 1995).

In *rheumatoid arthritis*, an autoimmune disease, levels of soluble CD44 in synovial fluid are increased in relation to the degree of inflammation (HAYNES et al. 1991b). Expression of surface CD44 is strongly increased on synovial fluid lymphocytes compared to peripheral blood lymphocytes (KELLEHER et al. 1995). Not only CD44 is increased but also hyaluronan in the inflamed joints. In a murine model of arthritis treatment with anti-CD44s antibodies rapidly reduces tissue swelling and leukocyte infiltration, which are the two major components of inflammation (MIKECZ et al. 1995). Treatment with anti-CD44 antibodies abrogates the highly increased levels of soluble CD44, of which at least two apparently variant isoforms have been detected in the synovial fluid (MIKECZ et al. 1995). Although the antibody used (IM7) also inhibits the accumulation of hyaluronan in the joints, it cannot block the interaction betwen CD44 and hyaluronan because it does not interfere with the hyaluronan binding region (ZHENG et al. 1995). Thus CD44 function in rheumatoid inflammation may relate to distinct properties of the infiltrated leukocytes and the inflamed synovial cells, in addition to the CD44-hyaluronan interaction. Various cytokines secreted by activated infiltrating lymphocytes during inflammatory processes are known inducers of CD44 variant isoforms (HAEGEL et al. 1993; MACKAY et al. 1994).

Inflammatory bowel diseases manifest with specific immune abnormalities, including autoimmune disorders. In ulcerative colitis the expression of v3- and v6-containing CD44 isoforms is highly increased in the epithelial cells of the crypts, the main focus of the inflammation (ROSENBERG et al. 1995). The upregulation is restricted to the inflamed area; more distal regions without signs of inflammation are negative for v3 and v6 immunoreactivity. Other inflammatory colonic diseases, such as Crohn's disease, radiation colitis, infective colitis, and coeliac disease, do not show the upregulation of v3 and v6 isoforms (ROSENBERG et al. 1995). It remains to be clarified which specific conditions in ulcerative colitis lead to the induction of variant isoforms. Although the strong expression of CD44 variant isoforms on the inflamed crypt epithelium identifies this tissue as the principal target of the immune-mediated inflammation, the role of the infiltrating lymphocytes in this area has not been addressed so far.

Disruption of the interleukin (IL)-2 genes by gene targeting has unexpectedly generated mice with ulcerative colitis (SADLACK et al. 1993). Similar phenotypes evolved in T cell receptor (TCR)-α, TCR-β, TCR-$\beta \times \delta$, and class II major histocompatibility complex mutant mice, as well as in mice with a disrupted IL-10 gene (reviewed in STROBER and EHRHARDT 1993). The phenotype of these gene-targeted mice suggests that chronic intestinal inflammation is a result of specific immunological defects, such as a dysregulted B cell activity leading to the production of auto-antibodies against gut constituents. The gut infiltrating T cells in IL-2 deficient mice are not only greatly increased in number but are highly activated and proliferative (SADLACK et al. 1993). These T cell abnormalities manifest preferentially as chronic, noninfectious intestinal inflammation. It is tempting to speculate that the infiltrating T lymphocytes in human ulcerative colitis upregulate CD44v3- and CD44v6-containing longer isoforms, resulting in an impaired interaction with B cells.

In the analysis of gastric mucosae, expression of CD44 variant isoforms, particularly those containing v9, is strongly associated leukocyte infiltration (MAYER et al. 1993), as outlined in the previous section. Whether this inflammatory reaction is the predisposition for neoplastic metaplasia of the epithelium requires further investigation. All patients showing dysplasia associated with ulcerative colitis have mild to moderate inflammation and exhibit v6 positivity in the dysplastic area (ROSENBERG et al. 1995).

Although only a preliminary study, the observation that CD44 variant isoforms are upregulated during the *alloimmune response* in rejecting renal tissue of sheep should be noted (KRISHNAN et al. 1995). The same, as yet not further characterized isoforms can also be induced in vitro by concanavalin A stimulation of peripheral blood mononuclear cells. Lymph draining of the allograft contains increased levels of interferon-γ and infiltrating lymphocytes express concomitantly the larger isoform (KRISHNAN et al. 1995).

While in the normal bronchial epithelium CD44s-and CD44v9-containing isoforms are moderately expressed, considerably more have been found in *asthmatic epithelium* (LACKIE et al., submitted). The increased expression of CD44 isoforms is confined to areas of epithelial damage, supporting the view that higher levels of epithelial damage and repair processes are associated with asthmatic disease. CD44 may play a significant role in the inflammatory and repair processes of the asthmatic epithelium (LACKIE et al., submitted).

In summary, the most recent data on the involvement of CD44 variant isoforms in inflammatory processes point out the apparent importance of this family of adhesion molecules in immunological disorders such as autoimmune diseases and lymphocyte activation processes. Whether distinct CD44 variant isoforms are directly involved in lymphocyte interactions must await refined analyses, such as gene targeting of specific exons from the variant region.

The initial infiltration by activated lymphocytes into areas of inflammation and the secretion of cytokines and other growth-/angiogenesis-inducing factors may start the upregulation of variant CD44 isoforms in the surface epithelium. When this upregulation becomes constituent, i.e., alternative splicing of the CD44 nuclear RNA is permanent, the cell-cell and cell-matrix interactions may be altered so effectively that neoplastic transformation is started.

5 Conclusion

Five years after the discovery of the CD44 variant isoforms and their involvement in rat metastasis formation it is still uncertain whether the most promising data in this animal model can be adopted to understand human tumor progression and be used for diagnosis and, more importantly, for treatment of patients. The data accumulated so far in clinical studies do not verify a causative role of CD44 variant isoforms in carcinoma progression in general. Depending on the tissue of origin, an up- or downregulation of variant isoforms may be induced during the development of the epithelial disorder. More convincing are the data on hematological disorders, such as NHL and multiple myeloma, and in autoimmune diseases, such as rheumatoid arthritis, ulcerative colitis and asthma.

A major aspect of the functional analysis of the isoforms in tumor progression and immunological disorders is still missing, namely the identification of specific ligands for the variable region. Apart from a rather unspecific attachment of bFGF and HB-EGF to the heparan sulfates of v3-containing isoforms, ligands specifically acting with the highly hydrophilic protein backbone of the variant region remain to be determined. We cannot exclude that it is not the gain of function, i.e., the variant region and its specific ligands, but a loss of function, i.e., inability of the variant isoforms to bind to hyaluronan, which is more decisive for tumor development. Overexpression of the hyaluronan receptor RHAMM, in fact, is transforming and causes spontaneous metastasis formation in the lungs of mice (HALL et al. 1995). The interrelation between the two hyaluronan receptors RHAMM and CD44 may require more attention, also in the light of the fact that both act apparently downstream of Ras (HALL et al. 1995; HOFMANN et al. 1993) and may compensate each other.

Adhesion receptors play pivotal roles in tumor progression, mostly during the invasive phase of the primary tumor and the organ-specific settlement of the secondary lesion, as well as during migration through the extracelllur matrix. These adhesive interactions involve cell-cell and cell-matrix contacts mediated by cadherins, members of the immunoglobulin superfamily and integrins. While the cell-matrix function of the standard CD44 isoform has been described (hyaluronan, fibronectin, collagen, laminin, proteoglycans), cell-cell interactions have not yet been detected but may be of importance for the function of the variant isoforms.

Acknowledgments. The author thanks Drs. Reinhard Stauder (Basel) and Patricia Ruiz (Berlin) for helpful comments and suggestions on the manuscript, and Viviane Anquez for technical assistance in the contribution to the work which is presented here. The Basel Institute was founded and is supported by Hoffmann-La Roche Ltd., Basel, Switzerland.

References

Bartolazzi A, Jackson DG, Bennett K, Aruffo A, Dickinson R, Shields J, Whittle N, Stamenkovic I (1995) Regulation of growth and dissemination of a human lymphoma by CD44 splice variants. J Cell Sci 108: 1723–1733

Bennett KL, Jackson DG, Simon JC, Tanczos E, Peach R, Modrell B, Stamenkovic I, Plowman G, Aruffo A (1995) CD44 isoforms containing exon v3 are responsible for the presentation of heparin-binding growth factor. J Cell Biol 128: 687–698

Bracke ME, van Roy FM, Mareel MM (1996) The E-cadherin/catenin complex in invasion and metastasis. Curr Top Microbiol Immunol 213 I: 123–162

Brown TA, Bouchard T, St John T, Wayner E, Carter WG (1991) Human keratinocytes express a new CD44 core protein (CD44E) as a heparan-sulfate intrinsic membrane proteoglycan with additional exons. J Cell Biol 113: 207–221

Christiansen H, Sahin K, Berthold F, Hero B, Terpe H-J, Lampert F (1995) Comparison of DNA aneuploidy, chromosome 1 abnormalities, MYCN amplification and CD44 expression as prognostic factors in neuroblastoma. Eur J Cancer 31A: 541–544

Combaret V, Lasset C, Frappaz D, Bouvier R, Thiesse P, Rebillard A-C, Philip T, Favrot MC (1995) Evaluation of CD44 prognostic value in neuroblastoma: comparison with the other prognostic factors. Eur J Cancer 31A: 545–549

Cooper DL, Dougherty GJ (1995) To metastasize or not? Selection of CD44 splice sites. Nat Med 1: 635–637

Dall P, Heider K-H, Sinn H-P, Skroch-Angel P, Adolf G, Kaufmann M, Herrlich P (1995) Comparison of immunohistochemistry and RT-PCR for detection of CD44v-expression, a new prognostic factor in human breast cancer. Int J Cancer 60: 471–477

Diamond MS, Springer TA (1994) The dynamic regulation of integrin adhesiveness. Curr Biol 4: 506–517

Droll A, Dougherty ST, Chiu RK, Dirks JF, McBride WH, Cooper DL, Dougherty GJ (1995) Adhesive interactions between alternatively spliced CD44 isoforms. J Biol Chem 270: 11567–11573

East JA, Hart IA (1993) CD44 and its role in tumour progression and metastasis. Eur J Cancer 29A: 1921–1922

Finke LH, Terpe H-J, Zörb C, Haensch W, Schlag PM (1995) Colorectal cancer prognosis and expression of exon-v6-containing CD44 proteins. Lancet 345: 583

Fox SB, Fawcett J, Jackson DG, Collins I, Gatter KC, Harris AL, Gearing A, Simmons DL (1994) Normal human tissues, in addition to some tumors, express multiple different CD44 isoforms. Cancer Res 54: 4539–4546

Friedrichs K, Kügler G, Franke F, Terpe H-J, Arlt J, Regidor P-A, Günthert U (1995a) CD44 isoforms in prognosis of breast cancer. Lancet 345: 1237

Friedrichs K, Franke F, Lisbon B-W, Kügler G, Gille J, Terpe H-J, Hölzel F, Maass H, Günthert U (1995b) CD44 isoforms correlate with cellular differentiation but not with prognosis in human breast cancer. Cancer Res 55: 5424–5433

Fujimoto K, Ichimori Y, Kakizoe T, Okajima E, Sakamoto H, Sugimura T, Terada M (1991) Increased serum levels of basic fibroblast growth factor in patients with renal cell carcinoma. Biochem Biophys Res Comm 180: 386–392

Gallagher JT (1992) The protein and proteoglycan guises of Hermes/CD44. Glycobiology 2: 93–94

Günthert U (1993) CD44: a multitude of isoforms with diverse functions. Curr Top Microbiol Immunol 184: 47–63

Günthert U, Hofmann M, Rudy W, Reber S, Zöller M, Haussmann I, Matzku S, Wenzel A, Ponta H, Herrlich P (1991) A new variant of glycoprotein CD44 confers metastatic potential to rat carcinoma cells. Cell 65: 13-24

Günthert U, Stauder R, Mayer B, Terpe H-J, Finke L, Friedrichs K (1995) Are CD44 variant isoforms involved in human tumor progression? Cancer Surv 24: 19–42

Haegel H, Tölg C, Hofmann M, Ceredig R (1993) Activated mouse astrocytes and T cells express similar CD44 variants. Role of CD44 in astrocyte/T cell binding. J Cell Biol 122: 1067–1077

Hall CL, Yang B, Yang X, Zhang S, Turley M, Samuel S, Lange LA, Wang C, Curpen GD, Savani RC, Greenberg AH, Turley EA (1995) Overexpression of the hyaluronan receptor RHAMM is transforming and is also required for H-*ras* transformation. Cell 82: 19–28

Haynes BF, Liao HX, Patton KL (1991a) The transmembrane hyaluronate receptor (CD44): multiple functions, multiple forms. Cancer Cells 3: 347–350

Haynes BF, Hale LP, Patton KL, Martin ME, McCallum RM (1991b) Measurement of an adhesion molecule as an indicator of inflammatory disease activity. Arthritis Rheum 34: 1434–1443

Herrlich P, Zöller M, Pals S, Ponta H (1993a) CD44 splice variants: metastases meet lymphocytes. Immunol Today 14: 395–399

Herrlich P, Rudy P, Hofmann M, Arch R, Zöller M, Zawadzki V, Tölg C, Hekele A, Koopman G, Pals S, Heider K-H, Sleeman J, Ponta H (1993b) CD44 and splice variants of CD44 in normal differentiation and tumor progression. In: Hemler ME, Mihich E (eds) Cell adhesion molecules. Plenum, New York, pp 265–288

Hofmann M, Rudy W, Zöller M, Tölg C, Ponta H, Herrlich P, Günthert U (1991) CD44 splice variants confer metastatic behaviour in rats: homologous sequences are expressed in human tumor cell lines. Cancer Res 51: 5292–5297

Hofmann M, Rudy W, Günthert U, Zimmer SG, Zawadzki, Zöller M, Lichtner RB, Herrlich P (1993) A link between ras and metastatic behavior of tumor cells: ras induces CD44 promoter activity and leads to low-level expression of metastasis-specific variants of CD44 in CREF cells. Cancer Res 53: 1516–1521

Hong R-L, Lee WJ, Shun CT, Shu JS, Chen YC (1995a) Expression of CD44 and its clinical implication in diffuse-type and intestinal-type gastric adenocarcinomas. Oncology 52: 334–339

Hong R-L, Pu Y-S, Hsieh T-S, Chu J-S, Lee W-J (1995b) Expressions of E-cadherin and exon v6 containing isoforms of CD44 and their prognostic values in human transitional cell carcinoma. J Urol 153: 2025–2028

Hudson DL, Sleeman J, Watt FM (1995) CD44 is the major peanut lectin-binding glycoprotein of human epidermal keratinocytes and plays a role in intercellular adhesion. J Cell Sci 108: 1959–1970

Isacke CM (1994) The role of the cytoplasmic domain in regulating CD44 function. J Cell Sci 107: 2353–2359

International Non-Hodgkin's Lymphoma Prognostic Factors Project (1993) A predictive model for aggressive non-Hodgkin's lymphoma. N Engl J Med 329: 987–994

Jackson DG, Bell JI, Dickinson R, Timans J, Shields J, Whittle N (1995) Proteoglycan forms of the lymphocyte homing receptor CD44 are alternatively spliced variants containing the V3 exon. J Cell Biol 128: 673–685

Katoh S, Zheng Z, Oritani K, Shimozato T, Kincade PW (1995) Glycosylation of CD44 negatively regulates its recogonition of hyaluronan. J Exp Med 182: 419–429

Kaufmann M, Heider K-H, Sinn H-P, von Minckwitz G, Ponta H, Herrlich P (1995) CD44 variant exon epitopes in primary breast cancer and length of survival. Lancet 345: 615–619

Kelleher D, Murphy A, Hall N, Omary MB, Kearns G, Long A, Casey EB (1995) Expression of CD44 on rheumatoid synovial fluid lymphocytes. Ann Rheum Dis 54: 566–570

Kincade PW (1993) Cell adhesion mechanisms utilized for lympho-hematopoiesis. In: Shimizu Y (ed) Lymphocyte adhesion molecules. Landes, Austin, pp 249–279

Kincade PW, He Q, Ishihara K, Miyake K, Lesley J, Hyman R (1993) CD44 and other cell interaction molecules contributing to B lymphopoiesis. Curr Top Microbiol Immunol 184: 215–222

Koopman G, Griffioen AW, Ponta H, Herrlich P, van den Berg F, Manten-Horst E, Pals ST (1993a) CD44 splice variants: expression on lymphocytes and in neoplasia. Res Immunol 144: 750–754

Koopman G, Heider K-H, Horst E, Adolf GR, van den Berg F, Ponta H, Herrlich P, Pals ST (1993b) Activated human lymphocytes and aggressive non-Hodgkin's lymphomas express a homologue of the rat metastasis-associated variant of CD44. J Exp Med 177: 897–904

Koretz K, Möller P, Lehnert T, Hinz U, Otto HF, Herfarth C (1995) Effect of CD44v6 on survival in colorectal carcinoma. Lancet 345: 327–328

Krishnan R, Kireta S, Carter JK, Grooby WL, Rao MM, Russ GR (1995) Expression of alternatively spliced CD44 mRNA in sheep renal allografts. Transplant Proc 27: 2170–2171

Labarriere N, Piau JP, Otry C, Denis M, Lustenberger P, Meflah K, Le Pendu J (1994) H-blood group antigen carried by CD44v modulates tumorigenicity of rat colon carcinoma cells. Cancer Res 54: 6275–6281

Lazaar AL, Puré E (1995) CD44: a model for regulated adhesion function. Immunologist 3: 19–25

Lesley J, Hyman R, Kincade PW (1993) CD44 and its interactions with the extracellular matrix. Adv Immunol 54: 271–335

Lesley J, English N, Perschl A, Gregoroff J, Hyman R (1995) Variant cell lines selected for alterations in the function of the hyaluronan receptor CD44 show differences in glycosylation. J Exp Med 182: 431–437

Mackay CR, Terpe HJ, Stauder R, Marston WL, Stark H, Günthert U (1994) Expression and modulation of CD44 variant isoforms in humans. J Cell Biol 124: 71–82

Mayer B, Jauch KW, Günthert U, Figdor CG, Schildberg FW, Funke I, Johnson JP (1993) De-novo expression of CD44 and survival in gastric cancer. Lancet 342: 1019–1022

Mikecz K, Brennan FR, Kim JH, Glant TT (1995) Anti-CD44 treatment abrogates tissue oedema and leukocyte infiltration in murine arthritis. Nat Med 1: 558–563

Moll J, Sleeman J, Kondo K, Hekele A, Plug R, Sherman L, Schmidt A, Zöller M, Horst E, Ponta H, Herrlich P, Pals St (1994) Analysis of molecular functions of the tumor metastasis promoting surface molecule CD44v4-v7 using transgenic mice. In: Hirohashi S, Moses HL, Ruoslahti E, Sugimura T, Takeichi M, Terada M (eds) Molecular and cellular basis for cell to cell interaction: its significance in cancer. Princeton Scientific, Princeton, pp 142–151

Mulder JWR, Kruyt PM, Sewnath M, Oosting J, Seldenrijk CA, Weidema WF, Offerhaus GJA, Pals ST (1994) Colorectal cancer prognosis and expression of exon-v6-containing CD44 proteins. Lancet 344: 1470–1472

Mulder JWR, Wielenga VJM, Polak MM, van den Berg FM, Adolf GR, Herrlich P, Pals ST, Offerhaus GJA (1995) Expression of mutant p53 protein and CD44 variant proteins in colorectal tumorigenesis. Gut 36: 76–80

Nanus DM, Schmitz-Dräger BJ, Motzer RJ, Lee AC, Vlamis V, Cordon-Cardo C, Albino AP, Reuter VE (1993) Expression of basic fibroblast growth factor in primary human renal tumors: correlation with poor survival. J Natl Cancer Inst 85: 1597–1599
Nguyen M, Watanabe H, Budson AE, Richie JP, Hayes DF, Folkman J (1994) Elevated levels of an angiogenic peptide, basic fibroblast growth factor, in the urine of patients with a wide spectrum of cancers. J Natl Cancer Inst 86: 356–361
Pals ST, Koopman G, Griffioen A, Ponta H, Herrlich P, van den Berg F, Horst E (1993a) CD44: a variety of isoforms with functions in cell adhesion and tumor metastasis. In: Shimizu Y (ed) Lymphocyte adhesion molecules. Landes, Austin, pp 135–150
Pals ST, Koopman G, Heider KH, Griffioen A, Adolf GR, van den Berg F, Ponta H, Herrlich P, Horst E (1993b) CD44 splice variants: expresssion during lymhpocyte activation and tumor progression. Behring Inst Mitt 92: 273–277
Peach RJ, Hollenbaugh D, Stamenkovic I, Aruffo A (1993) Identification of hyaluronic acid binding sites in the extracellular domain of CD44. J Cell Biol 122: 257–264
Perschl A, Lesley J, English N, Trowbridge I, Hyman R (1995) Role of CD44 cytoplasmic domain in hyaluronan binding. Eur J Immunol 25: 495–501
Ponta H, Sleeman J, Herrlich P (1994) Tumor metastasis formation: cell-surface proteins confer metastasis-promoting or -suppressing properties. Biochim Biophys Acta 1198: 1–10
Ponta H, Hekele A, Heider K-H, Kaufmann M, von Minckwitz G, Sinn H-P, Pals S, Wielenga VJM, Herrlich P (1995) Relevance of CD44 variant epitopes for diagnosis and therapy of human breast cancer. In: Jonat W (ed) Contribution to oncology. Karger, Basel (in press)
Ristamäki R, Hoensuu H, Söderström K-O, Jalkanen S (1995) CD44v6 expression in non-Hodgkin's lymphoma: an association with low histological grade and poor prognosis. J Pathol 176: 259–267
Ristamäki R, Joensuu H, Jalkanen S (1996) Does soluble CD44 reflect the clinical behavior of human cancer? Curr Top Microbiol Immunol 213 III: 139–150
Rosenberg WMC, Prince C, Kaklamanis L, Fox SB, Jackson DG, Simmons DL, Chapman RW, Trowell JM, Jewell DP, Bell JI (1995) Increased expression of CD44v6 and CD44v3 in ulcerative colitis but not colonic Crohn's disease. Lancet 345: 1205–1209
Rudy W, Hofmann M, Schwartz-Albiez R, Zöller M, Heider KH, Ponta H, Herrlich P (1993) The two major CD44 proteins expressed on a metastatic rat tumor cell line are derived from different splice variants: each one individually suffices to confer metastatic behavior. Cancer Res 53: 1262–1268
Ruiz P, Günthert U (1995) The cellular basis of metastasis. World J Urol (in press)
Ruiz P, Schwärzler C, Günthert U (1995) CD44 isoforms during differentiation and development. Bioessays 17: 17–24
Sadlack B, Merz H, Schorle H, Schimpl A, Feller AC, Horvak I (1993) Ulcerative colitis-like disease in mice with a disrupted interleukin-2 gene. Cell 75: 253–261
Seiter S, Arch R, Reber S, Komitowski D, Hofmann M, Ponta H, Herrlich P, Matzku S, Zöller M (1993) Prevention of tumor metastasis formation by anti-variant CD44. J Exp Med 177: 443–455
Sherman L, Sleeman J, Herrlich P, Ponta H (1994) Hyaluronate receptors: key players in growth, differentiation, migration and tumor progression. Curr Opin Cell Biol 6: 726–733
Sherman L, Sleeman J, Dall P, Hekele A, Moll J, Ponta H, Herrlich P (1996) The CD44 proteins in embryonic development and in cancer. Curr Top Microbiol Immunol 213 I: 249–269
Sleeman J, Moll J, Sherman L, Dall P, Pals ST, Ponta H, Herrlich P (1995) The role of CD44 splice variants in human metastatic cancer. Ciba Found Symp 189: 142–156
Stauder R, Günthert U (1995) CD44 isoforms – impact on lymphocyte activation and differentiation. Immunologist 3: 78–83
Stauder R, Eisterer W, Thaler J, Günthert U (1995) CD44 variant isoforms in non-Hodgkin's lymphoma: a new independent prognostic factor. Blood 85: 2885–2899
Störkel S, Thoenes W, Jacobi GH, Lippold R (1989) Prognostic parameters in renal cell carcinoma – a new approach. Eur Urol 16: 416–422
Strober W, Ehrhardt RO (1993) Chronic intestinal inflammation: an unexpected outcome in cytokine or T cell receptor mutant mice. Cell 75: 203–205
Sy MS, Guo YJ, Stamenkovic I (1991) Distinct effects of two CD44 isofoms on tumor growth in vivo. J Exp Med 174: 859–866
Sy MS, Guo YJ, Stamenkovic I (1992) Inhibition of tumor growth in vivo with a soluble CD44-immunoglobulin fusion protein. J Exp Med 176: 623–627
Sy M-S, Liu D, Schiavone R, Ma J, Mori H, Guo Y (1996) Interactions between CD44 and hyaluronic acid: their roles in tumor growth and metastasis. Curr Top Microbiol Immunol 213 III: 113–137
Tanabe KK, Saya H (1994) The CD44 adhesion molecule and metastasis. Crit Rev Oncog 5: 201–212

Tanaka Y, Adams DH, Shaw S (1993) Proteoglycans on endothelial cells present adhesion-inducing cytokines to leukocytes. Immunol Today 14: 111–115
Tarin D, Matsumura Y (1993) Deranged CD44 gene activity in malignancy. J Pathol 171: 249–250
Terpe H-J, Stark H, Prehm P, Günthert U (1994a) CD44 variant isoforms are preferentially expressed in basal epithelia of non-malignant human fetal and adult tissues. Histochemistry 101: 79–89
Terpe H-J, Koopman R, Imhof BA, Günthert U (1994b) Expression of integrins and CD44 isoforms in non-Hodgkin's lymphomas: CD44 variant isoforms are preferentially expressed in high grade malignant lymphomas. J Pathol 174: 89–100
Terpe H-J, Christiansen H, Gonzales M, Berthold F, Lampert F (1995a) Differentiation and prognosis of neuroblastoma in correlation to the expression of CD44s. Eur J Cancer 31A: 549–552
Terpe H-J, Störkel S, Zimmer U, Anquez V, Fischer C, Pantel K, Günthert U (1995b) Expression of CD44 isoforms in renal cell tumors: positive correlation to tumor progression. Am J Pathol (in press)
Thoenes W, Störkel S, Rumpelt HJ (1986) Histopathology and classification of renal cell tumors (adenomas, oncocytomas and carcinomas). Pathol Res Pract 181: 125–143
Underhill C (1992) CD44: the hyaluronan receptor. J Cell Sci 103: 293–298
Yang B, Yang BL, Savani RC, Turley EA (1994) Identification of a common hyaluronan binding motif in the hyaluronan binding proteins RHAMM, CD44 and link protein. EMBO J 13: 286–296
Zheng Z, Katoh S, He Q, Oritani K, Miyake K, Lesley J, Hyman R, Hamik A, Parkhouse RME, Farr AG, Kincade PW (1995) Monoclonal antibodies to CD44 and their influence on hyaluronan recognition. J Cell Biol 130: 485–495
Zöller M (1995) CD44 variant expession, lymphocyte homing and lymphogenic metastasis. In: Adhesion molecules and cell signalling: biology and clinical applications, vol 1. Raven, New York, pp 201–218
Zöller M (1996) Joint features of metastasis formation and lymphocyte maturation and activation. Curr Top Microbiol Immunol 213 I: 215–247
Zöller M, Kaufmann M (1994) CD44 and metastasis. Onkologie 17: 114–122

Subject Index

Current Topics in Microbiology and Immunology

Volumes published since 1989 (and still available)

Vol. 173: **Pfeffer, Klaus; Heeg, Klaus; Wagner, Hermann; Riethmüller, Gert (Eds.):** Function and Specificity of γ/δ TCells. 1991. 41 figs. XII, 296 pp. ISBN 3-540-53781-3

Vol. 174: **Fleischer, Bernhard; Sjögren, Hans Olov (Eds.):** Superantigens. 1991. 13 figs. IX, 137 pp. ISBN 3-540-54205-1

Vol. 175: **Aktories, Klaus (Ed.):** ADP-Ribosylating Toxins. 1992. 23 figs. IX, 148 pp. ISBN 3-540-54598-0

Vol. 176: **Holland, John J. (Ed.):** Genetic Diversity of RNA Viruses. 1992. 34 figs. IX, 226 pp. ISBN 3-540-54652-9

Vol. 177: **Müller-Sieburg, Christa; Torok-Storb, Beverly; Visser, Jan; Storb, Rainer (Eds.):** Hematopoietic Stem Cells. 1992. 18 figs. XIII, 143 pp. ISBN 3-540-54531-X

Vol. 178: **Parker, Charles J. (Ed.):** Membrane Defenses Against Attack by Complement and Perforins. 1992. 26 figs. VIII, 188 pp. ISBN 3-540-54653-7

Vol. 179: **Rouse, Barry T. (Ed.):** Herpes Simplex Virus. 1992. 9 figs. X, 180 pp. ISBN 3-540-55066-6

Vol. 180: **Sansonetti, P. J. (Ed.):** Pathogenesis of Shigellosis. 1992. 15 figs. X, 143 pp. ISBN 3-540-55058-5

Vol. 181: **Russell, Stephen W.; Gordon, Siamon (Eds.):** Macrophage Biology and Activation. 1992. 42 figs. IX, 299 pp. ISBN 3-540-55293-6

Vol. 182: **Potter, Michael; Melchers, Fritz (Eds.):** Mechanisms in B-Cell Neoplasia. 1992. 188 figs. XX, 499 pp. ISBN 3-540-55658-3

Vol. 183: **Dimmock, Nigel J.:** Neutralization of Animal Viruses. 1993. 10 figs. VII, 149 pp. ISBN 3-540-56030-0

Vol. 184: **Dunon, Dominique; Mackay, Charles R.; Imhof, Beat A. (Eds.):** Adhesion in Leukocyte Homing and Differentiation. 1993. 37 figs. IX, 260 pp. ISBN 3-540-56756-9

Vol. 185: **Ramig, Robert F. (Ed.):** Rotaviruses. 1994. 37 figs. X, 380 pp. ISBN 3-540-56761-5

Vol. 186: **zur Hausen, Harald (Ed.):** Human Pathogenic Papillomaviruses. 1994. 37 figs. XIII, 274 pp. ISBN 3-540-57193-0

Vol. 187: **Rupprecht, Charles E.; Dietzschold, Bernhard; Koprowski, Hilary (Eds.):** Lyssaviruses. 1994. 50 figs. IX, 352 pp. ISBN 3-540-57194-9

Vol. 188: **Letvin, Norman L.; Desrosiers, Ronald C. (Eds.):** Simian Immunodeficiency Virus. 1994. 37 figs. X, 240 pp. ISBN 3-540-57274-0

Vol. 189: **Oldstone, Michael B. A. (Ed.):** Cytotoxic T-Lymphocytes in Human Viral and Malaria Infections. 1994. 37 figs. IX, 210 pp. ISBN 3-540-57259-7

Vol. 190: **Koprowski, Hilary; Lipkin, W. Ian (Eds.):** Borna Disease. 1995. 33 figs. IX, 134 pp. ISBN 3-540-57388-7

Vol. 191: **ter Meulen, Volker; Billeter, Martin A. (Eds.):** Measles Virus. 1995. 23 figs. IX, 196 pp. ISBN 3-540-57389-5

Vol. 192: **Dangl, Jeffrey L. (Ed.):** Bacterial Pathogenesis of Plants and Animals. 1994. 41 figs. IX, 343 pp. ISBN 3-540-57391-7

Vol. 193: **Chen, Irvin S. Y.; Koprowski, Hilary; Srinivasan, Alagarsamy; Vogt, Peter K. (Eds.):** Transacting Functions of Human Retroviruses. 1995. 49 figs. IX, 240 pp. ISBN 3-540-57901-X

Vol. 194: **Potter, Michael; Melchers, Fritz (Eds.):** Mechanisms in B-cell Neoplasia. 1995. 152 figs. XXV, 458 pp. ISBN 3-540-58447-1

Vol. 195: **Montecucco, Cesare (Ed.):** Clostridial Neurotoxins. 1995. 28 figs. XI., 278 pp. ISBN 3-540-58452-8

Vol. 196: **Koprowski, Hilary; Maeda, Hiroshi (Eds.):** The Role of Nitric Oxide in Physiology and Pathophysiology. 1995. 21 figs. IX, 90 pp. ISBN 3-540-58214-2

Vol. 197: **Meyer, Peter (Ed.):** Gene Silencing in Higher Plants and Related Phenomena in Other Eukaryotes. 1995. 17 figs. IX, 232 pp. ISBN 3-540-58236-3

Vol. 198: **Griffiths, Gillian M.; Tschopp, Jürg (Eds.):** Pathways for Cytolysis. 1995. 45 figs. IX, 224 pp. ISBN 3-540-58725-X

Vol. 199/I: **Doerfler, Walter; Böhm, Petra (Eds.):** The Molecular Repertoire of Adenoviruses I. 1995. 51 figs. XIII, 280 pp. ISBN 3-540-58828-0

Vol. 199/II: **Doerfler, Walter; Böhm, Petra (Eds.):** The Molecular Repertoire of Adenoviruses II. 1995. 36 figs. XIII, 278 pp. ISBN 3-540-58829-9

Vol. 199/III: **Doerfler, Walter; Böhm, Petra (Eds.):** The Molecular Repertoire of Adenoviruses III. 1995. 51 figs. XIII, 310 pp. ISBN 3-540-58987-2

Vol. 200: **Kroemer, Guido; Martinez-A., Carlos (Eds.):** Apoptosis in Immunology. 1995. 14 figs. XI, 242 pp. ISBN 3-540-58756-X

Vol. 201: **Kosco-Vilbois, Marie H. (Ed.):** An Antigen Depository of the Immune System: Follicular Dendritic Cells. 1995. 39 figs. IX, 209 pp. ISBN 3-540-59013-7

Vol. 202: **Oldstone, Michael B. A.; Vitković, Ljubiša (Eds.):** HIV and Dementia. 1995. 40 figs. XIII, 279 pp. ISBN 3-540-59117-6

Vol. 203: **Sarnow, Peter (Ed.):** Cap-Independent Translation. 1995. 31 figs. XI, 183 pp. ISBN 3-540-59121-4

Vol. 204: **Saedler, Heinz; Gierl, Alfons (Eds.):** Transposable Elements. 1995. 42 figs. IX, 234 pp. ISBN 3-540-59342-X

Vol. 205: **Littman, Dan. R. (Ed.):** The CD4 Molecule. 1995. 29 figs. XIII, 182 pp. ISBN 3-540-59344-6

Vol. 206: **Chisari, Francis V.; Oldstone, Michael B. A. (Eds.):** Transgenic Models of Human Viral and Immunological Disease. 1995. 53 figs. XI, 345 pp. ISBN 3-540-59341-1

Vol. 207: **Prusiner, Stanley B. (Ed.):** Prions Prions Prions. 1995. 42 figs. VII, 163 pp. ISBN 3-540-59343-8

Vol. 208: **Farnham, Peggy J. (Ed.):** Transcriptional Control of Cell Growth. 1995. 17 figs. IX, 141 pp. ISBN 3-540-60113-9

Vol. 209: **Miller, Virginia L. (Ed.):** Bacterial Invasiveness. 1996. 16 figs. IX, 115 pp. ISBN 3-540-60065-5

Vol. 210: **Potter, M.; Rose, N. R. (Eds.):** Immunology of Silicones. 1996. 136 figs. XX, 430 pp. ISBN 3-540-60272-0

Vol. 211: **Wolff, L.; Perkins, A. S. (Eds.):** Molecular Aspects of Myeloid Stem Cell Development. 1996. 98 figs. XIV, 298 pp. ISBN 3-540-60414-6

Vol. 212: **Vainio, O.; Imhof, B. A. (Eds.):** Immunology and Developmental Biology of the Chicken. 1996. 43 figs. IX, 281 pp. ISBN 3-540-60585-1

Vol. 213/II: **Günthert, U.; Birchmeier, W. (Eds.):** Attempts to Understand Metastasis Formation II. 1996. 33 figs. XV, 288 pp. ISBN 3-540-60681-5

Vol. 213/III: **Günthert, U.; Schlag, P. M.; Birchmeier, W. (Eds.):** Attempts to Understand Metastasis Formation III. 1996. 14 figs. XV, 262 pp. ISBN 3-540-60682-3

MIX
Papier aus verantwortungsvollen Quellen
Paper from responsible sources
FSC® C105338

If you have any concerns about our products,
you can contact us on
ProductSafety@springernature.com

In case Publisher is established outside the EU,
the EU authorized representative is:
Springer Nature Customer Service Center GmbH
Europaplatz 3, 69115 Heidelberg, Germany

Printed by Libri Plureos GmbH
in Hamburg, Germany